Lecture Notes in Artificial Intelligence 10757

Subseries of Lecture Notes in Computer Science

More information about this series at http://www.springer.com/series/1244

Elizabeth Black · Sanjay Modgil
Nir Oren (Eds.)

Theory and Applications
of Formal Argumentation

4th International Workshop, TAFA 2017
Melbourne, VIC, Australia, August 19–20, 2017
Revised Selected Papers

 Springer

Editors
Elizabeth Black
Department of Informatics
King's College London
London
UK

Sanjay Modgil
King's College London
London
UK

Nir Oren
Department of Computing Science
University of Aberdeen
Aberdeen
UK

ISSN 0302-9743   ISSN 1611-3349   (electronic)
Lecture Notes in Artificial Intelligence
ISBN 978-3-319-75552-6   ISBN 978-3-319-75553-3   (eBook)
https://doi.org/10.1007/978-3-319-75553-3

Library of Congress Control Number: 2018933181

LNCS Sublibrary: SL7 – Artificial Intelligence

Printed on acid-free paper

This Springer imprint is published by the registered company Springer International Publishing AG part of Springer Nature
The registered company address is: Gewerbestrasse 11, 6330 Cham, Switzerland

Preface

Recent years have witnessed a rapid growth of interest in formal models of argumentation and their application in diverse sub-fields and domains of application of AI, including reasoning in the presence of inconsistency, non-monotonic reasoning, decision-making, inter-agent communication, the Semantic Web, grid applications, ontologies, recommender systems, machine learning, neural networks, trust computing, normative systems, social choice theory, judgement aggregation and game theory, and law and medicine. Argumentation thus shows great promise as a theoretically grounded tool for a wide range of applications. The 4th International Workshop on the Theory and Applications of Formal Argumentation (TAFA 2017) aimed to contribute to the realization of this promise, by promoting and fostering the uptake of argumentation as a viable AI paradigm with wide-ranging application, and providing a forum for further development of ideas and the initiation of new and innovative collaborations.

Co-located with the International Joint Conference on Artificial Intelligence (IJCAI 2017) in Melbourne, Australia, TAFA 2017 built on the success of TAFA 2011, TAFA 2013, and TAFA 2015 with a range of strong papers submitted by authors from across Europe, Israel, China, and the USA. For the first time this year, TAFA 2017 included a systems track for short papers presenting argumentation solvers, algorithms, implementation details, and empirical evaluations. The track included submissions from participants of the Second International Competition on Computational Models of Argumentation,[1] the results of which were presented at TAFA 2017.

TAFA 2017 received 20 submissions, of which 15 were accepted and presented at the workshop after a rigorous review process. We would like to thank the authors of this volume's papers for their high-quality contributions, and acknowledge the reviewers' efforts for their in-depth feedback to authors. We also thank the participants of the workshop for their lively and thought-provoking discussions. The papers included here point to not only the exciting work taking place in the field today, but also to challenges and exciting opportunities for further research in the area, which will no doubt lead to future volumes in this series of proceedings.

January 2018

Elizabeth Black
Sanjay Modgil
Nir Oren

---

[1] http://argumentationcompetition.org/2017/

Organization

TAFA 2017 took place at the RMIT University in Melbourne, Australia, during August 19–20, 2017, as a workshop at IJCAI 2017, the 26th International Joint Conference on Artificial Intelligence.

Workshop Chairs

Elizabeth Black	King's College London, UK
Sanjay Modgil	King's College London, UK
Nir Oren	University of Aberdeen, UK

Program Committee

Leila Amgoud	IRIT, CNRS, France
Pietro Baroni	DII, University of Brescia, Italy
Stefano Bistarelli	Università di Perugia, Italy
Elizabeth Black	King's College London, UK
Elise Bonzon	LIPADE, Université Paris Descartes, France
Katarzyna Budzynska	Polish Academy of Sciences, Poland, and University of Dundee, UK
Martin Caminada	Cardiff University, UK
Federico Cerutti	Cardiff University, UK
Carlos Chesñevar	Universidad Nacional del Sur, Argentina
Cosmina Croitoru	Max Planck Institute for Informatics, Germany
Sylvie Doutre	IRIT, University of Toulouse 1, France
Sarah Alice Gaggl	TU Dresden, Germany
Massimiliano Giacomin	University of Brescia, Italy
Anthony Hunter	University College London, UK
Souhila Kaci	LIRMM, France
João Leite	Universidade Nova de Lisboa, Portugal
Beishui Liao	Zhejiang University, China
Thomas Linsbichler	Vienna University of Technology, Austria
Marco Maratea	University of Genoa, Italy
Nicolas Maudet	Université Pierre et Marie Curie, France
Sanjay Modgil	King's College London, UK
Nir Oren	University of Aberdeen, UK
Sylwia Polberg	University College London, UK
Henry Prakken	University of Utrecht and University of Groningen, The Netherlands
Odinaldo Rodrigues	King's College London, UK

Contents

An Investigation of Argumentation Framework Characteristics

Josh Murphy$^{(\boxtimes)}$, Isabel Sassoon, Michael Luck, and Elizabeth Black

Department of Informatics, King's College London, London, UK
{josh.murphy,isabel.sassoon,michael.luck,elizabeth.black}@kcl.ac.uk

Abstract. We investigate the relationship between the structural properties of argumentation frameworks and their argument-based characteristics, examining the characteristics of structures of Dung-style frameworks and two generalisations: extended argumentation frameworks and collective-attack frameworks. Our results show that the structural properties of frameworks have an impact on the size of extensions produced, on the proportion of subsets of arguments that determine some topic argument to be acceptable, and on the likelihood that the addition of some new argument will affect the acceptability of an existing argument, all characteristics that are known to affect the performance of argumentation-based technologies. We demonstrate the applicability of our results with two case studies.

1 Introduction

Argumentation is a key sub-field of AI that provides an intuitive reasoning mechanism for dealing with inconsistent, uncertain and incomplete knowledge. A set of arguments and the relationships between them can be represented as a directed graph (referred to as an *argumentation framework*) to which one of a number of *semantics* can be applied to determine which arguments it is coherent to accept [12]. While progress has been made in the development of argumentation-based technologies (*e.g.*, argument solvers [5] and real-world applications [18]) realistic evaluations of such technologies is difficult, due to the shortage of repositories of argumentation frameworks that are representative of real-world domains [10]; typically, argument technologies are evaluated on randomly generated frameworks, with little understanding of how the *structure* of such frameworks impacts on performance. It has been shown that structural differences in argumentation frameworks can affect the performance of argumentation-based technologies, such as dialogue systems [3] and argument solvers [2]. We argue here that a better understanding of these effects can not only allow for a more thorough evaluation, but can also inform development of technologies that are optimised for specific framework structures.

In order to explore the characteristics of argumentation frameworks with different structural properties, we consider the classic Dung-style argumentation frameworks (which represent attacks between arguments) [12] and two generalisations of these that each have their own particular structural traits: extended

© Springer International Publishing AG, part of Springer Nature 2018
E. Black et al. (Eds.): TAFA 2017, LNAI 10757, pp. 1–16, 2018.
https://doi.org/10.1007/978-3-319-75553-3_1

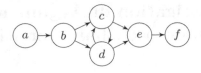

Fig. 1. An example DAF.

argumentation frameworks (which allow an argument to attack the attack between two arguments) [17] and collective-attack frameworks (which allow attacks from sets of arguments) [22]). We investigate three key characteristics.

1. The size of the set of acceptable arguments generated by the grounded and preferred sceptical semantics. This characteristic is known to affect performance of argument solvers [8];
2. The proportion of argument subsets of the framework that determine some topic argument to be acceptable. This characteristic is known to be a factor in the effectiveness of strategies for persuasion [3];
3. Whether the addition of a new argument to the framework results in a change of acceptability of some topic argument. This is a type of dynamic argumentation, another factor in the effectiveness of dialogue strategies [1], and also may be a key property for improving the computational efficiency of a variety of other argument technologies [16].

We demonstrate applicability of our results with two case studies: a Dung-style framework from a decision-making tool for aggregating the effects of medical treatment [15], and an extended framework from a statistical model selection tool in a clinical domain [25].

2   Argumentation Frameworks (AFs)

Since Dung's seminal work [12], the dominant approach to argumentation-based reasoning is to represent arguments as abstract entities in an argumentation framework that captures the relationships between them, and then to apply one of several argumentation semantics to determine which subsets of arguments it is rational to present as a coherent set. We now define *Dung-style argumentation frameworks* [12] (DAFs), which capture attacks between arguments.

Definition 1. *A **Dung-style argumentation framework (DAF)** is a pair $\langle A, R \rangle$ s.t. A is a finite set of arguments and $R \subseteq A \times A$ is a set of attacks. $(x, y) \in R$ means x attacks y.*

Argumentation semantics are based on the intuitive principles that it is not rational to accept any two conflicting arguments, and that an argument which is attacked can only be accepted if all of its attacking arguments are themselves attacked by an accepted argument [12].

Definition 2. *Let $\langle A, R \rangle$ be a DAF and $S \subseteq A$.*

- *S is **conflict-free** iff $\forall a, b \in S$: $(a, b) \notin R$.*
- *$a \in A$ is **acceptable** w.r.t. S iff $\forall b$ s.t. $(b, a) \in R$: $\exists c \in S$ s.t. $(c, b) \in R$.*
- *S is **admissible** iff S is conflict-free and each argument in S is acceptable w.r.t. S.*

There are a range of different semantics that build on these principles and determine sets of arguments that can rationally be presented as coherent, known as **extensions**. Here we consider two semantics: an argument is acceptable under the **preferred sceptical semantics** if it is part of all maximal admissible sets; an argument is acceptable under the **grounded semantics** if it is in the smallest set S such that every argument that is acceptable w.r.t. S is in S. In the DAF shown in Fig. 1, a and f are the only arguments that are acceptable under the preferred sceptical semantics, while a is the only argument acceptable under the grounded semantics.

Example 1. Considering the DAF in Fig. 1, the only argument acceptable under the grounded semantics is a, whereas the arguments a and f are acceptable under the preferred sceptical semantics.

Though Dung-style argumentation frameworks are expressive, many generalisations have been proposed which provide explicit representation of relationships other than attacks between arguments, seeking to intuitively capture particular aspects of argumentation [4]. *Extended argumentation frameworks* (EAFs) allow the representation of arguments that attack attack relations [17]. Thus, given an argument a which attacks b, an argument c may attack the attack between a and b. In this way, an EAF may be used to capture (possibly conflicting) preference relations between arguments. For example, see Fig. 2 in which c represents a preference for a over b, which conflicts with d representing a preference for b over a. EAFs are an especially expressive model as they represent preferences as defeasible arguments, allowing agents to argue about their preferences and, powerfully, about preferences over other preferences.

Definition 3. *An **extended argumentation framework (EAF)** is a tuple $\langle A, R, D \rangle$ s.t. A is a finite set of arguments, $R \subseteq A \times A$ is a set of attacks,*

- *$D \subseteq A \times R$ is a set of attacks on attacks, and*
- *if $(z, (x, y)), (z', (y, x)) \in D$ then $(z, z'), (z', z) \in R$.*

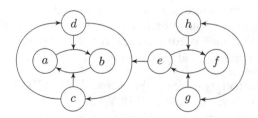

Fig. 2. An example EAF, which is also a HEAF.

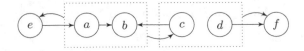

Fig. 3. An example CAF.

We especially consider here *hierarchical EAFs* (HEAFs), a particularly inter-esting class of EAFs that can be used to formalise practical reasoning [17].

Definition 4. *An EAF* $\langle A, R, D \rangle$ *is a **hierarchical extended argu-mentation framework (HEAF)** iff there exists a partition* $P = [\langle\langle A_1, R_1\rangle, D_1\rangle, ..., \langle\langle A_j, R_j\rangle, D_j\rangle, ...]$ *s.t.:*

- $A = \cup_{i=1}^{\infty} A_i, R = \cup_{i=1}^{\infty} R_i, D = \cup_{i=1}^{\infty} D_i$, *and for* $i = 1, ..., \infty$, $\langle A_i, R_i \rangle$ *is a DAF.*
- *If* $(z, (x, y)) \in D_i$ *then* $(x, y) \in R_i, z \in A_{i+1}$.

We refer to an argument a as being in a lower *partition than an argument b if* $a \in A_p$, $b \in A_q$, *and* $p < q$.

The arguments in Fig. 2 can be partitioned into 4 levels: $\{a, b\}$, $\{c, d\}$, $\{e, f\}$, and $\{g, h\}$, where $\{a, b\}$ is the lowest partition and $\{g, h\}$ is the highest.

EAF argumentation semantics are defined equivalently as for DAFs, with the following adjustments [17].

Definition 5. *Let* $\langle A, R, D \rangle$ *be an EAF and* $S \subseteq A$.

- *a* **defeats**$_S$ *b (also written as a* $\rightarrow^S b$*) iff* $(a, b) \in R$ *and* $\nexists c \in S$ *s.t.* $(c, (a, b)) \in D$.
- *S is* **conflict-free** *iff* $\forall a, b, \in S$*: if* $(a, b) \in R$ *then* $(b, a) \notin R$ *or* $\exists c \in S$ *s.t.* $(c, (a, b)) \in D$.
- $R_S = \{x_1 \rightarrow^S y_1, ..., x_n \rightarrow^S y_n\}$ *is a* **reinstatement set** *for* $c \rightarrow^S b$ *iff: (i)* $c \rightarrow^S b \in R_S$*; (ii)* $\forall i \in \{1, ..., n\}$*:* $x_i \in S$*, and (iii)* $\forall x \in R_s$*,* $\forall y'$ *s.t.* $(y', (s, y)) \in D$*:* $\exists x' \rightarrow^S y' \in R_S$.
- $a \in A$ *is* **acceptable** *w.r.t. S iff* $\forall b$ *s.t.* $b \rightarrow^S a$*:* $\exists c \in S$ *s.t.* $c \rightarrow^S b$ *and there is a reinstatement set for* $c \rightarrow^S b$.

Collective-attack frameworks (CAFs) allow the representation of sets of argu-ments that attack an argument [22]. They can allow for a more intuitive represen-tation of common-sense reasoning and human dialogues and have been shown to be useful in practical applications of argumentation [21,23]. See Fig. 3, in which there are three collective attacks: the set of arguments $\{a, b\}$ attacks the argument c, $\{a, b\}$ attacks e, and $\{c, d\}$ attacks f.

Definition 6. *A **collective-attack framework (CAF)** is a pair* $\langle A, R \rangle$ *s.t. A is a finite set of arguments, and* $R \subseteq (2^A \setminus \{\emptyset\}) \times A$ *is a set of attacks where* $(X, y) \in R$ *is an attack from the set of arguments X to the argument y.*

Similarly to EAFs, CAF argumentation semantics are defined equivalently as for DAFs but with the following adjustments [22].

Definition 7. *Let* $\langle A, R \rangle$ *be a CAF and* $S \subseteq A$.

- S *is* **conflict-free** *iff* $\nexists a \in S$ *s.t.* $\exists S' \subseteq S$ *s.t.* $(S', a) \in R$.
- $a \in A$ *is* **acceptable** *w.r.t.* S *iff* $\forall B \subseteq A$ *s.t.* $(B, a) \in R$: $\exists b \in B$, $\exists S' \subseteq S$ *s.t.* $(S', b) \in R$.

3 Structural Properties of AFs

There are many different structural properties of DAFs, HEAFs and CAFs we could investigate. Here we consider the DAF attack density, the distribution of arguments across the different levels of a HEAF, and the restriction on the number of arguments that may appear in a CAF collective-attack set. Our analysis of the characteristics of these different structural properties (Sect. 4) provides valuable insights for understanding their impact on the performance of argument technologies such as argument solvers or dialogue systems, particularly for domains or applications in which the structural properties we consider here are typical. Our case studies (Sect. 5) demonstrate the applicability of two of the structural classes we consider. More generally, our results show there is significant difference in the characteristics of different structural classes of AFs, which it can be important to consider when developing argument technologies or selecting the most appropriate AF representation (*e.g.*, [9]).

3.1 DAF Attack Density

Attack density of a DAF is the ratio of attack relations to the number of arguments. A framework with many attacks with respect to the number of arguments is *dense*, while a framework with fewer attacks is *sparse*.

Definition 8. *An* ***n-sparse DAF (n-DAF)*** *is a DAF* $\langle A, R \rangle$ *s.t.* $|R| = \frac{|A|}{n}$, *where* $n \in [0, 1]$.

We investigate 0.25-DAFs, 0.5-DAFs and 0.75-DAFs. Note that as n increases, the framework becomes more sparse. Note also that the number of attacks in the framework is linearly related to the number of arguments in the frameworks. We found in initial testing that if the number of attacks is tied instead to the number of possible attacks in the graph (which increases exponentially with the number of arguments) small changes in sparseness value produce very sharp changes in the characteristics of that structural class of DAF; linearly relating the number of attacks to arguments allows us to explore this relationship more finely.

We also consider a class of DAFs that correspond to *minimum-spanning trees* (mst-DAFs), which is a fully connected DAF in which the number of attacks is linearly related to the number of arguments ($|R| = |A| - 1$).

Definition 9. *A* **mst-DAF** *is a DAF* $\langle A, R \rangle$ *such that* $\langle A, R \rangle$ *is a minimum-spanning tree of* $\langle A, A \times A \rangle$.

3.2   Distributed HEAFs

In some domains, particularly human dialogues, it seems reasonable to assume that the number of arguments will be higher than the number of preferences over those arguments, which will be higher than the number of preferences over preferences, *etc.* We consider two different distributions of the proportion of arguments that appear in the different HEAF partitions. For *normally-distributed HEAFs* (nEAFs), we use the binomial coefficient to approximate the normal distribution (continuous) over a finite number of partitions (discrete), and thus the proportions with which to assign arguments to each partition. We use the number of partitions relative to the number of arguments in the graph that allows for the best fit with the normal distribution (computed with Sturges' formula [26]). The choice of normal distribution provides the desired trend of decreasing proportions, and is somewhat common in nature [14].

Definition 10. *The discrete normal distribution over l partitions is given by the formula* norm_dist$(l) = [d_0, d_1, ..., d_{l-1}]$ *s.t.:*

- $n = 2l - 1$
- $d_k = \frac{n!}{k!(n-k)!}$

The proportional weights of the partitions are thus given by the formula norm_prop$(l) = [p_0, p_1, ..., p_{l-1}]$ *such that* $p_i = 2(d_i) \div 2^n$.

We can then use this definition of a normal distribution over partitions to define normally-distributed HEAFs. Note that the HEAF in Fig. 2 us a normally-distributed HEAF.

Definition 11. *A **normally-distributed HEAF (nEAF)** is a HEAF $\langle A, R, D \rangle$ with a partition $P = [\langle\langle A_1, R_1 \rangle, D_1 \rangle, ..., \langle\langle A_m, R_m \rangle, D_m \rangle]$ s.t.:*

- $A = \cup_{i=1}^m A_i, R = \cup_{i=1}^m R_i, D = \cup_{i=1}^m D_i$, *and for* $i = 1, ..., m$, $\langle A_i, R_i \rangle$ *is a DAF,*
- *if* $(z, \langle x, y \rangle) \in D_i$ *then* $(x, y) \in R_i, z \in A_{i+1}$,
- $m = \lfloor \log_2 |A| \rfloor + 1$ *(Sturges' formula), and*
- $|A_j| = \lfloor (p_{l-j} \times |A|) + 1 \rfloor$ *where* norm_prop$(m) = [p_0, p_1, ..., p_{l-1}]$.

We also consider *evenly-distributed HEAFs* (eEAFs), in which each level of the partition has an equal number of arguments. We consider eEAFs to be an interesting corner-case to investigate. Again, we use Sturges' formula to compute an appropriate number of partitions.

Definition 12. *An **evenly-distributed HEAF (eEAF)** is a HEAF $\langle A, R, D \rangle$ with a partition $P = [\langle\langle A_1, R_1 \rangle, D_1 \rangle, ..., \langle\langle A_m, R_m \rangle, D_m \rangle]$ such that:*

- $A = \cup_{i=1}^m A_i, R = \cup_{i=1}^m R_i, D = \cup_{i=1}^m D_i$, *and for* $i = 1, ..., m$, $\langle A_i, R_i \rangle$ *is a DAF.*
- *If* $(z, (x, y)) \in D_i$ *then* $(x, y) \in R_i, z \in A_{i+1}$.
- $m = \lfloor \log_2 |A| \rfloor + 1$.
- *For* $i = 0, ..., m$, $|A_i| = \lceil (|A| \div m \pm 1) \rceil$.

3.3 Capped CAFs

We consider two structures of CAF: those in which the size of any collective-attack set is no greater than (*capped at*) 3 and CAFs in which there is no restriction on the size of collective-attacks sets. We refer to capped frameworks as *cCAFS*, and those which are uncapped as *uCAFs*.

Definition 13. *A* **capped collective-attack framework (cCAF)** *is a CAF* $\langle A, R \rangle$ *s.t.* $\forall (S, a) \in R : |S| \leq 3$.

Note, in the rest of this paper, to emphasise the distinction with capped collective-attack frameworks, we refer to collective-attack frameworks where it is not necessarily the case that there is an upper bound of 3 on the size of the attacking sets as **uncapped collective-attack frameworks, (uCAFs)**.

4 Characteristics of Structural Classes of AF

We ran experiments with the following structural classes of AF: 0.25-DAF, 0.5-DAF, 0.75-DAF, mst-DAF, eEAF, nEAF, cCAF and uCAF. We consider specifically the size of the grounded and the preferred sceptical extensions (known to affect the performance of argument solvers [8]), the proportion of argument subsets that determine some topic argument to be acceptable (a factor in the effectiveness of dialogue strategies for persuasion [3]), and whether the addition of a new argument to the framework results in a change of acceptability of some topic argument (also a factor in the effectiveness of dialogue strategies [1] and intrinsic to a variety of other argument technologies [16]). To investigate these properties empirically, we generate random instances of the specified structures.

When generating DAFs, we ensure that each possible weakly-connected DAF with the specified density is equally likely to be generated, only excluding frameworks that contain self-attacking arguments. For EAFs, we begin by generating each partition as a 0.5-DAF (in the same manner as described above), where the number of arguments in the partitions depends on the distribution of the EAF (e.g. whether it is a eEAF or nEAF). Then, we add one random preference relation from each argument (excluding those in the lowest partition), to a random attack relation in the partition directly below it; preference relations are generated one at a time, ensuring that the EAF has a valid HEAF structure (specifically maintaining the property in Definition 3, bullet 2). Finally, for CAFs, we begin by generating a random 0.25-DAF (in the manner described above); the attacks generated form the singleton attacks of the CAF. We then add attacks from sets of more than one argument so that the total number of attacks in the resulting CAF is the same as the number of attacks in a 0.5-DAF with the same number of arguments. We begin by first randomly selecting the size of the attacking set (for cCAFs either 2 or 3, for uCAFs from 2 to $|A| - 1$), we then randomly select a set of arguments of that size and then randomly select an argument to be attacked by that set; we repeat until we have the required number of attacks.

Our experiments were implemented in Java, partly using the Tweety library [27]. Our code is available at github.com/joshlmurphy. Experiments were run on an Intel i5 3.20 GHz CPU, with 4 GB RAM.

4.1 Size of Extension

The argument solver competition [8], in which argument solvers attempt to complete a set of tasks related to computational argumentation as efficiently as possible (such as computing an extension, or determining whether a particular argument is acceptable) used three benchmark sets of DAFs. Two of these benchmarks were characterised by the size of their extensions: frameworks with large grounded extensions and frameworks with large preferred extensions. Most solvers were slower when tasked with frameworks with a large preferred extension compared to those frameworks with a large grounded extension. This indicates that the size of the extensions of a framework is an important consideration when employing an argument solver for certain tasks. We investigated how the average size of both the grounded and the preferred sceptical extensions differs between our chosen framework classes.

For each framework class, we randomly generated at least 1,000 frameworks with n number of arguments, where $n = 12, 24, 36$. Figure 4 shows the average size of both the grounded and the preferred sceptical extension of the frameworks we generated. For DAFs, we observe a trend for both semantics that the more dense the DAF, the smaller the size of the extension. We also observe that the larger the framework, the larger the extension.

Interestingly, CAFs reverse this trend when using the grounded semantics: the larger a uCAF/cCAF, the smaller (on average) the grounded extension. This surprising result can be explained by the intuition that as you increase the number of arguments in a CAF, this increases the *proportion* of group attacks, and thus the more arguments that are part of a collective-attack relation, leading to a higher number of attack cycles (the more arguments in a set S that collectively attack an argument a, the higher the chance that a will attack at least one argument in S, causing a cycle) and the more attack cycles in a framework the smaller the grounded extension is likely to be. This is supported by the observation that uCAFs have on average a smaller grounded extension than cCAFs, which, we conclude, is due to more arguments being part of a collective-attack relation in uCAFs (as there is no cap on the number of arguments in the attack relation). When using the preferred semantics, cycles are less of a factor in the size of the extension, and so we observe that the size of the preferred sceptical extension increases as the size of CAF increases.

We find that eEAFs are more likely to have a larger grounded extension than nEAFs, but have similar sized preferred sceptical extensions. We reason that in EAFs, the more arguments that attack an attack between two arguments that exist in a framework, the more likely attack relations in the partition *below* will be defeated. This effectively lowers the attack density in lower partitions. So in the frameworks with more preferences on average (eEAFs) there will be a lower overall attack density. As we observe in DAFs, the lower the attack density of

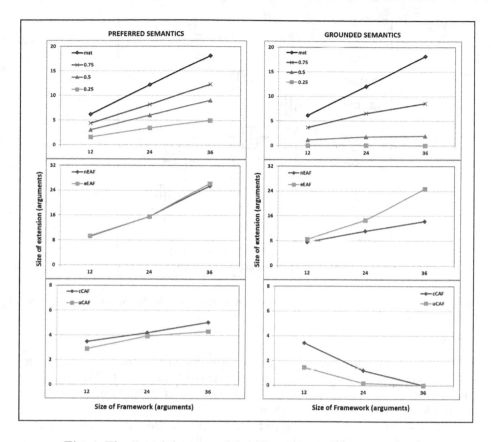

Fig. 4. The size of the grounded and preferred sceptical extensions.

the framework, the larger the extension—this is reflected in the results for the grounded extension.

4.2 Subsets that Determine a Topic Acceptable

Some particular *topic* argument will be determined acceptable by some subsets of arguments, but not others. Any topic argument will be acceptable in at most 50% of the subsets, since it will not exist in half of the subsets of the power set (an argument is deemed unacceptable in a framework it is not a part of). We refer to the proportion of subsets in which the topic argument is determined to be acceptable as *SA*. This property has been found to be an important factor in the success of persuasion dialogues [3]: the lower SA, the more difficult it is to persuade an agent that the topic argument is acceptable.

We investigated whether average SA differs between the selected framework classes. Our implementation is naive, exhaustively checking whether some topic is acceptable in every set in the power set of arguments. The time for these experiments is very high due to the exponential growth in the size of the power

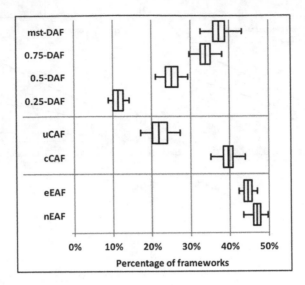

Fig. 5. The percentage of argument subsets in which a particular argument is acceptable under the grounded semantics.

set. To feasibly compute the results we used the grounded semantics (which are faster to compute) and limited the framework size to 12 arguments. We generated at least 1,000 random instances of each framework class with 12 arguments, each time randomly selecting a topic argument.

Using the analysis of variance test (ANOVA, a collective of tests used to analyse the difference between the means of multiple groups [13]) we find that the different argumentation framework classes have significantly different SA (apart from nEAF and eEAF which are distinct from other classes but not from each other), and thus that each class is a distinct population ($p < 0.05$ for each class); this implies that the framework class is a significant factor in determining SA. The largest difference between two classes is between *mst-DAF* and *nEAF*, with a 36.06% points difference between means.

In the different classes of DAF, we observe a clear trend that the more dense a framework class, the lower SA is for that class (see Fig. 5). This follows the trend observed in Fig. 4, where the more dense a DAF, the smaller the grounded extension. Similarly, uCAF frameworks typically have a smaller grounded extension than cCAF frameworks, and this trend is repeated for SA. For nEAF and eEAF frameworks of 12 arguments, there is little difference between the size of grounded extensions, and this trend is again shown for SA, in which eEAF and nEAF were not found to have significantly different SA. When using the grounded semantics it appears that the size of the extension and SA are linked.

Table 1. The percentage of frameworks that are resistant.

Framework class	12 args	24 args	36 args
mst-DAF	80.9	85.8	90.7
0.75-DAF	81.9	89.5	93.4
0.5-DAF	87.5	90.8	95.1
0.25-DAF	92.3	93.1	96.0
nEAF	87.2	91.7	96.8
eEAF	89.9	93.4	98.5
cCAF	76.7	84.1	88.8
uCAF	69.2	74.3	79.6

4.3 Resistance of AFs

Argumentation is an inherently dynamic process, with arguments and attack relations changing as new knowledge becomes available. The dynamic nature of argumentation can potentially be exploited for computational efficiency [16] as well as for strategic advantage [6]. Amgoud and Vesic consider whether the addition of a new argument to a framework changes the acceptability of a specific argument (termed the *topic argument*) [1]. If the addition of a new argument does not cause a change in the topic argument's acceptability we say the framework is *resistant*, otherwise it is *susceptible*. To investigate whether there is a difference in their resistance, for each framework class, we randomly generated at least 1,000 frameworks with n number of arguments, where $n = 12, 24, 36$, selecting both a topic argument and a test argument at random, determining the AF to be resistant if the acceptability of the topic argument is unaffected by the inclusion of the test argument. Table 1 shows the percentage of the framework instances we generated that are resistant.

For all classes we observe that the larger the framework, the more likely it is to be resistant. Intuitively, the more arguments in a framework, the more likely it is that an argument is topographically further away from the topic, and therefore the less likely the test argument will change the acceptability of the topic (this relationship can be used as a heuristic to inform an argument dialogue strategy [20]).

In a cCAF, a new argument can alter the acceptability of arguments both through introducing new argument-argument attacks as well as new collective-attacks. This is also true in uCAFs, though they have a greater chance of introducing collective-attacks: since the size of a collective attack is uncapped, each argument is in more collective attack relations on average. Thus, when we add a new argument to a uCAF it is likely to result in more changes in the acceptability of arguments, and this is why cCAFs are more resistant.

We see that eEAFs are more resistant than nEAFs, indicating that the higher the proportion of preference arguments to arguments, the more resistant the EAF will be. This is because an argument cannot alter the acceptability of an argument in a partition higher than its own partition, since all attack relations are either to arguments in the same partition or to arguments in the partition directly below. Therefore, if the topic argument is in a higher partition than the test argument, the framework is guaranteed to be resistant. In eEAFs it is more likely that the topic will be in a higher partition (since it is randomly selected and there are more arguments in higher partitions than in a nEAF) and thus the less likely it is that the test argument will affect the topic's acceptability.

5   Case Studies

We examine two case study frameworks, obtained from argument technologies deployed on real-world data. These motivate the relevance of the classes of framework structure we investigate (showing that the results of our experiments on randomly generated AFs map to the properties of our case-study frameworks) and also allow us to demonstrate how our results can be used to inform argument technologies.

5.1   Trial Aggregation

As evidence-based decision-making becomes increasingly important, clinical trials can provide an important source of information to inform healthcare professionals. Hunter and Williams propose an argument-based approach for aggregating the positive and negative effects of potential treatments, which has been shown to produce recommendations that align with published clinical guidelines [15]. The approach performs a type of meta-analysis on a range of clinical literature, producing a DAF (very sparse, almost a mst-DAF in structure) on which reasoning about possible treatment options is done. We use such a framework as our first case study.

Table 2 shows the number of arguments present in our trial aggregation case study DAF ($|Args|$), the size of its grounded and preferred sceptical extensions ($|Gr|$ and $|Pr|$), the average SA over all possible topic arguments (SA) and the percentage of cases that were resistant over each possible topic argument with a randomly selected test argument (Res). We see that the results correlate with the results obtained from mst-DAF presented earlier in this paper, with the size of extensions, SA, and resistance being within the expected ranges of mst-DAFs. This evidences the relevance of the structures we investigate.

We consider particularly the resistance of this case study framework to demonstrate how our results may be used to inform a specific application. The resistance of the trial aggregation framework is exceptionally high (97.2%). This indicates that new arguments added in the future, in this case by the addition of new clinical studies, are unlikely to change the acceptability of other arguments in the framework. This implication of this is that new studies are unlikely to have an affect on the recommended treatment, meaning there can be confidence

Table 2. Case study results.

	|Args|	|Gr|	|Pr|	SA	Res
Trial aggregation	34	9	9	41.9	97.2
Model selection	13	7	7	46.5	89.1

in the current recommendation. If a framework produced by the trial aggregation approach had a low resistance, new studies would be likely to change the recommended treatment, and this would imply that the recommendation is not yet reliable.

5.2   Statistical Model Selection

Clinicians without statistical training often need support to select a suitable model to correctly analyse and reason about their data. Sassoon *et al.* propose a tool that uses argumentation to aid in the process of deciding which statistical model is most suited to a user's research question, data and preferences [25]. The requirements and preferences of the user, as well as preferences from applicable context domains, are captured in an EAF, which can then inform the user of the most suitable model to use. We use a framework produced by using this tool with real-world data from a study involving clinicians (originally presented in [25]) as our second case study. The framework is an eEAF, being a HEAF with the same number of arguments at each level of the hierarchy.

Table 2 shows the number of arguments present in our statistical model selection case study eEAF (column \|Args\|), the size of its grounded and the preferred sceptical extensions (column \|Gr\| and column \|Pr\|), the average SA over all possible topic arguments (column SA) and the percentage of cases that were resistant over each possible topic argument with a randomly selected test argument (column Res). We see that the results correlate with our experiments over randomly generated eEAFs. Perhaps the most interesting result from this case study is the high SA of the framework (*46.5%*). Empirical investigations have demonstrated that the higher SA, the easier it is for a persuader to convince a persuadee of the acceptability of some argument [3,19]. Therefore, we would expect the persuasion of a user to use a particular statistical model to be likely to be successful when the underlying AF is an eEAF, as in this case study we consider here.

6   Discussion

We have shown that the type of AF and its structural properties have a significant effect on the size of the grounded and preferred sceptical extensions, on the proportion of subsets that determine some topic argument to be acceptable, and on the resistance of the framework; these characteristics are known to be important factors in the performance of different argument technologies.

Understanding these relationships is therefore important when considering how to evaluate such technologies. Furthermore, it can allow technologies to be optimised for specific domains in which certain structures of AF are known to be typical (such as our case study domains). For example, solvers can be developed to be faster for particular classes of framework, or a dialogue strategy can be effective for particular knowledge domains.

Related work considers how graph-theoretic properties of DAFs can be used to predict the "best" argument solver for a particular DAF [7], specifically the work considers how fast solvers are for DAFs with structures based on social networks. In contrast, we consider a range of general argument-based characteristics that are known to impact on various argumentation-technologies, including argument solvers. The structures we investigate are based on those derived from generalised argumentation frameworks commonly used in argument-technology, and our case-studies demonstrate the relevance of these structures.

We could also examine structures of framework derived from natural human-style argumentation (such as recent work by Rosenfield and Kraus [24]). Argument mining offers the possibility of obtaining large datasets of frameworks from real-world human-based argumentation, and can be applied to a vast array of domains, providing a range of framework structures related to human-reasoning. However, representing human reasoning in a formal argumentation framework is a challenging task; detecting arguments can be difficult in human dialogues because conflict tends to be hidden [11]. Nevertheless, investigating the properties of structural patterns that may be emerging in representations of human reasoning is a possible direction for future work.

References

1. Amgoud, L., Vesic, S.: On revising argumentation-based decision systems. In: Sossai, C., Chemello, G. (eds.) ECSQARU 2009. LNCS (LNAI), vol. 5590, pp. 71–82. Springer, Heidelberg (2009). https://doi.org/10.1007/978-3-642-02906-6_8
2. Bistarelli, S., Rossi, F., Santini, F.: A comparative test on the enumeration of extensions in abstract argumentation. Fundam. Inform. **140**(3–4), 263–278 (2015)
3. Black, E., Coles, A., Bernardini, S.: Automated planning of simple persuasion dialogues. In: Bulling, N., van der Torre, L., Villata, S., Jamroga, W., Vasconcelos, W. (eds.) CLIMA 2014. LNCS (LNAI), vol. 8624, pp. 87–104. Springer, Cham (2014). https://doi.org/10.1007/978-3-319-09764-0_6
4. Brewka, G., Polberg, S., Woltran, S.: Generalizations of dung frameworks and their role in formal argumentation. IEEE Intell. Syst. **29**(1), 30–38 (2014)
5. Brochenin, R., Linsbichler, T., Maratea, M., Wallner, J.P., Woltran, S.: Abstract solvers for dung's argumentation frameworks. In: Black, E., Modgil, S., Oren, N. (eds.) TAFA 2015. LNCS (LNAI), vol. 9524, pp. 40–58. Springer, Cham (2015). https://doi.org/10.1007/978-3-319-28460-6_3
6. Cayrol, C., de Saint-Cyr, F., Lagasquie-Schiex, M.: Change in abstract argumentation frameworks: adding an argument. J. Artif. Intell. Res. **38**, 49–84 (2010)
7. Cerutti, F., Giacomin, M., Vallati, M.: Algorithm selection for preferred extensions enumeration. In: 5th International Conference on Computational Models of Argument. IOS Frontiers in AI and Applications, vol. 266, pp. 221–232 (2014)

8. Cerutti, F., Oren, N., Strass, H., Thimm, M., Vallati, M.: A benchmark framework for a computational argumentation competitions. In: 5th International Conference on Computational Models of Argument. IOS Frontiers in AI and Applications, vol. 266, pp. 459–460 (2014)

9. Cerutti, F., Palmer, A., Rosenfeld, A., Snajder, J., Toni, F.: A pilot study in using argumentation frameworks for online debates. In: 1st International Workshop on Systems and Algorithms for Formal Argumentation, pp. 63–74 (2016)

10. Charwat, G., Dvorak, W., Gaggl, S., Wallner, J., Woltran, S.: Methods for solving reasoning problems in abstract argumentation: a survey. Artif. Intell. **220**, 28–63 (2015)

11. Concannon, S., Healey, P., Purver, M.: How natural is argument in natural dialogue? In: 16th International Workshop on Computational Models of Natural Argument (2016)

12. Dung, P.: On the acceptability of arguments and its fundamental role in nonmonotonic reasoning, logic programming, and n-person games. Artif. Intell. **77**(2), 321–357 (1995)

13. Fisher, R.: Statistical Methods for Research Workers. Genesis Publishing, Guildford (1925)

14. Frank, S.: The common patterns of nature. J. Evol. Biol. **22**(8), 1563–1585 (2009)

15. Hunter, A., Williams, M.: Aggregating evidence about the positive and negative effects of treatments. Artif. Intell. Med. **56**(3), 173–190 (2012)

16. Liao, B., Jin, L., Koons, R.: Dynamics of argumentation systems: a division-based method. Artif. Intell. **175**(11), 1790–1814 (2011)

17. Modgil, S.: Reasoning about preferences in argumentation frameworks. Artif. Intell. **173**(9), 901–934 (2009)

18. Modgil, S., et al.: The added value of argumentation. In: Ossowski, S. (ed.) Agreement Technologies, vol. 8, pp. 357–403. Springer, Dordrecht (2013). https://doi.org/10.1007/978-94-007-5583-3_21

19. Murphy, J., Black, E., Luck, M.: Arguing from similar positions: an empirical analysis. In: Black, E., Modgil, S., Oren, N. (eds.) TAFA 2015. LNCS (LNAI), vol. 9524, pp. 177–193. Springer, Cham (2015). https://doi.org/10.1007/978-3-319-28460-6_11

20. Murphy, J., Black, E., Luck, M.: A heuristic strategy for persuasion dialogues. In: 6th International Conference on Computational Models of Argument. IOS Frontiers in AI and Apps, vol. 287, pp. 411–417 (2016)

21. Nielsen, S., Parsons, S.: An application of formal argumentation: fusing Bayes nets in MAS. In: 1st International Conference on Computational Models of Argument. IOS Frontiers in AI and Applications, vol. 144, pp. 33–44 (2006)

22. Nielsen, S.H., Parsons, S.: A generalization of Dung's abstract framework for argumentation: arguing with sets of attacking arguments. In: Maudet, N., Parsons, S., Rahwan, I. (eds.) ArgMAS 2006. LNCS (LNAI), vol. 4766, pp. 54–73. Springer, Heidelberg (2007). https://doi.org/10.1007/978-3-540-75526-5_4

23. Oren, N., Norman, T., Preece, A.: Loose lips sink ships: a heuristic for argumentation. In: 3rd International Workshop on Argumentation in Multi-Agent Systems. LNAI, vol. 4049, pp. 121–134. Springer (2006)

24. Rosenfield, A., Kraus, S.: Strategical argumentative agent for human persuasion. In: 22nd European Conference on Artificial Intelligence. IOS Frontiers in AI and Applications, vol. 285, pp. 320–328 (2016)

25. Sassoon, I., Keppens, J., McBurney, P.: Preferences in argumentation for statistical model selection. In: 6th International Conference on Computational Models of Argument. IOS Frontiers in AI and Applications, vol. 287, pp. 53–60 (2016)

26. Sturges, H.: The choice of a class interval. J. Am. Stat. Assoc. **21**(153), 65–66 (1926)
27. Thimm, M.: Tweety - a comprehensive collection of Java libraries for logical aspects of artificial intelligence and knowledge representation. In: 14th International Conference on Principles of Knowledge Representation and Reasoning, pp. 528–537. AAAI Press (2014)

Hypersequent-Based Argumentation: An Instantiation in the Relevance Logic RM

AnneMarie Borg[1(✉)], Ofer Arieli[2], and Christian Straßer[1]

[1] Institute of Philosophy II, Ruhr University Bochum, Bochum, Germany
{annemarie.borg,christian.strasser}@rub.de
[2] School of Computer Science, The Academic College of Tel-Aviv, Tel-Aviv, Israel
oarieli@mta.ac.il

Abstract. In this paper we introduce *hypersequent-based* frameworks for the modeling of defeasible reasoning by means of logic-based argumentation. These frameworks are an extension of sequent-based argumentation frameworks, in which arguments are represented not only by sequents, but by more general expressions, called *hypersequents*. This generalization allows us to overcome some of the weaknesses of logical argumentation reported in the literature and to prove several desirable properties, stated in terms of rationality postulates. For this, we take the relevance logic RM as the deductive base of our formalism. This logic is regarded as "by far the best understood of the Anderson-Belnap style systems" (Dunn and Restall, Handbook of Philosophical Logic, vol. 6). It has a clear semantics in terms of Sugihara matrices, as well as sound and complete Hilbert- and Gentzen-type proof systems. The latter are defined by hypersequents and admit cut elimination. We show that hypersequent-based argumentation yields a robust defeasible variant of RM with many desirable properties (e.g., rationality postulates and crash-resistance).

1 Introduction

Argumentation theory has been described as "*a core study within artificial intelligence*" [11]. Among others, it is a standard method for modeling defeasible reasoning. Logical argumentation (sometimes called deductive or structural argumentation) is a branch of argumentation theory in which arguments have a specific structure. This includes rule-based argumentation, such as the ASPIC⁺ framework [26] and methods that are based on Tarskian logics, like Besnard and Hunter's approach [12], in which classical logic is the deductive base (the so-called *core logic*).

The latter approach was generalized to *sequent-based argumentation* in [4], where Gentzen-style sequents [19], extensively used in proof theory, are incorporated for representing arguments, and attacks are formulated by special inference

---

The first two authors are supported by the Israel Science Foundation (grant 817/15). The first and the third author are supported by the Alexander von Humboldt Foundation and the German Ministry for Education and Research.

rules called *sequent elimination rules*. The result is a generic and modular app-roach to logical argumentation, in which any logic with a corresponding sound and complete sequent calculus can be used as the underlying core logic. A dynamic proof theory as a computational tool for sequent-based argumenta-tion was introduced in [6]. This allows for reasoning with these argumentation frameworks in an automatic way.

In this paper we further extend sequent-based argumentation to *hyperse-quents* [7,22,24]. This is a powerful generalization of Gentzen's sequents which was used for providing cut-free Gentzen-type systems for the relevance logic RM, its 3-valued version RM_3 and the modal logic S5. It allows a high degree of par-allelism in constructing proofs and has some applications in the proof theory of fuzzy logics (see, e.g., [21]). In the context of argumentation, the incorpora-tion of hypersequents enables to split sequents into different components, and so different rationality postulates [1,13] can be satisfied, some of which are not available otherwise.

The usefulness of logical argumentation with hypersequents is demonstrated here on frameworks whose core logic is RM. This logic was introduced by Dunn and McCall and later extensively studied by Dunn and Meyer [17] and Avron [7,9] (see also [3,18]). In [18, p. 81], RM is regarded as *"by far the best understood of the Anderson-Belnap style systems"*. The basic idea behind this logic (and relevance logics in general) is that the set of premises should be 'rele-vant' to its conclusion. This way some problematic phenomena of classical logic, such as the paradoxes of material implication, are avoided. In addition, it was shown that RM is semi-relevant, paraconsistent, decidable and has the Scroggs' property [3, Sect. 29.4]. Furthermore, RM has a clear semantics in terms of Sugi-hara matrices [3, Sect. 29.3] and sound and complete Hilbert- and Gentzen-type proof systems are available for it (see, e.g., [7,9]). The latter admit cut elim-ination and are expressed in terms of hypersequents, a fact which makes RM particularly suitable for our purpose.

We will show that hypersequent-based frameworks, with RM as the core logic, satisfy the logic-based rationality postulates from [1] and non-interference and crash-resistance from [14]. In particular, this proves that such formalisms avoid the problem of logical argumentation raised in [15], and further discussed in [2] (to which we shall refer below). A byproduct of our approach is a defeasible variant of RM with many desirable properties.

The rest of the paper is organized as follows. The next two sections contain some preliminary material: in Sect. 2 we recall some basic notions of sequent-based argumentation, and in Sect. 3 we review the notion of hypersequents and the logic RM. Then, in Sect. 4 we extend sequent-based argumentation frame-works to hypersequent-based ones, and in Sect. 5 we consider some properties of these frameworks, instantiated in RM. Finally, in Sect. 6 we conclude.

2   Sequent-Based Argumentation

We start by recalling the setting of sequent-based argumentation [4]. Throughout the paper we consider propositional languages, denoted by $\mathcal{L}$, that may contain connectives in $\{\neg, \wedge, \vee, \supset, \leftrightarrow\}$. Sets of formulae are denoted by $\mathcal{S}, \mathcal{T}$, finite sets of formulae are denoted by Γ, Δ, formulae are denoted by ϕ, ψ and atomic formulae are denoted by p, q, all of which can be primed or indexed. We denote by $\bigwedge \Gamma$ (respectively, by $\bigvee \Gamma$), the conjunction (respectively, the disjunction) of all the formulae in Γ. Furthermore, we let $\neg \mathcal{S} = \{\neg \phi \mid \phi \in \mathcal{S}\}$.

Definition 1. *A logic for a language $\mathcal{L}$ is a pair $\mathsf{L} = \langle \mathcal{L}, \vdash \rangle$, where $\vdash$ is a (Tarskian) consequence relation for $\mathcal{L}$, satisfying, for all sets $\mathcal{T}, \mathcal{T}'$ of $\mathcal{L}$-formulas and every $\mathcal{L}$-formula ϕ, the following properties:*

- reflexivity: *if $\phi \in \mathcal{T}$, then $\mathcal{T} \vdash \phi$;*
- transitivity: *if $\mathcal{T} \vdash \phi$ and $\mathcal{T}', \phi \vdash \psi$, then $\mathcal{T}, \mathcal{T}' \vdash \psi$;*
- monotonicity: *if $\mathcal{T}' \vdash \phi$ and $\mathcal{T}' \subseteq \mathcal{T}$, then $\mathcal{T} \vdash \phi$.*

As usual in logical argumentation (see, e.g., [12, 23, 25, 27]), arguments have a specific structure based on the underlying formal language. In the current setting arguments are represented by the well-known proof theoretical notion of a *sequent*.

Definition 2. *Let $\mathsf{L} = \langle \mathcal{L}, \vdash \rangle$ be a logic and let $\mathcal{S}$ be a set of formulae in $\mathcal{L}$.*

- *An $\mathcal{L}$-sequent (sequent for short) is an expression of the form $\Gamma \Rightarrow \Delta$, where Γ and Δ are finite sets of formulae in $\mathcal{L}$ and $\Rightarrow$ is a symbol that does not appear in $\mathcal{L}$.*
- *An L-argument (argument for short) is an $\mathcal{L}$-sequent $\Gamma \Rightarrow \psi$,[1] where $\Gamma \vdash \psi$. Γ is called the support set of the argument and ψ is its conclusion.*
- *An L-argument based on $\mathcal{S}$ is an L-argument $\Gamma \Rightarrow \psi$, where $\Gamma \subseteq \mathcal{S}$. We denote by $Arg_\mathsf{L}(\mathcal{S})$ the set of all the L-arguments based on $\mathcal{S}$.*

The formal systems used for the constructions of sequents (and so of arguments) for a logic $\mathsf{L} = \langle \mathcal{L}, \vdash \rangle$, are called *sequent calculi* [19]. In what follows we shall assume that a sequent calculus C is sound and complete for its logic (i.e., $\Gamma \Rightarrow \psi$ is provable in C iff $\Gamma \vdash \psi$). One of the advantages of sequent-based argumentation is that any logic with a corresponding sound and complete sequent calculus can be used as the core logic. Furthermore, unlike other logic-based approaches to argumentation (see, e.g., [2]), it is not required that the support set is minimal, nor that it is consistent.[2] The construction of arguments from simpler arguments is done by the *inference rules* of the sequent calculus [19].

Argumentation systems contain also attacks between arguments. In our case, attacks are represented by *sequent elimination rules*. Such a rule consists of an attacking argument (the first condition of the rule), an attacked argument (the last condition of the rule), conditions for the attack (the conditions in between) and a conclusion (the eliminated attacked sequent). The outcome of an application of such a rule is that the attacked sequent is 'eliminated'. The elimination of a sequent $s = \Gamma \Rightarrow \Delta$ is denoted by $\overline{s}$ or $\Gamma \not\Rightarrow \Delta$.

---

[1] Set signs in arguments are omitted.

[2] See [4] for further advantages of this approach.

Definition 3. A sequent elimination rule *(or attack rule) is a rule $\mathcal{R}$ of the form:*

$$\frac{\Gamma_1 \Rightarrow \Delta_1 \ldots \Gamma_n \Rightarrow \Delta_n}{\Gamma_n \not\Rightarrow \Delta_n} \quad \mathcal{R} \tag{1}$$

Let $L = \langle \mathcal{L}, \vdash \rangle$ be a logic with corresponding sequent calculus C, $\Gamma \Rightarrow \psi, \Gamma' \Rightarrow \psi' \in Arg_L(\mathcal{S})$ and let $\mathcal{R}$ be an elimination rule as above. If $\Gamma \Rightarrow \psi$ is an instance of $\Gamma_1 \Rightarrow \Delta_1$, $\Gamma' \Rightarrow \psi'$ is an instance of $\Gamma_n \Rightarrow \Delta_n$ and all the other conditions of $\mathcal{R}$ (i.e., $\Gamma_i \Rightarrow \Delta_i$ for $i = 2, \ldots, n-1$) are provable in C, then we say that $\Gamma \Rightarrow \psi$ $\mathcal{R}$-attacks $\Gamma' \Rightarrow \psi'$.

Example 1. We refer to [4, 29] for a definition of many sequent elimination rules. Below are three of them (assuming that $\Gamma_2 \neq \emptyset$):

$$\text{Defeat:} \quad \frac{\Gamma_1 \Rightarrow \psi_1 \quad \Rightarrow \psi_1 \supset \neg \bigwedge \Gamma_2 \quad \Gamma_2 \Rightarrow \psi_2}{\Gamma_2 \not\Rightarrow \psi_2} \quad \text{Def}$$

$$\text{Undercut:} \quad \frac{\Gamma_1 \Rightarrow \psi_1 \quad \Rightarrow \psi_1 \leftrightarrow \neg \bigwedge \Gamma_2 \quad \Gamma_2, \Gamma_2' \Rightarrow \psi_2}{\Gamma_2, \Gamma_2' \not\Rightarrow \psi_2} \quad \text{Ucut}$$

$$\text{Consistency undercut} \quad \frac{\Rightarrow \neg \bigwedge \Gamma \quad \Gamma, \Gamma' \Rightarrow \psi}{\Gamma, \Gamma' \not\Rightarrow \psi} \quad \text{ConUcut}$$

Note that the attacker and the attacked argument must be elements of $Arg_L(\mathcal{S})$.[3] A sequent-based argumentation framework is now defined as follows:

Definition 4. *A sequent-based argumentation framework for a set of formulae $\mathcal{S}$ based on a logic $L = \langle \mathcal{L}, \vdash \rangle$ and a set AR of sequent elimination rules, is a pair $\mathcal{AF}_L(\mathcal{S}) = \langle Arg_L(\mathcal{S}), \mathcal{A} \rangle$, where $\mathcal{A} \subseteq Arg_L(\mathcal{S}) \times Arg_L(\mathcal{S})$ and $(a_1, a_2) \in \mathcal{A}$ iff there is an $\mathcal{R} \in AR$ such that a_1 $\mathcal{R}$-attacks a_2.*

Example 2. Suppose that $\{p, \neg p\} \subseteq \mathcal{S}$. When classical logic (CL) is the core logic, the sequents $p \Rightarrow p$ and $\neg p \Rightarrow \neg p$ attack each other according to defeat and undercut (see Example 1). The tautological sequent $\Rightarrow \psi \vee \neg \psi$ is not defeated or undercut by any argument in $Arg_{CL}(\mathcal{S})$, since it has an empty support set.

Given a (sequent-based) argumentation framework $\mathcal{AF}_L(\mathcal{S})$, Dung-style semantics [16] can be applied to it, to determine what combinations of arguments (called *extensions*) can collectively be accepted from it.

Definition 5. *Let $\mathcal{AF}_L(\mathcal{S}) = \langle Args_L(\mathcal{S}), \mathcal{A} \rangle$ be an argumentation framework and let $S \subseteq Args_L(\mathcal{S})$ be a set of arguments. It is said that:*

- *S attacks an argument a if there is an $a' \in S$ such that $(a', a) \in \mathcal{A}$;*
- *S defends an argument a if S attacks every attacker of a;*
- *S is conflict-free if there are no arguments $a_1, a_2 \in S$ such that $(a_1, a_2) \in \mathcal{A}$;*
- *S is admissible if it is conflict-free and it defends all of its elements.*

---

[3] By requiring that both the attacking and the attacked argument should be in $Arg_L(\mathcal{S})$ we prevent "irrelevant attacks", that is: situations in which, e.g., $\neg p \Rightarrow \neg p$ attacks $p \Rightarrow p$ (by Undercut), although $\mathcal{S} = \{p\}$.

An admissible set that contains all the arguments that it defends is a complete
extension *of* $\mathcal{AF}_L(\mathcal{S})$. *Below are definitions of some other extensions of* $\mathcal{AF}_L(\mathcal{S})$:

- *the* grounded extension *of* $\mathcal{AF}_L(\mathcal{S})$ *is the minimal (with respect to* $\subseteq$*) complete extension of* $Arg_L(\mathcal{S})$;[4]
- *a* preferred extension *of* $\mathcal{AF}_L(\mathcal{S})$ *is a maximal (with respect to* $\subseteq$*) admissible subset of* $Arg_L(\mathcal{S})$;
- *a* stable extension *of* $\mathcal{AF}_L(\mathcal{S})$ *is an admissible subset of* $Arg_L(\mathcal{S})$ *that attacks every argument not in it.*

In what follows we shall refer to either complete (cmp), grounded (gr), preferred (prf) or stable (stb) semantics as *completeness-based semantics*. We denote
by $\mathsf{Ext}_{\mathsf{sem}}(\mathcal{AF}_L(\mathcal{S}))$ the set of all the extensions of $\mathcal{AF}_L(\mathcal{S})$ under the semantics
sem $\in \{\mathsf{cmp}, \mathsf{gr}, \mathsf{prf}, \mathsf{stb}\}$. The subscript is omitted when this is clear from the
context.

Example 3. Let $\mathcal{AF}_{\mathsf{CL}}(\mathcal{S})$ be a sequent-based argumentation framework for
$\mathcal{S} = \{p, \neg p, q\}$, based on CL, with Ucut as the sole attack rule. Then, as noted
in Example 2, the sequent $\Rightarrow p \vee \neg p$ belongs to every complete extension of
$\mathcal{AF}_{\mathsf{CL}}(\mathcal{S})$, since it cannot be undercut-attacked. Similarly, $q \Rightarrow q$ also belongs
to every complete extension of $\mathcal{AF}_L(\mathcal{S})$, since $\Rightarrow p \vee \neg p$ counter-attacks any
attacker of $q \Rightarrow q$ that belongs to $Arg_{\mathsf{CL}}(\mathcal{S})$.[5] This implies that both $\Rightarrow p \vee \neg p$
and $q \Rightarrow q$ are in the grounded extension of $\mathcal{AF}_{\mathsf{CL}}(\mathcal{S})$.

Definition 6. *Given a sequent-based argumentation framework* $\mathcal{AF}_L(\mathcal{S})$, *the
semantics as defined in Definition 5 induces corresponding (nonmonotonic)
entailment relations:* $\mathcal{S} \mathrel{|\!\!\!\sim}^{\cap}_{\mathsf{sem}} \phi$ ($\mathcal{S} \mathrel{|\!\!\!\sim}^{\cup}_{\mathsf{sem}} \phi$) *iff for every (some) extension* $\mathcal{E} \in$
$\mathsf{Ext}_{\mathsf{sem}}(\mathcal{AF}_L(\mathcal{S}))$ *there is an argument* $\Gamma \Rightarrow \phi \in \mathcal{E}$ *for some* $\Gamma \subseteq \mathcal{S}$.

Example 4. Note that, since the grounded extension is unique, $\mathrel{|\!\!\!\sim}^{\cap}_{\mathsf{gr}}$ and $\mathrel{|\!\!\!\sim}^{\cup}_{\mathsf{gr}}$ coin-
cide (so both can be denoted by $\mathrel{|\!\!\!\sim}_{\mathsf{gr}}$). For instance, in Example 3, $p, \neg p, q \mathrel{|\!\!\!\sim}_{\mathsf{gr}} q$,
while $p, \neg p, q \mathrel{|\!\!\!\not\sim}_{\mathsf{gr}} p$ and $p, \neg p, q \mathrel{|\!\!\!\not\sim}_{\mathsf{gr}} \neg p$.

3   Hypersequents and RM

Ordinary sequent calculi do not capture all the interesting logics. For some logics,
which have a clear and simple semantics, no standard cut-free sequent calculus is
known. Notable examples are the Gödel–Dummett intermediate logic LC, the relevance logic RM and the modal logic S5. A large range of extensions of Gentzen's
original sequent calculi have been introduced for providing decent proof systems for different non-classical logics. Here we consider a natural extension of
sequent calculi, called *hypersequent calculi*. Hypersequents were independently
introduced by Mints [22], Pottinger [24] and Avron [7], nowadays Avron's notation is mostly used (see, e.g., [8]). Intuitively, a hypersequent is a finite set (or

---

[4] It is well-known (see [16]) that the grounded extension of a framework is unique.
[5] This follows since any attacker of $q \Rightarrow q$ has an inconsistent support.

sequence) of sequents, which is valid if and only if at least one of its component sequents is valid. This allows to define new inference (and elimination) rules for "multi-processing" different sequents. These types of rules increase the expressive power of hypersequents compared to ordinary sequent calculi, and as a result the corresponding argumentation systems have some desirable properties that are not available for ordinary sequent-based frameworks.

To illustrate the application of hypersequents in argumentation, we take RM as the core logic and use a hypersequent calculus for it, as well as extended versions of the attack rules for standard sequents. In this section we formally define what a hypersequent is and present a hypersequent calculus for RM.

3.1   Hypersequents and Inference Rules for Them

Definition 7. *An $\mathcal{L}$-hypersequent is a finite multiset of sequents: $\Gamma_1 \Rightarrow \Delta_1 \mid \ldots \mid \Gamma_n \Rightarrow \Delta_n$, where $\Gamma_i \Rightarrow \Delta_i$ $(1 \leq i \leq n)$ are $\mathcal{L}$-sequents and $\mid$ is a new symbol, not appearing in $\mathcal{L}$.*[6] *Each $\Gamma_i \Rightarrow \Delta_i$ is called a* component *of the hypersequent.*

Note that every ordinary sequent is a hypersequent as well. In what follows, hypersequents are denoted by $\mathcal{G}, \mathcal{H}$, primed or indexed if needed. Given a hypersequent $\mathcal{H} = \Gamma_1 \Rightarrow \Delta_1 \mid \ldots \mid \Gamma_n \Rightarrow \Delta_n$, the *support* of $\mathcal{H}$ is the set $\mathsf{Supp}(\mathcal{H}) = \{\Gamma_1, \ldots, \Gamma_n\}$ and the *consequent* of $\mathcal{H}$ is the formula $\mathsf{Conc}(\mathcal{H}) = \bigvee \Delta_1 \vee \ldots \vee \bigvee \Delta_n$. Given a set Λ of hypersequents, we let $\mathsf{Concs}(\Lambda) = \{\mathsf{Conc}(\mathcal{H}) \mid \mathcal{H} \in \Lambda\}$.

Example 5. Like in Gentzen's sequent calculi, hypersequent axioms have the form $A \Rightarrow A$. Consider the right implication rule of Gentzen's calculus LK for classical logic (on the left below). The corresponding hypersequent rule is similar, now with added components (on the right below):

$$\frac{\Gamma, A \Rightarrow \Delta, B}{\Gamma \Rightarrow \Delta, A \supset B} \Rightarrow \supset \qquad \frac{\mathcal{G} \mid \Gamma, A \Rightarrow \Delta, B \mid \mathcal{H}}{\mathcal{G} \mid \Gamma \Rightarrow \Delta, A \supset B \mid \mathcal{H}} \Rightarrow \supset$$

As noted in [8], many sequent rules can be translated like this. However, it can be that there are two versions (an additive form and a multiplicative form), which are equivalent if contraction, exchange and weakening are all part of the system. Take for example the right conjunction rule of LK. The dual hypersequent rule in an additive form is:

$$\frac{\mathcal{G} \mid \Gamma \Rightarrow \Delta, A \mid \mathcal{H} \quad \mathcal{G} \mid \Gamma \Rightarrow \Delta, B \mid \mathcal{H}}{\mathcal{G} \mid \Gamma \Rightarrow \Delta, A \wedge B \mid \mathcal{H}} \Rightarrow \wedge$$

and the multiplicative form of the same rule is:

$$\frac{\mathcal{G}_1 \mid \Gamma_1 \Rightarrow \Delta_1, A \mid \mathcal{H}_1 \quad \mathcal{G}_2 \mid \Gamma_2 \Rightarrow \Delta_2, B \mid \mathcal{H}_2}{\mathcal{G}_1 \mid \mathcal{G}_2 \mid \Gamma_1, \Gamma_2 \Rightarrow \Delta_1, \Delta_2, A \wedge B \mid \mathcal{H}_1 \mid \mathcal{H}_2} \Rightarrow \wedge$$

---

[6] The common, intuitive interpretation of the sign "$\mid$" is disjunction.

3.2   The Logic **RM** and the hypersequent calculus **GRM**

As noted previously, we will demonstrate hypersequent-based argumentation by the core logic RM. This is the best understood and researched logic among the relevance logics from the Anderson-Belnap approach [3].[7] Moreover, it is paraconsistent, decidable [9], has a simple semantics [3, Sect. 29] and is characterized by a Hilbert-style system [3, Sect. 27] (see also [9]). Like other relevance logics (such as R), RM does not satisfy the classical implication paradoxes $\phi \supset (\psi \supset \phi)$, $\neg\phi \supset (\phi \supset \psi)$, $(\phi \wedge \neg\phi) \supset \psi$ and $\phi \supset (\psi \supset \psi)$.[8] This makes RM suitable for the modeling of defeasible reasoning and hence an appropriate core logic for argumentation-based reasoning.

An ordinary cut-free sequent calculus for RM is not known. Figure 1 presents a hypersequent proof system for RM, called GRM.

Axioms: $\mathcal{G} \mid \psi \Rightarrow \psi$

Logical rules:

$$[\neg\Rightarrow] \quad \frac{\mathcal{G} \mid \Gamma \Rightarrow \Delta, \varphi}{\mathcal{G} \mid \neg\varphi, \Gamma \Rightarrow \Delta} \qquad\qquad [\Rightarrow\neg] \quad \frac{\mathcal{G} \mid \varphi, \Gamma \Rightarrow \Delta}{\mathcal{G} \mid \Gamma \Rightarrow \Delta, \neg\varphi}$$

$$[\supset\Rightarrow] \quad \frac{\mathcal{G} \mid \Gamma_1 \Rightarrow \Delta_1, \varphi \qquad \mathcal{G} \mid \psi, \Gamma_2 \Rightarrow \Delta_2}{\mathcal{G} \mid \Gamma_1, \Gamma_2, \varphi \supset \psi \Rightarrow \Delta_1, \Delta_2} \qquad [\Rightarrow\supset] \quad \frac{\mathcal{G} \mid \Gamma, \varphi \Rightarrow \Delta, \psi}{\mathcal{G} \mid \Gamma \Rightarrow \Delta, \varphi \supset \psi}$$

$$[\wedge\Rightarrow] \quad \frac{\mathcal{G} \mid \Gamma, \varphi \Rightarrow \Delta}{\mathcal{G} \mid \Gamma, \varphi \wedge \psi \Rightarrow \Delta} \quad \frac{\mathcal{G} \mid \Gamma, \psi \Rightarrow \Delta}{\mathcal{G} \mid \Gamma, \varphi \wedge \psi \Rightarrow \Delta} \qquad [\Rightarrow\wedge] \quad \frac{\mathcal{G} \mid \Gamma \Rightarrow \Delta, \varphi \qquad \mathcal{G} \mid \Gamma \Rightarrow \Delta, \psi}{\mathcal{G} \mid \Gamma \Rightarrow \Delta, \varphi \wedge \psi}$$

$$[\vee\Rightarrow] \quad \frac{\mathcal{G} \mid \Gamma, \varphi \Rightarrow \Delta \qquad \mathcal{G} \mid \Gamma, \psi \Rightarrow \Delta}{\mathcal{G} \mid \Gamma, \varphi \vee \psi \Rightarrow \Delta} \qquad [\Rightarrow\vee] \quad \frac{\mathcal{G} \mid \Gamma \Rightarrow \Delta, \varphi}{\mathcal{G} \mid \Gamma \Rightarrow \Delta, \varphi \vee \psi} \quad \frac{\mathcal{G} \mid \Gamma \Rightarrow \Delta, \psi}{\mathcal{G} \mid \Gamma \Rightarrow \Delta, \varphi \vee \psi}$$

Structural rules:

$$[\text{EC}] \quad \frac{\mathcal{G} \mid s \mid s}{\mathcal{G} \mid s} \qquad\qquad [\text{EW}] \quad \frac{\mathcal{G}}{\mathcal{G} \mid s}$$

$$[\text{Sp}] \quad \frac{\mathcal{G} \mid \Gamma_1, \Gamma_2 \Rightarrow \Delta_1, \Delta_2}{\mathcal{G} \mid \Gamma_1 \Rightarrow \Delta_1 \mid \Gamma_2 \Rightarrow \Delta_2} \qquad [\text{Mi}] \quad \frac{\mathcal{G} \mid \Gamma_1 \Rightarrow \Delta_1 \qquad \mathcal{G} \mid \Gamma_2 \Rightarrow \Delta_2}{\mathcal{G} \mid \Gamma_1, \Gamma_2 \Rightarrow \Delta_1, \Delta_2}$$

$$[\text{Cut}] \quad \frac{\mathcal{G} \mid \Gamma_1 \Rightarrow \Delta_1, \varphi \qquad \mathcal{G} \mid \varphi, \Gamma_2 \Rightarrow \Delta_2}{\mathcal{G} \mid \Gamma_1, \Gamma_2 \Rightarrow \Delta_1, \Delta_2}$$

Fig. 1. The proof system GRM [7]

---

[7] Strictly speaking, RM is a *semi-relevance logic*: it does satisfy the basic relevance criterion (introduced in [3]) and the minimal semantic relevance criterion [9], but it does not have the variable sharing property (introduced in [3]), see, e.g., [9].

[8] Unlike R, RM does satisfy the *mingle axiom* $\phi \supset (\phi \supset \phi)$.

In [7] it is shown that GRM admits cut-elimination and that it satisfies the following soundness and completeness result for RM:

Theorem 1. *Let* $\mathcal{H} = \Gamma_1 \Rightarrow \Delta_1 \mid \ldots \mid \Gamma_n \Rightarrow \Delta_n$ *be a hypersequent, where for each* $1 \leq i \leq n$, $\Gamma_i = \{\gamma_1^i, \ldots, \gamma_{m_i}^i\}$ *and* $\Delta_i = \{\delta_1^i, \ldots, \delta_{l_i}^i\}$. *We denote:*

$$\tau(\mathcal{H}) = \left(\neg\gamma_1^1 \vee \ldots \vee \neg\gamma_{m_1}^1 \vee \delta_1^1 \vee \ldots \vee \delta_{l_1}^1\right) \vee$$
$$\ldots \vee \left(\neg\gamma_1^n \vee \ldots \vee \neg\gamma_{m_n}^n \vee \delta_1^n \vee \ldots \vee \delta_{l_n}^n\right). \tag{2}$$

Then $\mathcal{H}$ *is derivable in* GRM *if and only if* $\tau(\mathcal{H})$ *is a theorem* RM, *that is, the sequent* $\Rightarrow \tau(\mathcal{H})$ *is derivable in a (complete) sequent calculus for* RM *[7].*

To define hypersequent-based argumentation frameworks, it is not enough to simply take the hypersequent inference rules to create arguments. A new definition of arguments is required and sequent elimination rules should be turned into hypersequent elimination rules. This is what we will do in the next section.

4   Hypersequent-Based Argumentation

Given a logic $\mathsf{L} = \langle \mathcal{L}, \vdash \rangle$ with a sound and complete hypersequent calculus H, from now on, an *argument* (or an L-hyperargument) is an $\mathcal{L}$-hypersequent (i.e., whose components are $\mathcal{L}$-sequents) that is provable in H.[9] In the remainder of the paper, an argument based on a set $\mathcal{S}$ (of formulae in $\mathcal{L}$), is an L-hyperargument $\mathcal{H}$ such that $\Gamma \subseteq \mathcal{S}$ for every $\Gamma \in \mathsf{Supp}(\mathcal{H})$. We shall continue to denote by $\mathsf{Arg}_\mathsf{L}(\mathcal{S})$ the set of arguments that are based on $\mathcal{S}$.

As before, arguments are constructed by the inference rules of the hypersequent calculus under consideration (see Sect. 3). For the elimination rules, we continue to use the same notation: $\overline{\mathcal{H}}$ denotes the elimination of the hypersequent $\mathcal{H}$. The structure of such rules remains the same as before as well: the first hypersequent in the conditions of the rule is the attacking argument, the last hypersequent in the conditions is the attacked argument and the rest of the conditions are to be satisfied for the attack to take place.

Example 6. The elimination rules Def_H, Ucut_H and $\mathsf{ConUcut}_H$ are the hypersequent counterparts of the rules in Example 1. Let $\mathcal{G}, \mathcal{H}$ be two arguments, where $\mathsf{Supp}(\mathcal{H}) = \{\Delta_1, \ldots, \Delta_m\}$. We also assume that $\Delta_j \neq \emptyset$ for Def_H, $\emptyset \neq \Delta_j' \subseteq \Delta_j$ for Ucut_H, and $\bigcup_{i=1}^m \Delta_i \neq \emptyset$ for $\mathsf{ConUcut}_H$.

$$\frac{\mathcal{G} \quad \Rightarrow \mathsf{Conc}(\mathcal{G}) \supset \neg \bigwedge \Delta_j \quad \mathcal{H}}{\overline{\mathcal{H}}} \ \mathsf{Def}_H$$

$$\frac{\mathcal{G} \quad \Rightarrow \mathsf{Conc}(\mathcal{G}) \leftrightarrow \neg \bigwedge \Delta_j' \quad \mathcal{H}}{\overline{\mathcal{H}}} \ \mathsf{Ucut}_H$$

$$\frac{\Rightarrow \neg \bigwedge \bigcup_{i=1}^m \Delta_i \quad \mathcal{H}}{\overline{\mathcal{H}}} \ \mathsf{ConUcut}_H$$

---

[9] Since a sequent is a particular case of a hypersequent and hypersequent calculi generalize sequent calculi, arguments in the sense of the previous sections are particular cases of the arguments according to the new definition.

The notion of attack between hypersequents is the same as in Definition 3, except that sequents are replaced by hypersequents and the sequent calculus C is replaced by a hypersequent calculus H. Now, a hypersequent-based argumentation framework can be defined in a similar way as that of a sequent-based argumentation framework (cf. Definition 4).

Definition 8. *A hypersequent-based argumentation framework for a set of formulae S based on a logic $L = \langle \mathcal{L}, \vdash \rangle$ and a set AR of hypersequent elimination rules, is a pair $\mathcal{AF}_L(S) = \langle Arg_L(S), \mathcal{A} \rangle$, where $\mathcal{A} \subseteq Arg_L(S) \times Arg_L(S)$ and $(a_1, a_2) \in \mathcal{A}$ iff there is an $\mathcal{R} \in$ AR such that a_1 $\mathcal{R}$-attacks a_2.*

Given a hypersequent-based argumentation framework $\mathcal{AF}_L(S)$, Dung-style semantics are defined in an equivalent way to those in Definition 5.

Example 7. Let $\mathcal{AF}_{RM}(S)$ be a hypersequent-based argumentation framework for $S = \{p, q, \neg p \vee \neg q\}$, based on RM, with Ucut_H as the sole attack rule. From the axioms $p \Rightarrow p$ and $q \Rightarrow q$, by the Mingle Rule [Mi] (see Fig. 1) the sequent $p, q \Rightarrow p, q$ can be derived in GRM and by the Splitting Rule [Sp] the hypersequent $p \Rightarrow q \mid q \Rightarrow p$ is derivable in GRM as well. The hypersequent $p, q \Rightarrow p, q$ is Ucut_H-attacked by the axiom $\neg p \vee \neg q \Rightarrow \neg p \vee \neg q$, but the hypersequent $p \Rightarrow q \mid q \Rightarrow p$ is not Ucut_H-attacked by this axiom. However, both hypersequents are Ucut_H-attacked by the hypersequents $p, \neg p \vee \neg q \Rightarrow \neg q$ and $q, \neg p \vee \neg q \Rightarrow \neg p$.

Definition 9. *Given a hypersequent-based argumentation framework $\mathcal{AF}_L(S)$, the following entailment relations can be defined as in Definition 6: $S \hspace{0.5mm}\vdash^\cap_{H,\text{sem}} \phi$ $(S \hspace{0.5mm}\vdash^\cup_{H,\text{sem}} \phi)$ iff for every (some) extension $\mathcal{E} \in \text{Ext}_{\text{sem}}(\mathcal{AF}_L(S))$ there is an argument $\mathcal{H} \in \mathcal{E}$ such that $\text{Conc}(\mathcal{H}) = \phi$ and $\bigcup \text{Supp}(\mathcal{H}) \subseteq S$. The subscript H is omitted when this is clear from the context.*

5 Discussion of Some Properties

In this section we consider some useful properties of hypersequent-based argumentation. We begin by showing that in some cases this kind of argumentation overcomes a shortcoming of some other frameworks (including sequent-based ones) that under some completeness-based semantics (Definition 5) extensions may not always be consistent [2,15].

Example 8 (Based on Example 2 in [2]). Let $\mathcal{AF}_{CL}(S) = \langle Arg_{CL}(S), \mathcal{A} \rangle$, where $S = \{p, q, \neg p \vee \neg q, t\}$ and the attack rules are Def and/or Ucut. The following sequents are in $Arg_{CL}(S)$:

$a_1 = t \Rightarrow t \qquad a_2 = p \Rightarrow p \qquad a_3 = q \Rightarrow q \qquad a_4 = \neg p \vee \neg q \Rightarrow \neg p \vee \neg q$

$a_5 = p \Rightarrow \neg((\neg p \vee \neg q) \wedge q) \qquad a_6 = q \Rightarrow \neg((\neg p \vee \neg q) \wedge p)$

$a_7 = p, q \Rightarrow p \wedge q \qquad a_8 = \neg p \vee \neg q, q \Rightarrow \neg p \qquad a_9 = \neg p \vee \neg q, p \Rightarrow \neg q$

It can be shown that $\mathcal{E} = \{a_1, a_2, a_3, a_4, a_5, a_6\}$ is admissible in $\mathcal{AF}_{CL}(S)$, for either of the attack rules Def or Ucut. However, $\text{Concs}(\mathcal{E})$ is inconsistent.

Next, we show that the problem of the last example may be avoided by using a hypersequent-based framework[10].

Example 9 (Example 8, continued). Let $\mathcal{AF}'_{\mathsf{L}}(\mathcal{S}) = \langle \mathrm{Arg}'_{\mathsf{L}}(\mathcal{S}), \mathcal{A}' \rangle$ be a hypersequent-based argumentation framework (Definition 8) for $\mathsf{L} \in \{\mathsf{CL}, \mathsf{RM}\}$, the attack rules Def_H and Ucut_H, and $\mathcal{S}$ as in Example 8. With the possibility of splitting components, we get $\mathrm{Arg}'_{\mathsf{L}}(\mathcal{S}) \supseteq \mathrm{Arg}_{\mathsf{L}}(\mathcal{S}) \cup \{a_{10}, a_{11}, a_{12}\}$ where:

$$a_{10} = \neg p \lor \neg q \Rightarrow \neg p \mid q \Rightarrow \neg p \qquad a_{11} = \neg p \lor \neg q \Rightarrow \neg q \mid p \Rightarrow \neg q$$
$$a_{12} = p \Rightarrow p \land q \mid q \Rightarrow p \land q$$

See Fig. 2 for a graphical representation of the situation in CL (the graph for RM is similar). The dashed graph (nodes and arrows) represents Example 8, the ordinary sequent-based argumentation graph. When generalizing to hypersequents, the three solid nodes and all solid arrows are added.

The following three sets of arguments are part of different complete extensions: $\mathcal{E}_1 = \{a_1, a_2, a_3, a_5, a_6, a_7, a_{12}\}$, $\mathcal{E}_2 = \{a_1, a_3, a_4, a_6, a_8, a_{10}\}$ and $\mathcal{E}_3 = \{a_1, a_2, a_4, a_5, a_9, a_{11}\}$. Furthermore, although $\mathcal{E} = \{a_1, a_2, a_3, a_4, a_5, a_6\}$ is conflict-free, a_2, for example, is attacked by a_{10}. In order to defend a_2, $\mathcal{E}$ must be extended with a hypersequent like a_7, a_9, a_{11} or a_{12}, however, then the new extension is not conflict-free anymore. Hence $\mathcal{E}$ is not part of a complete extension. Additionally, each extension contains a_1, therefore, the system $\mathcal{AF}'_{\mathsf{L}}(\mathcal{S})$ does not only avoid inconsistent extensions, it provides extensions from which the free arguments follow[11].

In the next subsection it will be shown, among others, that the outcome of the last example is not a coincidence.

5.1   Rationality Postulates

In this section we show that, for a hypersequent-based argumentation framework with the attack rules Def_H and Ucut_H, and core logic RM, the logic-based rationality postulates in [1] hold.

Definition 10. *Let $\mathsf{L} = \langle \mathcal{L}, \vdash \rangle$ be a logic and let $\mathcal{T}$ be a set of $\mathcal{L}$-formulae, where $\mathcal{L}$ contains the connectives $\neg$ and $\land$.*

- *The closure of $\mathcal{T}$ is denoted by $\mathsf{CN}_{\mathsf{L}}(\mathcal{T})$ (thus, $\mathsf{CN}_{\mathsf{L}}(\mathcal{T}) = \{\phi \mid \mathcal{T} \vdash \phi\}$).*
- *$\mathcal{T}$ is consistent (for $\vdash$), if there are no formulae $\phi_1, \ldots, \phi_n \in \mathcal{T}$ such that $\vdash \neg \bigwedge_{i=1}^{n} \phi_i$[12].*

---

[10] Intuitively, this is so due to the possibility of *splitting* hypersequents into different components. A formal justification will be given in the next subsection.

[11] Where free arguments are those arguments that are based only on premises that are not involved in minimally inconsistent subsets of $\mathcal{S}$ (see Definition 10).

[12] Note that if $\mathcal{T}$ is consistent, then so are $\mathsf{CN}_{\mathsf{L}}(\mathcal{T})$ and $\mathcal{T}'$ for every $\mathcal{T}' \subseteq \mathcal{T}$. If $\mathcal{T}$ is inconsistent, then so is every superset of $\mathcal{T}$.

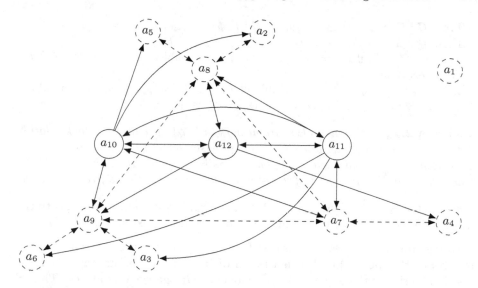

Fig. 2. Part of the hypersequent-based argumentation graph for $\mathcal{S} = \{p, q, \neg p \vee \neg q, t\}$, with defeat as attack rule. The dashed graph is part of the ordinary sequent-based graph, the solid nodes and arrows become available when generalizing to hypersequents.

– *A subset $\mathcal{C}$ of $\mathcal{T}$ is a* minimal conflict *of $\mathcal{T}$ (w.r.t. $\vdash$), if $\mathcal{C}$ is inconsistent and for any $c \in \mathcal{C}$, the set $\mathcal{C} \setminus \{c\}$ is consistent. We denote by $\mathsf{Free}(\mathcal{T})$ the set of formulae in $\mathcal{T}$ that are not part of any minimal conflict of $\mathcal{T}$.*

Let $\mathcal{AF}_{\mathsf{L}}(\mathcal{S}) = \langle \mathrm{Arg}_{\mathsf{L}}(\mathcal{S}), \mathcal{A} \rangle$ be a hypersequent-based argumentation framework and let $\mathcal{H}, \mathcal{H}' \in \mathrm{Arg}_{\mathsf{L}}(\mathcal{S})$ such that $\mathcal{H} = \Gamma_1 \Rightarrow \phi_1 \mid \ldots \mid \Gamma_n \Rightarrow \phi_n$ and $\mathcal{H}' = \Gamma_1' \Rightarrow \phi_1' \mid \ldots \mid \Gamma_m' \Rightarrow \phi_m'$. Then $\mathcal{H}'$ is a *sub-argument* of $\mathcal{H}$ if for each $i \in \{1, \ldots, m\}$ there is a $j \in \{1, \ldots, n\}$ such that $\Gamma_i' \subseteq \Gamma_j$. The set of sub-arguments of $\mathcal{H}$ is denoted by $\mathsf{Sub}(\mathcal{H})$.

Definition 11. *Let $\mathcal{AF}_{\mathsf{L}}(\mathcal{S}) = \langle \mathrm{Arg}_{\mathsf{L}}(\mathcal{S}), \mathcal{A} \rangle$ be an argumentation framework for the logic $\mathsf{L} = \langle \mathcal{L}, \vdash \rangle$, the set $\mathcal{S}$ of $\mathcal{L}$-formulae and a fixed (set of) semantics* sem. *We say that $\mathcal{AF}_{\mathsf{L}}(\mathcal{S})$ has the properties below (for* sem*), if they are satisfied for every $\mathcal{E} \in \mathsf{Ext}_{\mathsf{sem}}(\mathcal{AF}_{\mathsf{L}}(\mathcal{S}))$.*

– closure of extensions: $\mathsf{Concs}(\mathcal{E}) = \mathsf{CN}_{\mathsf{L}}(\mathsf{Concs}(\mathcal{E}))$.
– closure under sub-arguments: *if $\mathcal{H} \in \mathcal{E}$ then* $\mathsf{Sub}(\mathcal{H}) \subseteq \mathcal{E}$.
– consistency: $\mathsf{Concs}(\mathcal{E})$ *is consistent.*
– exhaustiveness: *For every $\mathcal{H} \in Arg_{\mathsf{L}}(\mathcal{S})$ such that $\bigcup \mathsf{Supp}(\mathcal{H}) \cup \{\mathsf{Conc}(\mathcal{H})\} \subseteq \mathsf{Concs}(\mathcal{E})$, $\mathcal{H} \in \mathcal{E}$.*
– free precedence: $Arg_{\mathsf{L}}(\mathsf{Free}(\mathcal{S})) \subseteq \mathcal{E}$.

Note 1. For proving the above postulates, we shall use (sometimes implicitly) the following admissible rules of GRM:

– Transitivity: if $\mathcal{G}_1 \mid \Gamma \Rightarrow \phi_1 \mid \mathcal{H}_1$ and $\mathcal{G}_2 \mid \phi_1 \Rightarrow \phi_2 \mid \mathcal{H}_2$ are derivable, then $\mathcal{G}_1 \mid \mathcal{G}_2 \mid \Gamma \Rightarrow \phi_2 \mid \mathcal{H}_1 \mid \mathcal{H}_2$ is derivable.

- From $\mathcal{G} \mid \Gamma \Rightarrow \phi \supset \psi, \Delta \mid \mathcal{H}$ derive $\mathcal{G} \mid \Gamma, \phi \Rightarrow \psi, \Delta \mid \mathcal{H}$.
- From $\mathcal{G} \mid \Delta \Rightarrow \phi \mid \mathcal{H}$ derive $\mathcal{G} \mid \ \Rightarrow \neg\phi \supset \neg \bigwedge \Delta \mid \mathcal{H}$.
- For any $\Gamma' \subseteq \Gamma$, if $\mathcal{G} \mid \ \Rightarrow \phi \supset \neg \bigwedge \Gamma' \mid \mathcal{H}$ is derivable then $\mathcal{G} \mid \ \Rightarrow \phi \supset \neg \bigwedge \Gamma \mid$ $\mathcal{H}$ is derivable.
- $\Gamma_1 \Rightarrow \phi_1 \mid \ldots \mid \Gamma_n \Rightarrow \phi_n$ is derivable iff $\Gamma_1, \ldots, \Gamma_n \Rightarrow \phi_1, \ldots, \phi_n$ is derivable.
- $\phi_1 \vee \ldots \vee \phi_n \Rightarrow \phi_1 \mid \ldots \mid \phi_1 \vee \ldots \vee \phi_n \Rightarrow \phi_n$ is derivable.

Theorem 2. *Any argumentation framework $\mathcal{AF}_{\mathsf{RM}}(\mathcal{S})$ with the attack relation Def_H or $Ucut_H$, and under any completeness-based semantics, satisfies closure of extensions, closure under sub-arguments, consistency and exhaustiveness. Moreover, when $ConUcut_H$ is part of the attack rules, it also satisfies free precedence.*

Proof. Let $\mathcal{AF}_{\mathsf{RM}}(\mathcal{S}) = \langle \mathrm{Arg}_{\mathsf{RM}}(\mathcal{S}), \mathcal{A} \rangle$ be an argumentation framework, with the attack rules Def_H and/or $Ucut_H$ and let $\mathcal{E}$ be a complete extension of $\mathcal{AF}_{\mathsf{RM}}(\mathcal{S})$.

Sub-argument closure: For both Def_H and $Ucut_H$ it can be shown that any attacker of $\mathcal{H}' \in \mathsf{Sub}(\mathcal{H})$ is also an attacker of $\mathcal{H}$. If $\mathcal{H} \in \mathcal{E}$, for any completeness-based extension $\mathcal{E}$ there is a $\mathcal{G} \in \mathcal{E}$ that defends $\mathcal{H}$ against this attack. Thus $\mathcal{E}$ defends $\mathcal{H}'$ as well. Therefore, $\mathcal{H}' \in \mathcal{E}$.

Closure of extensions: Showing that $\mathsf{Concs}(\mathcal{E}) \subseteq \mathsf{CN}_{\mathsf{RM}}(\mathsf{Concs}(\mathcal{E}))$ is trivial. For the other direction, assume that $\phi \in \mathsf{CN}_{\mathsf{RM}}(\mathsf{Concs}(\mathcal{E}))$. Then there are arguments $\mathcal{H}_1, \ldots, \mathcal{H}_n \in \mathcal{E}$ such that $\mathcal{H}_i = \Gamma_1^i \Rightarrow \psi_1^i \mid \ldots \mid \Gamma_{m_i}^i \Rightarrow \psi_{m_i}^i$, with $\phi_i = \psi_1^i \vee \ldots \vee \psi_{m_i}^i$ and $\phi_1, \ldots, \phi_n \vdash_{\mathsf{RM}} \phi$.

It can be shown that the argument $\mathcal{H}' = \bigwedge_{k=1}^{n} \bigwedge_{j=1}^{m_k} \bigwedge \Gamma_j^k \Rightarrow \phi_1 \wedge \ldots \wedge \phi_n$ is derivable in GRM. By transitivity and splitting we get that $\mathcal{H} = \Gamma_1^1 \Rightarrow \phi \mid \ldots \mid$ $\Gamma_{m_1}^1 \Rightarrow \phi \mid \ldots \mid \Gamma_1^n \Rightarrow \phi \mid \ldots \mid \Gamma_{m_n}^n \Rightarrow \phi$ is provable in GRM. For both attack rules Def_H and $Ucut_H$, any attacker of $\mathcal{H}$ is an attacker of one of the arguments $\mathcal{H}_1, \ldots, \mathcal{H}_n$. Hence $\mathcal{H} \in \mathcal{E}$, and so $\phi \in \mathsf{Concs}(\mathcal{E})$.

Consistency: Assume, towards a contradiction, that $\mathsf{Concs}(\mathcal{E})$ is not consistent. Then there are $\phi_1, \ldots, \phi_n \in \mathsf{Concs}(\mathcal{E})$ such that $\Rightarrow \neg \bigwedge_{j=1}^{n} \phi_j$ is derivable in GRM. Let $\psi = \phi_1 \wedge \ldots \wedge \phi_n$. Furthermore, like the proof of closure, there are arguments $\mathcal{H}_1, \ldots, \mathcal{H}_n \in \mathcal{E}$, such that $\mathcal{H}_i = \Gamma_1^i \Rightarrow \psi_1^i \mid \ldots \mid \Gamma_{m_i}^i \Rightarrow \psi_{m_i}^i$ and $\phi_i = \psi_1^i \vee \ldots \vee \psi_{m_i}^i$. From these, arguments $\mathcal{H}_i' = \Gamma_1^i, \ldots, \Gamma_{m_i}^i \Rightarrow \phi_i$, for each $i \in \{1, \ldots, n\}$, can be derived. By applying $(\Rightarrow\wedge)$ to the $\mathcal{H}_i'$s, we drive $\Gamma_1^1, \ldots, \Gamma_{m_1}^1, \ldots, \Gamma_1^n, \ldots, \Gamma_{m_n}^n \Rightarrow \psi$.

Then, for each $j \in \{1, \ldots, n\}$ and $k \in \{1, \ldots, m_j\}$, $\neg\psi, \Gamma_1^1, \ldots, \Gamma_{m_1}^1, \ldots, \Gamma_1^n$, $\ldots, \Gamma_{m_n}^n \Rightarrow \neg \bigwedge \Gamma_k^j$ is derivable, where Γ_k^j is taken out of $\Gamma_1^1, \ldots, \Gamma_{m_1}^1, \ldots, \Gamma_1^n$, $\ldots, \Gamma_{m_n}^n$. By transitivity from $\Rightarrow \neg\psi$ and splitting, $\mathcal{G} = \Gamma_1^1 \Rightarrow \neg \bigwedge \Gamma_k^j \mid \ldots \mid$ $\Gamma_{m_1}^1 \Rightarrow \neg \bigwedge \Gamma_k^j \mid \ldots \mid \Gamma_1^n \Rightarrow \neg \bigwedge \Gamma_k^j \mid \ldots \mid \Gamma_{m_n}^n \Rightarrow \neg \bigwedge \Gamma_k^j$ is derivable. Note that, for both attack rules Def_H and $Ucut_H$, any attacker of $\mathcal{G}$ is an attacker of one of the arguments $\mathcal{H}_1, \ldots, \mathcal{H}_n$, therefore $\mathcal{G} \in \mathcal{E}$. However, $\mathcal{G}$ attacks (defeats/undercuts) $\mathcal{H}_j$, a contradiction to the assumption that $\mathcal{E}$ is conflict-free. Thus $\mathsf{Concs}(\mathcal{E})$ is consistent.

Exhaustiveness: Let $\mathcal{H} \in \mathrm{Arg}_{\mathsf{RM}}(\mathcal{S})$ be an argument such that $\bigcup \mathsf{Supp}(\mathcal{H}) \cup \{\mathsf{Conc}(\mathcal{H})\} \subseteq \mathsf{Concs}(\mathcal{E})$. It easily follows that $\mathcal{E} \cup \{\mathcal{H}\}$ is conflict-free. Assume that some $\mathcal{G} = \Delta_1 \Rightarrow \psi_1 \mid \ldots \mid \Delta_n \Rightarrow \psi_n \in \mathsf{Arg}_{\mathsf{RM}}(\mathcal{S})$ defeats $\mathcal{H}$ (the case for undercut is similar and left to the reader). Then $\Rightarrow \mathsf{Conc}(\mathcal{G}) \supset \neg \bigwedge \Gamma$ is derivable in GRM, for some $\Gamma \in \mathsf{Supp}(\mathcal{H})$. Let $\Gamma = \{\gamma_1, \ldots, \gamma_m\}$. Then there are $\mathcal{H}_1, \ldots, \mathcal{H}_m \in \mathcal{E}$ such that $\mathsf{Conc}(\mathcal{H}_j) = \gamma_j$ and $\bigcup \mathsf{Supp}(\mathcal{H}_j) = \{\delta_1^j, \ldots \delta_{k_j}^j\}$ $(1 \leq j \leq m)$. By Theorem 1, $\delta_1^j, \ldots, \delta_{k_j}^j \vdash_{\mathsf{RM}} \gamma_j$, thus $\delta_1^1, \ldots, \delta_{k_1}^1, \ldots, \delta_1^m, \ldots,$ $\delta_{k_m}^m \vdash_{\mathsf{RM}} \gamma_1 \wedge \ldots \wedge \gamma_m$, and so $\neg \bigwedge \Gamma, \delta_1^1, \ldots, \delta_{k_1}^1, \ldots, \delta_1^m, \ldots, \delta_{k_{m-1}}^m \vdash_{\mathsf{RM}} \neg \delta_{k_m}^m$.

Now, by transitivity from $\mathsf{Conc}(\mathcal{G}) \Rightarrow \neg \bigwedge \Gamma$, Theorem 1, and splitting, we have that $\mathcal{G}' = \Delta_1 \Rightarrow \neg \delta_{k_m}^m \mid \ldots \mid \Delta_m \Rightarrow \neg \delta_{k_m}^m \mid \delta_1^1 \Rightarrow \neg \delta_{k_m}^m \mid \ldots \mid \delta_{k_{m-1}}^m \Rightarrow$ $\neg \delta_{k_m}^m \in \mathsf{Arg}_{\mathsf{RM}}(\mathcal{S})$. But then $\mathcal{G}'$ defeats $\mathcal{H}_m \in \mathcal{E}$, thus there is some $\mathcal{H}^* \in \mathcal{E}$ which defeats $\mathcal{G}'$. This attack has to be on some Δ_i, $i \in \{1, \ldots, n\}$, otherwise $\mathcal{E}$ would not be conflict-free. Hence $\mathcal{H}^*$ defeats $\mathcal{G}$ as well.

Since, by assumption, $\mathcal{E}$ is complete, $\mathcal{E} \cup \{\mathcal{H}\}$ is conflict-free and $\mathcal{E}$ defends $\mathcal{H}$, it follows that $\mathcal{H} \in \mathcal{E}$.

Free precedence: Suppose that $\mathsf{ConUcut}_H$ is among the attack rules in $\mathcal{AF}_{\mathsf{RM}}(\mathcal{S})$ as well. It can be shown that Def_H, Ucut_H and $\mathsf{ConUcut}_H$ are *conflict-dependent* in the sense of [1], that is: if $\mathcal{G}, \mathcal{H} \in \mathsf{Arg}_{\mathsf{RM}}(\mathcal{S})$ such that $\mathcal{G}$ attacks $\mathcal{H}$, then $\bigcup \mathsf{Supp}(\mathcal{G}) \cup \bigcup \mathsf{Supp}(\mathcal{H})$ is inconsistent.

Assume that some $\mathcal{G} \in \mathsf{Arg}_{\mathsf{RM}}(\mathcal{S})$ attacks an argument $\mathcal{H} \in \mathsf{Arg}_{\mathsf{RM}}(\mathsf{Free}(\mathcal{S}))$. Since each of the considered attack rules is conflict-dependent, $\bigcup \mathsf{Supp}(\mathcal{H}) \cup \bigcup \mathsf{Supp}(\mathcal{G})$ is inconsistent. However, $\bigcup \mathsf{Supp}(\mathcal{H}) \subseteq \mathsf{Free}(\mathcal{S})$, thus $\mathcal{G}$ has an inconsistent support set. Then there is an argument $\Rightarrow \neg \bigwedge \mathsf{Supp}(\mathcal{G}) \in \mathcal{E}$ that attacks $\mathcal{G}$ via the $\mathsf{ConUcut}_H$ rule. Since any attacker of $\mathcal{H}$ is counter-attacked by $\mathcal{E}$, it follows that $\mathcal{E}$ defends $\mathcal{H}$, and since $\mathcal{E}$ is complete, $\mathcal{H} \in \mathcal{E}$.

We have shown that $\mathcal{AF}_{\mathsf{RM}}(\mathcal{S})$, for the given attack rules, satisfies the five postulates under complete semantics. From this it follows (see, e.g., [1, Prop. 26]) that $\mathcal{AF}_{\mathsf{RM}}(\mathcal{S})$ satisfies the five postulates also under grounded, preferred and stables semantics. $\qquad \square$

Consider the following weakening of the definition of sub-arguments:

Definition 12. *We say that $\mathcal{H}'$ is a* weak sub-argument *of $\mathcal{H}$, if $\bigcup \mathsf{Supp}(\mathcal{H}') \subseteq \bigcup \mathsf{Supp}(\mathcal{H})$. We denote by $\mathsf{WSub}(\mathcal{H})$ the set of all weak sub-arguments of $\mathcal{H}$.*

Clearly, any sub-argument of $\mathcal{H}$ is in particular a weak sub-argument of $\mathcal{H}$. Interestingly, as the next proposition shows, closure of extensions and exhaustiveness imply closure under weak sub-arguments (and so closure under sub-argument).

Proposition 1. *Any argumentation framework $\mathcal{AF}_{\mathsf{RM}}(\mathcal{S})$ that satisfies closure of extensions and exhaustiveness also satisfies closure under weak sub-arguments.*

Proof. Let $\mathcal{AF}_{\mathsf{RM}}(\mathcal{S})$ be a hypersequent-based argumentation framework for the core logic RM and set of formulas $\mathcal{S}$ that satisfies closure of extensions and exhaustiveness. Suppose that $\mathcal{H} \in \mathcal{E}$ for some $\mathcal{E} \in \mathsf{Ext}_{\mathsf{sem}}(\mathcal{AF}_{\mathsf{RM}}(\mathcal{S}))$, and let $\mathcal{H}' \in \mathsf{WSub}(\mathcal{H})$. Then $\bigcup \mathsf{Supp}(\mathcal{H}') \subseteq \bigcup \mathsf{Supp}(\mathcal{H})$. Note that for each $\phi \in \bigcup \mathsf{Supp}(\mathcal{H})$, $\phi \Rightarrow \phi \in \mathcal{E}$ (since every attacker of $\phi \Rightarrow \phi$ is also an attacker of $\mathcal{H}$ and $\mathcal{E}$ is complete). Thus (†) $\bigcup \mathsf{Supp}(\mathcal{H}') \subseteq \mathsf{Concs}(\mathcal{E})$. Furthermore, since $\mathsf{Supp}(\mathcal{H}') \vdash_{\mathsf{RM}} \mathsf{Conc}(\mathcal{H}')$, by the monotonicity of $\vdash$ also $\mathsf{Supp}(\mathcal{H}) \vdash_{\mathsf{RM}} \mathsf{Conc}(\mathcal{H}')$ and by closure (‡) $\mathsf{Conc}(\mathcal{H}') \in \mathsf{Concs}(\mathcal{E})$. Thus $\mathcal{H}' \in \mathcal{E}$ by exhaustiveness in view of (†) and (‡). □

Note 2. Consider a hypersequent variant LK_H of the sequent calculus LK for classical propositional logic. This calculus would allow for internal weakening in addition to all the rules of GRM. Then all of the above proofs for the postulates hold also for classical logic with the calculus LK_H.

5.2 Crash-Resistance and Non-interference

Two additional postulates were introduced in [14] concerning crash-resistance, the problem that a system collapses when it is based on inconsistent information. For defining these postulates, some definitions and notations are necessary.

Let $\mathcal{AF}_{\mathsf{L}}(\mathcal{S}) = \langle \mathsf{Arg}_{\mathsf{L}}(\mathcal{S}), \mathcal{A} \rangle$ be an argumentation framework for the logic $\mathsf{L} = \langle \mathcal{L}, \vdash \rangle$ and a set $\mathcal{S}$ of $\mathcal{L}$-formulae.

- We denote by $\mathsf{Atoms}(\mathcal{S})$ the set of atoms that occur in the formulae in $\mathcal{S}$ and by $\mathsf{Atoms}(\mathcal{L})$ the set of all the atoms of the language.
- Sets $\mathcal{S}$, $\mathcal{T}$ of formulae are *syntactically disjoint*, if $\mathsf{Atoms}(\mathcal{S}) \cap \mathsf{Atoms}(\mathcal{T}) = \emptyset$.

Definition 13. *Let $\sim$ be an entailment relation for $\mathcal{L}$. A set $\mathcal{S}'$ of $\mathcal{L}$-formulae is called* contaminating *(with respect to $\sim$), if for any set $\mathcal{S}^* \subseteq \mathcal{L}$ such that $\mathcal{S}'$ and $\mathcal{S}^*$ are syntactically disjoint, and for every $\mathcal{L}$-formula ϕ, it holds that $\mathcal{S}' \hspace{-0.3em}\sim\hspace{-0.3em} \phi$ if and only if $\mathcal{S}' \cup \mathcal{S}^* \hspace{-0.3em}\sim\hspace{-0.3em} \phi$.*

Definition 14. *Let $\mathcal{L}$ be a propositional language and $\sim$ an entailment relation for $\mathcal{L}$. Then $\sim$ satisfies*

- non-interference: *if and only if for every syntactically disjoint sets $\mathcal{S}_1$, $\mathcal{S}_2$ of $\mathcal{L}$-formulae and any $\mathcal{L}$-formula ϕ such that $\mathsf{Atoms}(\phi) \subseteq \mathsf{Atoms}(\mathcal{S}_1)$, $\mathcal{S}_1 \hspace{-0.3em}\sim\hspace{-0.3em} \phi$ if and only if $\mathcal{S}_1 \cup \mathcal{S}_2 \hspace{-0.3em}\sim\hspace{-0.3em} \phi$;*
- crash-resistance: *if and only if there is no set $\mathcal{S}$ of $\mathcal{L}$-formulae that is contaminating w.r.t. $\sim$.*

For proving the above postulates, we need the next lemma. Its proof is partially based on [5, Lemma 1] and [16, Lemma 15], but omitted due to space restrictions.

Lemma 1. *Let $\mathcal{AF}_{\mathsf{RM}}(\mathcal{S})$ be a hypersequent-based argumentation framework for $\mathcal{S}$ (Definition 8) whose core logic is RM. The following are equivalent:*

(a) $\mathcal{E} \in \mathsf{Ext}_{\mathsf{prf}}(\mathcal{AF}_{RM}(\mathcal{S}))$;
(b) $\mathcal{E} \in \mathsf{Ext}_{\mathsf{stb}}(\mathcal{AF}_{RM}(\mathcal{S}))$;
(c) $\mathcal{E} = Arg_{RM}(\mathcal{S}')$, *where $\mathcal{S}'$ is a $\subseteq$-maximally consistent subset of $\mathcal{S}$.*

Theorem 3. *Let $\mathcal{AF}_{RM}(\mathcal{S})$ be a hypersequent-based argumentation framework for the logic RM, the attack rules Def_H and/or $Ucut_H$, and a set of formulae $\mathcal{S}$. Let also $\pi \in \{\cap, \cup\}$, and sem a completeness-based semantics. Then the induced entailment $\mathord{\vdash}^{\pi}_{H,\mathsf{sem}}$ (Definition 9) satisfies non-interference.*

Proof (outline). The structure of the proof is roughly based on the proofs in [30]. In what follows we abbreviate $\mathord{\vdash}^{\pi}_{H,\mathsf{sem}}$ by $\mathord{\vdash}^{\pi}$.

Let $\mathcal{AF}_{RM}(\mathcal{S})$ be some hypersequent-based argumentation framework for the logic RM, with the attack rules Def_H and/or $Ucut_H$ and a set of formulae $\mathcal{S}$. Consider two syntactically disjoint sets $\mathcal{S}_1, \mathcal{S}_2 \subseteq \mathcal{S}$ and let $\mathcal{S}' = \mathcal{S}_1 \cup \mathcal{S}_2$. For any $\mathsf{S} \subseteq \mathsf{Arg}_{RM}(\mathcal{S})$, let $\mathcal{D}_{\mathcal{AF}_{RM}(\mathcal{S})}(\mathsf{S}) = \{\mathcal{H} \in \mathsf{Arg}_{RM}(\mathcal{S}) \mid \mathsf{S} \text{ defends } \mathcal{H}\}$. Then, by Lemma 1 and the fact that RM satisfies the basic relevance criterion [9], the following points can be shown for complete, preferred and stable semantics (proofs are omitted due to space restrictions):

1. if $\mathcal{E} \in \mathsf{Ext}_{\mathsf{sem}}(\mathcal{AF}_{RM}(\mathcal{S}'))$, then $\mathcal{E} \cap \mathsf{Arg}_{RM}(\mathcal{S}_1) \in \mathsf{Ext}_{\mathsf{sem}}(\mathcal{AF}_{RM}(\mathcal{S}_1))$;
2. if $\mathcal{E}_1 \in \mathsf{Ext}_{\mathsf{sem}}(\mathcal{AF}_{RM}(\mathcal{S}_1))$ and $\mathcal{E}_2 \in \mathsf{Ext}_{\mathsf{sem}}(\mathcal{AF}_{RM}(\mathcal{S}_2))$, then $\mathcal{D}_{\mathcal{AF}_{RM}(\mathcal{S}')}(\mathcal{E}_1 \cup \mathcal{E}_2) \in \mathsf{Ext}_{\mathsf{sem}}(\mathcal{AF}_{RM}(\mathcal{S}'))$.

Let ϕ be a formula with $\mathsf{Atoms}(\phi) \subseteq \mathsf{Atoms}(\mathcal{S}_1)$. We show that $\mathcal{S}_1 \mathord{\vdash}^{\cap} \phi$ if and only if $\mathcal{S}' \mathord{\vdash}^{\cap} \phi$ (the proof for $\mathord{\vdash}^{\cup}$ is left to the reader).

$\Rightarrow$ Assume that $\mathcal{S}_1 \mathord{\vdash}^{\cap}\phi$ but $\mathcal{S}' \mathord{\not\vdash}^{\cap}\phi$. Thus, there is some $\mathcal{E} \in \mathsf{Ext}_{\mathsf{sem}}(\mathcal{AF}_{RM}(\mathcal{S}'))$, such that there is no argument $\mathcal{H} \in \mathcal{E}$ with $\mathsf{Conc}(\mathcal{H}) = \phi$. By Item 1 above $\mathcal{E} \cap \mathsf{Arg}_{RM}(\mathcal{S}_1) \in \mathsf{Ext}_{\mathsf{sem}}(\mathcal{AF}_{RM}(\mathcal{S}_1))$, a contradiction to $\mathcal{S}_1 \mathord{\vdash}^{\cap}\phi$.

$\Leftarrow$ Assume that $\mathcal{S}' \mathord{\vdash}^{\cap}\phi$ but $\mathcal{S}_1 \mathord{\not\vdash}^{\cap}\phi$. Thus, there is some $\mathcal{E} \in \mathsf{Ext}_{\mathsf{sem}}(\mathcal{AF}_{RM}(\mathcal{S}_1))$ such that there is no argument $\mathcal{H} \in \mathcal{E}$ with $\mathsf{Conc}(\mathcal{H}) = \phi$. By the basic relevance criterion [3], if $\mathcal{E}' \in \mathsf{Ext}_{\mathsf{sem}}(\mathcal{AF}_{RM}(\mathcal{S}_2))$ (in [16] it is shown that there is at least one such extension), there is no argument $\mathcal{H} \in \mathcal{E}'$ with $\mathsf{Conc}(\mathcal{H}) = \phi$ either. Thus, by Item 2 above, $\mathcal{D}_{\mathcal{AF}_{RM}(\mathcal{S}')}(\mathcal{E} \cup \mathcal{E}') \in \mathsf{Ext}_{\mathsf{sem}}(\mathcal{AF}_{RM}(\mathcal{S}'))$. By definition of $\mathcal{D}_{\mathcal{AF}_{RM}(\mathcal{S}')}$, there is no argument $\mathcal{H} \in \mathcal{D}_{\mathcal{AF}_{RM}(\mathcal{S}')}(\mathcal{E} \cup \mathcal{E}')$ with $\mathsf{Conc}(\mathcal{H}) = \phi$, a contradiction to $\mathcal{S}' \mathord{\vdash}^{\cap}\phi$.

It follows that $\mathord{\vdash}^{\cup}_{H,\mathsf{sem}}$ and $\mathord{\vdash}^{\cap}_{H,\mathsf{sem}}$, for sem $\in \{\mathsf{gr}, \mathsf{cmp}, \mathsf{prf}, \mathsf{stb}\}$, satisfy non-interference. $\square$

Theorem 4. *Let $\mathcal{AF}_{RM}(\mathcal{S})$ be a hypersequent-based argumentation framework for the logic RM, the attack rules Def_H and/or $Ucut_H$, and a set of formulae $\mathcal{S}$. Let also $\pi \in \{\cap, \cup\}$, and sem a completeness-based semantics. Then $\mathord{\vdash}^{\pi}_{H,\mathsf{sem}}$ satisfies crash-resistance.*

For the proof, we recall the following notion from [14]:

– Let AT be a set of atoms. We denote by $\mathcal{S}_{|\mathsf{AT}}$ the set of formulae in $\mathcal{S}$ that contain only atoms from AT. For a set $\mathcal{F}$ of sets of $\mathcal{L}$-formulae, we denote: $\mathcal{F}_{|\mathsf{AT}} = \{\mathcal{S}_{|\mathsf{AT}} \mid \mathcal{S} \in \mathcal{F}\}$.

– According to [14], a logic $\mathsf{L} = \langle \mathcal{L}, \vdash \rangle$ is called *non-trivial*, if for every non-empty set $\mathsf{AT} \subseteq \mathsf{Atoms}(\mathcal{L})$ there are sets $\mathcal{S}_1, \mathcal{S}_2$ of $\mathcal{L}$-formulae such that $\mathsf{Atoms}(\mathcal{S}_1) = \mathsf{Atoms}(\mathcal{S}_2) = \mathsf{AT}$ but $\mathsf{CN}_\mathsf{L}(\mathcal{S}_1)_{|\mathsf{AT}} \neq \mathsf{CN}_\mathsf{L}(\mathcal{S}_2)_{|\mathsf{AT}}$.

Proof (sketch). By Theorem 3, for every sem $\in \{\mathsf{gr}, \mathsf{cmp}, \mathsf{prf}, \mathsf{stb}\}$ the entailments $\vdash^{\cap}_{H,\mathsf{sem}}$ and $\vdash^{\cup}_{H,\mathsf{sem}}$ satisfy non-interference. Thus, since RM is non-trivial, crash-resistance follows from [14, Theorem 1].   □

Note 3. The basic relevance criterion [3] is a primary property of RM, used in the proof of Theorem 3 for showing non-interference (and so also for obtaining crash resistance in the proof of Theorem 4). We note that, although classical logic does not satisfy the basic relevance criterion, it is a uniform logic (i.e., for every two sets of formulae $\mathcal{S}, \mathcal{S}'$ and a formula ϕ, if $\mathcal{S}, \mathcal{S}' \vdash \phi$ and $\mathcal{S}'$ is a $\vdash$-consistent theory that is syntactically disjoint from $\mathcal{S} \cup \{\phi\}$, then $\mathcal{S} \vdash \phi$). By assuming that $\mathsf{ConUcut}_H$ is part of the attack rules, Items 1 and 2 in the proof of Theorem 3 still hold. In the $\Leftarrow$ direction of the proof the use of the basic relevance criterion can be replaced by the uniformity of the core logic and the fact that no arguments with inconsistent support set will be part of any extension. Hence, the proofs of Theorems 3 and 4 can be adjusted also for the case that, e.g., classical logic is the core logic.

6   Conclusion

Hypersequent-based argumentation, like sequent-based argumentation, avoids some limitations of other approaches to logic-based argumentation (e.g., [12]), where the support set of an argument has to be consistent and minimal. Furthermore, it incorporates any logic with a corresponding sound and complete (hyper)sequent calculus, and allows a great flexibility in the specification of the attack rules.

In this paper we have examined hypersequent frameworks that are based on the logic RM and with defeat and/or undercut as the attack rule. It was shown that such frameworks satisfy the logic-based rationality postulates from [1,13] and non-interference and crash-resistance from [14]. Moreover, a problem raised in [15] (and further discussed in [2]), in which complete extensions may not be consistent, is avoided.

A comparison to related literature has to be postponed. However, it is worth noting that our non-interference result is more general than the one in [30], where this is only proven for frameworks under complete semantics.

Future research directions include the extension of dynamic proof theory [6] from sequent-based frameworks to hypersequent-based ones. Moreover, we plan to investigate the integration of priorities among arguments and the use of assumptions, such as default assumptions [20] and assumptions taken in adaptive logics [10,28], for further extending the expressive power of hypersequent-based argumentation.

References

1. Amgoud, L.: Postulates for logic-based argumentation systems. Int. J. Approx. Reason. **55**(9), 2028–2048 (2014)
2. Amgoud, L., Besnard, P.: Logical limits of abstract argumentation frameworks. J. Appl. Non-Class. Logics **23**(3), 229–267 (2013)
3. Anderson, A., Belnap, N.: Entailment: The Logic of Relevance and Necessity, vol. 1. Princeton University Press, Princeton (1975)
4. Arieli, O., Straßer, C.: Sequent-based logical argumentation. Argum. Comput. **6**(1), 73–99 (2015)
5. Arieli, O., Straßer, C.: Argumentative approaches to reasoning with maximal consistency. In: Proceedings of KR 2016, pp. 509–512. AAAI Press (2016)
6. Arieli, O., Straßer, C.: Deductive argumentation by enhanced sequent calculi and dynamic derivations. Electron. Notes Theor. Comput. Sci. **323**, 21–37 (2016)
7. Avron, A.: A constructive analysis of RM. J. Symb. Logic **52**(4), 939–951 (1987)
8. Avron, A.: The method of hypersequents in the proof theory of propositional non-classical logics. In: Logic: Foundations to Applications, pp. 1–32. Oxford Science Publications, Oxford (1996)
9. Avron, A.: **RM** and its nice properties. In: Bimbó, K. (ed.) J. Michael Dunn on Information Based Logics. OCL, vol. 8, pp. 15–43. Springer, Cham (2016). https://doi.org/10.1007/978-3-319-29300-4_2
10. Batens, D.: A universal logic approach to adaptive logics. Log. Univers. **1**(1), 221–242 (2007)
11. Bench-Capon, T., Dunne, P.: Argumentation in artificial intelligence. Artif. Intell. **171**(10), 619–641 (2007)
12. Besnard, P., Hunter, A.: A logic-based theory of deductive arguments. Artif. Intell. **128**(1–2), 203–235 (2001)
13. Caminada, M., Amgoud, L.: On the evaluation of argumentation formalisms. Artif. Intell. **171**(5), 286–310 (2007)
14. Caminada, M., Carnielli, W., Dunne, P.: Semi-stable semantics. J. Logic Comput. **22**(5), 1207–1254 (2011)
15. Cayrol, C.: On the relation between argumentation and non-monotonic coherence-based entailment. In: Proceedings of the 14th International Joint Conference on Artificial Intelligence, pp. 1443–1448 (1995)
16. Dung, P.M.: On the acceptability of arguments and its fundamental role in non-monotonic reasoning, logic programming and n-person games. Artif. Intell. **77**(2), 321–357 (1995)
17. Dunn, M., Meyer, R.: Algebraic completeness results for Dummett's LC and its extensions. Zeitschrift für mathematische Logik und Grundlagen der Mathematik **17**, 225–230 (1971)
18. Dunn, M., Restall, G.: Relevance logic. In: Gabbay, D.M., Guenthner, F. (eds.) Handbook of Philosophical Logic, 2nd edn., vol. 6, pp. 1–136. Kluwer, Alphen aan den Rijn (2002)
19. Gentzen, G.: Untersuchungen über das logische Schließen I, II. Mathematische Zeitschrift **39**, 176–210, 405–431 (1934)
20. Makinson, D.: Bridges between classical and nonmonotonic logic. Logic J. IGPL **11**(1), 69–96 (2003)
21. Metcalfe, G., Olivetti, N., Gabbay, D.: Fundamental logics. In: Metcalfe, G., Olivetti, N., Gabbay, D. (eds.) Proof Theory for Fuzzy Logics, vol. 36, pp. 137–175. Springer, Dordrecht (2009). https://doi.org/10.1007/978-1-4020-9409-5_6

22. Mints, G.: Lewis' systems and system T (1965–1973). In: Feys, R. (ed.) "Modal Logic" (Russian Translation), pp. 422–501. Nauka (1974)
23. Pollock, J.: How to reason defeasibly. Artif. Intell. **57**(1), 1–42 (1992)
24. Pottinger, G.: Uniform, cut-free formulations of T, S4 and S5. J. Symb. Logic **48**, 900–901 (1983). Abstract
25. Prakken, H.: Two approaches to the formalisation of defeasible deontic reasoning. Stud. Logica. **57**(1), 73–90 (1996)
26. Prakken, H.: An abstract framework for argumentation with structured arguments. Argum. Comput. **1**(2), 93–124 (2010)
27. Simari, G., Loui, R.: A mathematical treatment of defeasible reasoning and its implementation. Artif. Intell. **53**(2–3), 125–157 (1992)
28. Straßer, C.: Adaptive Logics for Defeasible Reasoning: Applications in Argumentation. Trends in Logic, vol. 38. Normative Reasoning and Default Reasoning. Springer, Cham (2014). https://doi.org/10.1007/978-3-319-00792-2
29. Straßer, C., Arieli, O.: Normative reasoning by sequent-based argumentation. J. Logic Comput. (forthcoming). https://doi.org/10.1093/logcom/exv050
30. Wu, Y., Podlaszewski, M.: Implementing crash-resistance and non-interference in logic-based argumentation. J. Logic Comput. **25**(2), 303–333 (2014)

On the Interaction Between Logic and Preference in Structured Argumentation

Anthony P. Young$^{(\boxtimes)}$ (iD), Sanjay Modgil, and Odinaldo Rodrigues

Department of Informatics, King's College London, London, UK
{peter.young,sanjay.modgil,odinaldo.rodrigues}@kcl.ac.uk

Abstract. The *structure-preference* (SP) order is a way of defining argument preference relations in structured argumentation theory that takes into account how arguments are constructed. The SP order was first introduced in the context of endowing Brewka's prioritised default logic (PDL) with sound and complete argumentation semantics. In this paper, we further articulate the underlying intuitions of the SP order in terms of how an agent should construct arguments. We also compare the SP order to other argument preference relations and illustrate the different results one would obtain. Finally, we prove that the SP order allows for the original version of PDL to satisfy Brewka's and Eiter's postulates.

1 Introduction

Argumentation theory [1, 2, 10, 14] is a general framework for non-monotonic reasoning [3], where inference from an inconsistent knowledge base in a given non-monotonic logic (NML) can be expressed as the exchange of conflicting arguments with premises from that knowledge base, such that the inferred statements of the logic are the conclusions of justified arguments. As the study of how preferences are used to resolve conflicts has become a major topic in NML [5, 15, 18], argumentation theory has used preference relations to decide which arguments are justified. Such preferences over arguments could be taken as exogenously given, or be derived from more primitive concepts. Structured argumentation theories like ASPIC$^+$ [14], which treat arguments as structured objects made up of premises and inference rules, consider more primitive preferences that are given over argument components such as defeasible rules, such that these preferences over components are aggregated into an argument preference relation.

This paper makes the following contributions. We first motivate and define the *structure-preference* (SP) order. This is a rearrangement of the preference relations on the fallible components (i.e. the non-axiom premises and the non-deductive rules of inference) of a structured argumentation theory that takes into account the structure of arguments, understood as the actual order of applicability of the fallible components during argument construction. The SP order is an alternative preference relation that can also make use of the commonly-used

© Springer International Publishing AG, part of Springer Nature 2018
E. Black et al. (Eds.): TAFA 2017, LNAI 10757, pp. 35–50, 2018.
https://doi.org/10.1007/978-3-319-75553-3_3

aggregation techniques such as the elitist and democratic set-comparison relations, in accordance with the weakest-link and last-link principles [14, Sect. 5]. We also define the corresponding SP argument preference, which makes more certain arguments more preferred. The SP order was first devised to endow Brewka's prioritised default logic (PDL) [4] with argumentation semantics [21]. After recapping this result, our second contribution applies the insight of the SP order to show that Brewka and Eiter's principles for PDL [5,6] hold for the original version of PDL. We then discuss some related work, in particular the roots of the SP order in logic programming [8,18].

This paper is structured as follows. In Sect. 2 we review the relevant aspects of the ASPIC$^+$ framework for structured argumentation [10,14]. In Sect. 3 we define the SP order on the defeasible rules in the abstract context of ASPIC$^+$, establish its underlying intuitions, and compare it with different argument preference relations. In Sect. 4 we recap PDL [4] and its argumentation semantics in the case where the underlying priority is a total order [21]. We also recall Brewka and Eiter's claim that PDL "does not take seriously what they believe" [5,6], and their remedy by modifying PDL to satisfy two guiding principles. We prove that if PDL reasons with the SP order, then it will also satisfy these principles. Section 5 discusses related work [8,11–13,18] and Sect. 6 concludes.

2   Preferences in Structured Argumentation

To illustrate the idea of the SP order, we will use the ASPIC$^+$ framework for structured argumentation. However, it will become clear that any structured argumentation theory that considers preferences over the fallible components of arguments and has a well-defined notion of argument construction can accommodate the SP order.

2.1   The ASPIC$^+$ Framework for Structured Argumentation

We recap the relevant definitions of ASPIC$^+$ [10,14]. An *argumentation system* is a tuple $\langle \mathcal{L}, -, \mathcal{R}_s, \mathcal{R}_d, n \rangle$, where $\mathcal{L}$ is a set of well-formed formulae (wffs), $- : \mathcal{L} \to \mathcal{P}(\mathcal{L})$ is the contrary function[1] $\theta \mapsto \overline{\theta}$, where $\overline{\theta}$ is the set of wffs that disagree with θ. Let $m \in \mathbb{N}$ and $\theta_1, \ldots, \theta_m, \phi \in \mathcal{L}$.[2] $\mathcal{R}_s$ is *the set of strict inference rules*, where rules are denoted by $(\theta_1, \ldots, \theta_m \to \phi)$, meaning that if $\theta_1, \ldots, \theta_m$ are all true then ϕ is also true. $\mathcal{R}_d$ is *the set of defeasible inference rules*, where rules are denoted by $(\theta_1, \ldots, \theta_m \Rightarrow \phi)$, meaning that if $\theta_1, \ldots, \theta_m$ are all true then ϕ is tentatively true. We have that $\mathcal{R}_s \cap \mathcal{R}_d = \varnothing$. For $r = (\theta_1, \ldots, \theta_m \to / \Rightarrow \phi) \in \mathcal{R}_s \cup \mathcal{R}_d$ we define $Ante(r) := \{\theta_i\}_{i=1}^{m} \subseteq \mathcal{L}$ and $Cons(r) := \phi \in \mathcal{L}$.[3] Finally, $n : \mathcal{R}_d \to \mathcal{L}$ is a partial function that assigns a *name* to defeasible rules.

---

[1] If X is a set then $\mathcal{P}(X)$ is its power set.
[2] In this paper, $\mathbb{N} := \{0, 1, 2, 3, \ldots\}$ and $\mathbb{N}^+ := \{1, 2, 3, 4, \ldots\}$.
[3] If $m = 0$ then rules like $(\to \phi)$ and $(\Rightarrow \psi)$ are well-defined, with $Ante(r) = \varnothing$.

A *knowledge base* is a set $\mathcal{K} := \mathcal{K}_n \cup \mathcal{K}_p \subseteq \mathcal{L}$ where $\mathcal{K}_n \cap \mathcal{K}_p = \varnothing$. $\mathcal{K}_n$ is *the set of axioms*, and $\mathcal{K}_p$ is *the set of ordinary premises*. Given an argumentation system and knowledge base, ASPIC$^+$ arguments are constructed inductively:

1. (Base) $[\theta]$ is a *singleton argument* with $\theta \in \mathcal{K}$, with *conclusion* $Conc([\theta]) := \theta \in \mathcal{L}$, *premise set* $Prem([\theta]) := \{\theta\} \subseteq \mathcal{K}$, *top rule* $TopRule([\theta]) := *^4$ and *set of subarguments* $Sub([\theta]) := \{[\theta]\}$.
2. (Inductive) Let $n \in \mathbb{N}$ and $\{A_i\}_{i=1}^n$ be a set of arguments where for all $1 \le i \le n$, A_i has $Conc(A_i) \in \mathcal{L}$, $Prem(A_i) \subseteq \mathcal{L}$ and $Sub(A_i)$ well-defined. If we have a strict rule $r = (Conc(A_1), \ldots, Conc(A_n) \to \phi) \in \mathcal{R}_s$ and defeasible rule $s = (Conc(A_1), \ldots, Conc(A_n) \Rightarrow \psi) \in \mathcal{R}_d$, then $B := [A_1, \ldots, A_n \to \phi]$ and $C := [A_1, \ldots, A_n \Rightarrow \psi]$ are also arguments, with respective conclusions $Conc(B) := \phi$ and $Conc(C) := \psi$, premise sets $Prem(B)$, $Prem(C) := \bigcup_{i=1}^n Prem(A_i)$, top rules $TopRule(B) = r$ and $TopRule(C) = s$, and sets of subarguments $Sub(B) = \{B\} \cup \bigcup_{i=1}^n Sub(A_i)$ and $Sub(C) = \{C\} \cup \bigcup_{i=1}^n Sub(A_i)$.[5]

Let $\mathcal{A}$ denote the (unique) set of arguments constructed in this way.

Example 1. Working in propositional logic where $\mathcal{L}$ denotes the set of all propositional wffs and $\mathcal{R}_s$ contains all of the usual rules of proof. Let $a, b, c \in \mathcal{L}$ be propositional variables. Suppose we have $\mathcal{R}_d = \{(a \Rightarrow b), (b \Rightarrow c)\}$ and $\mathcal{K}_n = \{a\}$. Then we have arguments $A_0 := [a]$, $A_1 := [A_0 \Rightarrow b]$ and $A := [A_1 \Rightarrow c]$.

Two arguments are *equal* iff they are constructed identically in the above manner with syntactically identical formulae. For $A, B \in \mathcal{A}$ we say A is a *subargument of* B iff $A \in Sub(B)$ and write $A \subseteq_{\mathrm{arg}} B$; it is clear that $\subseteq_{\mathrm{arg}}$ is a preorder on $\mathcal{A}$. An argument A is *firm* iff $Prem(A) \subseteq \mathcal{K}_n$. For $A \in \mathcal{A}$ let $DR(A) \subseteq \mathcal{R}_d$ be the defeasible rules applied in constructing A. We say an argument is *strict* iff $DR(A) = \varnothing$; non-strict arguments are *defeasible*. We say that A attacks B on $B' \subseteq_{\mathrm{arg}} B$ iff at least one of the following hold.[6]

1. *Undermine* iff $(\exists \theta \in Prem(B) \cap \mathcal{K}_p)[B' = [\theta]$ and $Conc(A) \in \overline{\theta}]$.
2. *Rebut* iff $r := TopRule(B') \in \mathcal{R}_d$ and $Conc(A) \in \overline{Cons(r)}$.
3. *Undercut* iff $Conc(A) \in n(TopRule(B'))$.

Example 2. (Example 1 continued) Let $- : \mathcal{L} \to \mathcal{P}(\mathcal{L})$ be the contrary function representing *classical syntactic negation*, i.e. $\overline{\theta} := \{\psi\}$ where if θ is syntactically of the form $\neg\phi$, then $\psi = \phi$, else $\psi = \neg\theta$. Suppose a further defeasible rule $(a \Rightarrow \neg c) \in \mathcal{R}_d$, then $B := [A_0 \Rightarrow \neg c]$ is an argument such that $B \rightharpoonup A$, and $A \rightharpoonup B$, both rebutting each other at their conclusions.

ASPIC$^+$ also includes preferences on each arguments' fallible components, namely ordinary premises in $\mathcal{K}_p$ and defeasible rules in $\mathcal{R}_d$. These sets are

---

[4] In this paper, undefined quantities are denoted with $*$.

[5] From Footnote 3: when $n = 0$ then arguments like $[\to \phi]$ and $[\Rightarrow \psi]$ are well-defined, each with empty premises and only itself as a subargument.

[6] See [14] for why attacks are distinguished in this way.

equipped with the respective strict partial orders $<_K$ and $<_D$, assumed to be exogenously given, such that $\langle \mathcal{K}_p, <_K \rangle$ and $\langle \mathcal{R}_d, <_D \rangle$ are respectively lifted to relations $\lhd_K$ and $\lhd_D$ that compare finite subsets of $\mathcal{K}_p$ and $\mathcal{R}_d$ respectively.[7] This information is aggregated to an argument preference relation $\precsim$ on $\mathcal{A}$ (see Sect. 2.2). We can then use this preference relation to determine which attacks succeed as defeats. The *defeat* relation $\hookrightarrow$ on $\mathcal{A}$ is defined as: $A \hookrightarrow B$ on $B' \Leftrightarrow [A \rightharpoonup B$ on $B' \subseteq_{\mathrm{arg}} B$, $A \not\precsim B']$, in the cases where $A \rightharpoonup B$ is an undermine or rebut.[8] This gives us a directed graph called an *argumentation framework* $\langle \mathcal{A}, \hookrightarrow \rangle$ where the usual methods of calculating the justified arguments from abstract argumentation apply [10]. For our purposes we say that a set of arguments $S \subseteq \mathcal{A}$ is *justified* (i.e. a *stable extension*) iff it is conflict free, $\hookrightarrow \cap S^2 = \varnothing$, and $(\forall B \notin S)(\exists A \in S) A \hookrightarrow B$.

2.2 Principles for Argument Preferences

We now elaborate on how ASPIC$^+$ derives argument preference relations from the strict partially ordered set (poset) $\langle \mathcal{K}_p, <_K \rangle$ and $\langle \mathcal{R}_d, <_D \rangle$ [14, Sect. 5]. Let X be $\mathcal{K}_p$ or $\mathcal{R}_d$ and $<$ be either $<_K$ or $<_D$. To lift $\langle X, < \rangle$ to $\langle \mathcal{P}_{\mathrm{fin}}(X), \lhd \rangle$,[9] we have the following formulae [14,16]: for $\Gamma, \Gamma' \in \mathcal{P}_{\mathrm{fin}}(X)$ and $\lhd \in \{\lhd_{Eli}, \lhd_{Dem}\}$:

$$\Gamma \unlhd_{Eli} \Gamma' \Leftrightarrow [\Gamma = \Gamma' \text{ or } \Gamma \lhd_{Eli} \Gamma'], \tag{1}$$

$$\text{where } \Gamma \lhd_{Eli} \Gamma' \Leftrightarrow (\exists x \in \Gamma)(\forall y \in \Gamma') \, x <_D y, \text{ and} \tag{2}$$

$$\Gamma \unlhd_{Dem} \Gamma' \Leftrightarrow [\Gamma = \Gamma' \text{ or } \Gamma \lhd_{Dem} \Gamma'], \tag{3}$$

$$\text{where } \Gamma \lhd_{Dem} \Gamma' \Leftrightarrow \begin{cases} \text{true if } \Gamma \neq \varnothing, \, \Gamma' = \varnothing, \\ \text{false if } \Gamma = \Gamma' = \varnothing \text{ or } (\Gamma = \varnothing, \Gamma' \neq \varnothing), \\ \text{else, } (\forall x \in \Gamma)(\exists y \in \Gamma') \, x <_D y. \end{cases} \tag{4}$$

The relation $\lhd_{Eli}$ (Eq. 2) is the *strict elitist set-comparison relation*. The relation $\lhd_{Dem}$ (Eq. 4) is the *strict democratic set-comparison relation*. Notice in all cases $\varnothing$ is a maximal element to reflect that what should be most preferred should be arguments with no such fallible component.

To relate these set-comparison relations to arguments, we recall the *last link principle* (LLP) [14, Definition 20]. For $A, B \in \mathcal{A}$ and $\unlhd \in \{\unlhd_{Eli}, \unlhd_{Dem}\}$, define

$$A \precsim B \Leftrightarrow \begin{cases} Prem_p(A) \unlhd Prem_p(B) & LDR(A) = LDR(B), \\ LDR(A) \unlhd LDR(B) & \text{else,} \end{cases} \tag{5}$$

where for $A \in \mathcal{A}$: if A is singleton then $LDR(A) = \varnothing$, else if $A = [A_1, \ldots, A_n \Rightarrow Conc(A)]$ then $LDR(A) = \{(Conc(A_1), \ldots, Conc(A_n)) \Rightarrow Conc(A))\}$, else we have $LDR(A) = \bigcup_{i=1}^{n} LDR(A_i)$. Alternatively, one can use the *weakest link principle* (WLP) [14, Definition 21]. For $A, B, \unlhd$ as above,

---

[7] The subscript K stands for "knowledge" and D stands for "defeasible".

[8] This is adequate for our purposes. For a discussion of the subtleties of how this depends on the contrary function and for the case of undercutting attacks, see [14].

[9] If X is a set then $\mathcal{P}_{\mathrm{fin}}(X)$ is the set of all finite subsets of X.

$$A \precsim B \Leftrightarrow \begin{cases} Prem_p(A) \lhd Prem_p(B) \text{ if } A, B \text{ are strict,} \\ DR(A) \lhd DR(B) \text{ if } A, B \text{ are firm,} \\ Prem_p(A) \lhd Prem_p(B) \text{ and } DR(A) \lhd DR(B) \text{ else.} \end{cases} \quad (6)$$

Both the LLP and WLP are commonly used ways of defining preferences $\precsim$ on arguments from more primitive preferences on the fallible components of arguments.[10] The strict preference is defined as $A \prec B \Leftrightarrow [A \precsim B, B \not\precsim A]$.

3 The Structure-Preference Order

3.1 Guiding Intuition – How to Construct Arguments

We now articulate the guiding intuition of the SP order, which is related to how agents should construct and compare arguments. Preferences have long been used to guide reasoning in non-monotonic logics (NMLs) and logic programming. In [9], Delgrande et al. review the ways preferences are treated in NMLs. They distinguish between two types of preferences. *Prescriptive preferences* provide information on which of the applicable rules *should* be selected, i.e. "applicable" in the sense of having all of the antecedents of a rule known. *Descriptive preferences* specify the exact order of how the rules *are actually* applied. How do these ideas translate to structured argumentation theory?

Assume we have an inferentially ideal agent who, when constructing arguments, is able to apply all applicable strict rules in $\mathcal{R}_s$ when it is possible to do so. Such an agent would begin with all premises $\mathcal{K}$ (as singleton arguments) and deductively close under all possible strict rules to form a *core*. Of the applicable defeasible rules, the agent would choose the $<_D$-most preferred ones to be applied. The agent then continues deductively closing with respect to the strict rules, and then adding the $<_D$-most preferred defeasible rules... and so on. This view of argument construction gives a canonical enumeration of *how far* a given argument is from the agent's core, in terms of the number of times the agent has added a defeasible rule and closed under all possible strict rules. This canonical enumeration also creates a preference over the defeasible rules that is descriptive in the sense of Delgrande et al.

How can we define such a descriptive preference on $\mathcal{R}_d$ in structured argumentation theory? For simplicity, we will assume arguments are *firm* and that the agent has a prescriptive preference relation $<_D$ over $\mathcal{R}_d$ that is a strict *total* order [19, Chap. 3], and that $\mathcal{R}_d$ is a finite set. We define a preference $<_{SP}$ on $\mathcal{R}_d$, where SP stands for *structure-preference*, as follows. The most $<_{SP}$-preferred defeasible rule, a_1, is the most $<_D$-preferred *applicable* rule after all strict arguments are constructed, i.e. the core. The next most $<_{SP}$-preferred defeasible rule, a_2, is the next most $<_D$-preferred applicable rule after a_1... etc. and so on until all defeasible rules are added. If $<_D$ is a total order then $<_{SP}$ is also a total order. We will formalise this idea in Definition 2 below.

---

[10] The infallible components of arguments, i.e. the axiom premises and deductive rules of inference, are by convention incomparable because they are all true.

3.2   The SP Order - Definitions, Comparisons, Properties

We give the following definitions. We work with an arbitrary argumentation system and knowledge base with $\mathcal{K}_p = \varnothing$ (Sect. 2).

Definition 1. *Let* $R \subseteq \mathcal{R}_d$. *The set* $Args(R) \subseteq \mathcal{A}$ *is defined as:* $A \in Args(R) \Leftrightarrow DR(A) \subseteq R$. *This is* **the set of arguments freely constructed with defeasible rules restricted to those in** R.

The set $Args(R)$ has all arguments with premises in $\mathcal{K}$, strict rules in $\mathcal{R}_s$ and defeasible rules in R. Given R, $Args(R)$ exists and is unique. Further, we will assume that there are no *irrelevant rules*, i.e. there is no $r \in \mathcal{R}_d$ such that $r \notin DR(\mathcal{A}) := \bigcup_{A \in \mathcal{A}} DR(A)$. Therefore, all rules in $\mathcal{R}_d$ feature in some argument. Also, we generalise the conclusion function (Sect. 2) to sets of arguments. For $S \subseteq \mathcal{A}$, $Conc(S) := \bigcup_{A \in S} \{Conc(A)\}$.

We will define the SP order for argumentation systems where

1. $\mathcal{K}_p = \varnothing$ (i.e. all arguments are firm),
2. $\mathcal{R}_d$ is finite and
3. $<_D$ is a total order on $\mathcal{R}_d$.

We will briefly consider how assumptions 1 and 3 above might be lifted in Sect. 6. Definition 2 below formalises our discussion of Sect. 3.1. Each $<_D$ over $\mathcal{R}_d$ can be transformed into $<_{SP}$ that incorporates the logical relationship of the defeasible rules, which is determined by the order they are applied when constructing arguments.

Definition 2. *Let* $N := |\mathcal{R}_d|$ *and* $1 \leq i \leq N$. *We define a rearrangement of the defeasible rules* $r \in \mathcal{R}_d$ *to* $a_i \in \mathcal{R}_d$ *as follows:*

$$a_i := \max_{<_D} \left[\left\{ r \in \mathcal{R}_d \,\middle|\, Ante(r) \subseteq Conc\left[Args\left(\bigcup_{k=1}^{i-1} \{a_k\} \right) \right] \right\} - \bigcup_{j=1}^{i-1} \{a_j\} \right]. \quad (7)$$

The **(strict) structure-preference (SP) order** *on* $\mathcal{R}_d$, *denoted by* $<_{SP}$, *is:*

$$(\forall 1 \leq i, j \leq N)\, a_i <_{SP} a_j \Leftrightarrow j < i. \quad (8)$$

The **non-strict SP order** *is* $a_i \leq_{SP} a_j \Leftrightarrow [a_i = a_j \ or \ a_i <_{SP} a_j]$.[11]

As $<_D$ is a total order and $\mathcal{R}_d$ is finite, a_i exists and is unique. The agent first constructs all strict (and firm) arguments $Args(\varnothing)$, then adds the $<_D$-most preferred applicable rule $a_1 = \max_{<_D} \{r \in \mathcal{R}_d \,|\, Ante(r) \subseteq Conc(Args(\varnothing))\}$. Then the agent adds the next $<_D$-most preferred applicable rule a_2... etc. until all rules are exhausted. Note that the second union after the set difference in Eq. 7 ensures that each rule is only applied once. The result is such that $<_{SP}$-larger defeasible rules belong to smaller arguments or are more preferred. Clearly, $<_{SP}$ is also a strict total order on $\mathcal{R}_d$, and the transformation $<_D \mapsto <_{SP}$ is functional.

---

[11] This is well-defined because $i \mapsto a_i$ is bijective between $\mathcal{R}_d$ and $\{1, 2, 3, \ldots, |\mathcal{R}_d|\}$.

3.3   The SP Argument Preference

Inspired by preferences in NML, $<_{SP}$ provides a new way of defining argument preference relations, because it takes into account how arguments are constructed. We now lift $<_{SP}$ to its corresponding argument preference, $\prec_{SP}$. The guiding intuition is that arguments further away from the core should be less preferred, because (as $\mathcal{K}_p = \varnothing$) arguments in the core are certain (strict and firm); one might expect an agent to prefer arguments that are more certain (closer to the core) by virtue of it having less fallible elements and thus being less susceptible to attack. We formalise this as $A \sqsubseteq_{\text{arg}} B \Rightarrow A \precsim B$, and investigate through examples whether the other ASPIC$^+$ preferences satisfy this property.

Example 3. (Example 1 continued) Consider the arguments A_1 and A from Example 1. Clearly $LDR(A_1) = \{(a \Rightarrow b)\}$ and $LDR(A) = \{(b \Rightarrow c)\}$. Suppose $(a \Rightarrow b) <_D (b \Rightarrow c)$. By $\trianglelefteq_{Eli}$-LLP (Eqs. 1, 2 and 5), we have $A_1 \prec A$. By Eq. 7 $\lfloor a \rfloor \in Args(\varnothing)$ and $Ante(a \Rightarrow b) = \{a\} \subseteq Conc(Args(\varnothing))$ so $a_1 = (a \Rightarrow b)$. Similarly, $a_2 = (b \Rightarrow c)$, hence $(b \Rightarrow c) <_{SP} (a \Rightarrow b)$. Therefore, by $\trianglelefteq_{Eli}$-LLP under $<_{SP}$, we have that $A \prec A_1$.

Example 3 shows that under $<_D$, it is possible for $\trianglelefteq_{Eli}$-LLP to rank an argument A that is further from the core (because it has two defeasible rules composed in series) to be more preferred than an argument A_1 that is closer to the core.

The next example shows that $<_{SP}$ does not completely capture that arguments should be less preferred than their (smaller) subarguments under $\trianglelefteq_{Dem}$.

Example 4. (Examples 1 to 3 continued) Let $r_1 := (a \Rightarrow b)$, $r_2 := (b \Rightarrow c)$ and $r_3 := (a \Rightarrow \neg c)$. Suppose $r_1 <_D r_3 <_D r_2$. Applying $\trianglelefteq_{Dem}$-WLP, we have $A_1 \prec B \prec A \prec A_0$, which by Example 2 means that $A \hookrightarrow B$ on B, so c is a justified conclusion. From Eq. 7, we have $r_2 <_{SP} r_1 <_{SP} r_3$, hence the new preference is $A, A_1 \prec B \prec A_0$, with A and A_1 incomparable.

As $A_1 \sqsubseteq_{\text{arg}} A$, we would like $A \precsim A_1$. Does $\trianglelefteq_{Eli}$ fare any better?

Example 5. (Example 4 continued) Consider the same situation but with $\trianglelefteq_{Eli}$-WLP. From $r_1 <_D r_3 <_D r_2$ we have $A, A_1 \prec B \prec A_0$ with A and A_1 incomparable. However, from $r_2 <_{SP} r_1 <_{SP} r_3$ we have $A \prec A_1 \prec B \prec A_0$.

Example 5 makes the larger argument A less preferred than its subargument A_1. However, even when using $\trianglelefteq_{Eli}$, this does not generally hold true.

Example 6. Consider a different example where $a, b, c \in \mathcal{L}$, $\mathcal{K}_n := \{a\}$, $\mathcal{R}_d := \{r_1 := (a \Rightarrow b), r_2 := (b \Rightarrow c), r_3 := (b \Rightarrow \neg c)\}$. We can construct the arguments $A := [[[a] \Rightarrow b] \Rightarrow c]$ and $B := [[[a] \Rightarrow b] \Rightarrow \neg c]$, with $DR(A) = \{r_1, r_2\}$ and $DR(B) = \{r_1, r_3\}$. Suppose we have that $r_3 <_D r_2 <_D r_1$, which gives $r_3 <_{SP} r_2 <_{SP} r_1$ by Eq. 7. Under $\trianglelefteq_{Eli}$-WLP, both A and B are incomparable.

The $\prec_{SP}$-smaller argument should be that which has the $<_{SP}$-smallest rule, i.e. be further from the core. Example 6 shows that $\trianglelefteq_{Eli}$ does not behave well when comparing arguments with shared rules. We now define a set-comparison relation that compares arguments at their non-shared rules.

Definition 3. *Given $<_D$ and $<_{SP}$ from Eq. 8, the **structure preference set-comparison relation**, $\lhd_{SP}$ is the following binary relation on $\mathcal{P}_{fin}(\mathcal{R}_d)$:*

$$\Gamma \lhd_{SP} \Gamma' \Leftrightarrow (\exists x \in \Gamma - \Gamma')(\forall y \in \Gamma' - \Gamma) \, x <_{SP} y. \tag{9}$$

It can be shown that as $<_{SP}$ is a strict total order on $\mathcal{R}_d$, then $\lhd_{SP}$ is also a strict total order on $\mathcal{P}_{fin}(\mathcal{R}_d)$ [21, Lemma 4.2].[12] We specialise this relation to obtain the corresponding argument preference relation:

Definition 4. *Given $<_D$, $<_{SP}$ from Eq. 8 and $\lhd_{SP}$ from Eq. 9, the **(strict) structure-preference (SP) argument preference relation** is the relation $\prec_{SP}$, which is Eq. 9 specialised to WLP:*

$$A \prec_{SP} B \Leftrightarrow DR(A) \lhd_{SP} DR(B), \tag{10}$$

*with the **non-strict SP argument preference relation** defined as $A \precsim_{SP} B \Leftrightarrow [A \prec_{SP} B \text{ or } DR(A) = DR(B)]$.*

We use WLP to avoid situations like Example 3, where arguments further away from the core are more preferred. It follows that $\precsim_{SP}$ is a total preorder on $\mathcal{A}$. In particular, $\precsim_{SP}$ satisfies the following two properties that reflect how arguments further from the core are $\prec_{SP}$-less preferred.

1. Larger arguments are less preferred than smaller arguments, i.e. $A \subseteq_{\mathrm{arg}} B \Rightarrow B \precsim_{SP} A$ [21, Lemma 4.1].
2. Infallible arguments, in this case strict arguments, are $\precsim_{SP}$-maximal. It follows from the definition of the defeat relation that (e.g.) strict arguments concluding θ will defeat any defeasible argument concluding $\neg\theta$.

In summary, $\prec_{SP}$ is an ASPIC$^+$ argument preference relation based on $<_{SP}$, which captures the intuition that arguments further from the core are less certain and therefore less preferred. As shown in the preceding examples, these properties do not hold for LLP or $\unlhd_{Dem}$, and also fails for $\unlhd_{Eli}$ when there are shared defeasible rules.

4   Applications to Prioritised Default Logic

In this section, we remind the reader that $\prec_{SP}$ has been used to endow Brewka's prioritised default logic (PDL) [4] with sound and complete argumentation semantics [21]. Further, we show that whereas when reasoning according to $<_D$ in PDL does not satisfy Brewka and Eiter's two principles (articulated in Sect. 4.2 below, also see [5,6]), the principles are satisfied if $<_{SP}$ is used instead. We work in first order logic where the set of formulae is $\mathcal{FL}$, and the set of formulae without free variables is $\mathcal{SL} \subset \mathcal{FL}$, with the usual quantifiers and connectives. Classical entailment is denoted by $\models$. Given $S \subseteq \mathcal{FL}$, the *deductive closure* of S is $Th(S) \subseteq \mathcal{FL}$, and given $\theta \in \mathcal{FL}$, the *addition operator* is $S + \theta := Th(S \cup \{\theta\})$.

---

[12] Equation 9 has previously been considered in a different context [7].

4.1   Brewka's Prioritised Default Logic as Argumentation

In this paper, we assume *closed normal defaults* of the form $\frac{\theta:\phi}{\phi}$ read as: if the *antecedent* $\theta \in \mathcal{SL}$ is the case and the *consequent* ϕ is consistent with what we know, then ϕ is also the case. Given $S \subseteq \mathcal{SL}$, a default is *active (in S)* iff $[\theta \in S, \phi \notin S, \neg\phi \notin S]$. Active defaults are precisely those that can be applied, such that the information gained is new and consistent with what we know. A *finite prioritised default theory* (PDT) is a structure $T := \langle D, W, < \rangle$, where $W \subseteq \mathcal{SL}$ is a possibly infinite set of known facts and $\langle D, < \rangle$ is a *finite* strict poset of defaults, where $d < d' \Leftrightarrow$ means d' is *more prioritised than* d. Intuitively, D consists of the defaults that nonmonotonically extend W. The inferences of a PDT are defined by its extensions. Let $<^+ \supseteq <$ be a linearisation of $<$. A *prioritised default extension (with respect to $<^+$)* (PDE) is a set $E := \bigcup_{i\in\mathbb{N}} E_i \subseteq \mathcal{SL}$ built inductively as

$$E_0 := Th\,(W) \text{ and } E_{i+1} := \begin{cases} E_i + \phi\,, & \text{if property 1} \\ E_i\,, & \text{else} \end{cases} \tag{11}$$

where "property 1" iff "ϕ is the consequent of the $<^+$-greatest default d active in E_i". Intuitively, one first generates all classical consequences from the facts W, and then iteratively adds the nonmonotonic consequences from the most prioritised default to the least. The set of defaults thus added are called the *generating defaults of E*, denoted by $GD(<^+) \subseteq D$. Notice if W is inconsistent then $E_0 = E = \mathcal{FL}$. It can be shown that the ascending chain $E_i \subseteq E_{i+1}$ stabilises at some finite $i \in \mathbb{N}$ and that E is consistent provided that W is consistent. E does not have to be unique because there are many distinct linearisations $<^+$ of $<$. We say the PDT T *sceptically infers* $\theta \in \mathcal{SL}$ iff $\theta \in E$ for *all* PDEs E.

Henceforth, we will assume a linearised PDT (LPDT) $T = \langle D, W, < \rangle$ where $<$ is a strict total order unless otherwise stated. By Eq. 11, since $<$ is total, there is only one way to apply the defaults in D, hence the PDE is unique and all inferences are sceptical. We say that θ *follows from* T iff $\theta \in E$ where E is the PDE of T. Further, we will assume W is consistent.

Given an LPDT $T := \langle D, W, < \rangle$ we translate directly into an argumentation system and knowledge base. For the argumentation system, we have that $\mathcal{L} = \mathcal{FL}$, $-$ is classical syntactic negation (as in Example 2),

$$\mathcal{R}_s := \{(\theta_1, \ldots, \theta_n \to \phi) \,|\, \theta_1, \ldots, \theta_n \models \phi\}, \tag{12}$$

$$\mathcal{R}_d := \left\{ (\theta \Rightarrow \phi) \,\middle|\, \frac{\theta:\phi}{\phi} \in D \right\}, \quad (\theta \Rightarrow \phi) <_D (\theta' \Rightarrow \phi') \Leftrightarrow \frac{\theta:\phi}{\phi} < \frac{\theta':\phi'}{\phi'}. \tag{13}$$

Also, $n \equiv *$ (we do not need undercuts), $\mathcal{K}_n = W$ and $\mathcal{K}_p = \varnothing$. Arguments and attacks are defined as in Sect. 2. It has been shown that a sound and complete argumentation semantics for PDL is obtained if $<_{SP}$ is used rather than $<_D$ [21]. Further, it has been shown why the ASPIC$^+$ argument preferences (Sect. 2.2) cannot give a sound and complete argumentation semantics based on $<_D$ [20, Sect. 4.2.1]. Intuitively, the inference mechanism of PDL (Eq. 11) picks out those defaults that are most preferred *and active*. This requirement of being active is not a property of $<$, but rather a property of the way PDEs are defined (Eq. 11). When translating into argumentation, $<_D$ only contains the information from $<$. To achieve soundness and completeness, we must explicitly incorporate the idea for a default to be active, such that arguments containing rules corresponding to blocked defaults are defeated by being less preferred, which is what $<_{SP}$ captures. Further, common rules are ignored because either they are included in E or not, which is what $\prec_{SP}$ captures (Eq. 9).

Using $<_{SP}$ and the associated defeat relation $\hookrightarrow$, it can be shown that there is a unique stable extension of $\langle \mathcal{A}, \hookrightarrow \rangle$ [21, Theorem 5.2]. We then have the following soundness and completeness result.

Theorem 1. *Let T be an LPDT where W is consistent, and $DG(T) := \langle \mathcal{A}, \hookrightarrow \rangle$ be its defeat graph with $\hookrightarrow$ defined under $\prec_{SP}$.*

1. *Let E be the extension of T. Then there exists a unique stable extension $\mathcal{E} \subseteq \mathcal{A}$ of $DG(T)$ such that $Conc\,(\mathcal{E}) = E$.*
2. *If $\mathcal{E} \subseteq \mathcal{A}$ is the stable extension of $DG(T)$, then $Conc(\mathcal{E})$ is the extension of T.*

Proof. See [21, Theorem 5.3].   ∎

PDL is thus endowed with argumentation semantics. The following definition (Definition 5) is important for proving Theorem 1 because given the PDE E of T we can show that $Args\,(F\,(NBD(E)))$ is the unique stable extension $\mathcal{E}$ (Definition 1). The set $NBD(E)$ will become important in the next section.[13]

Definition 5. *(From [21, Equation 5.3 and Lemma 5.1]) Let $T = \langle D, W, < \rangle$ with extension E. Define* **the set of non-blocked defaults w.r.t.** E

$$NBD(E) := \left\{ \frac{\theta : \phi}{\phi} \in D \,\middle|\, \theta \in E \text{ and } \neg\phi \notin E \right\}. \qquad (14)$$

The corresponding set of defeasible rules is denoted by the obvious order isomorphism $F : D \to \mathcal{R}_d$ implicit in Eq. 13.

$$F\,(NBD(E)) := \{ (\theta \Rightarrow \phi) \in \mathcal{R}_d \,|\, \theta \in E \text{ and } \neg\phi \notin E \}. \qquad (15)$$

4.2   On Brewka and Eiter's Principles for Priorities

When PDL was defined, Prakken offered an alternative intuition of the preference that differed from Eq. 11 [4, Section 5]. Brewka modified PDL in order to accommodate Prakken's intuition, at the cost of a less intuitive, non-constructive inference mechanism. Brewka and Eiter later formalised this version of PDL with two intuitive principles which they argue all PDLs should satisfy (see below),

---

[13] Given T and its PDE E is generated by the total order $<$, we have $NBD(E) \neq GD\,(<)$ in general. See [21, Section 5.1] for an explanation.

which are satisfied by the non-constructive inference mechanism but not by Eq. 11. We will show that by importing $<_{SP}$ from the sound and complete argumentation semantics of PDL back into PDL (Definition 6), Eq. 11 satisfies both of these principles as well. This allows us to retain the constructive inference mechanism of PDL.

Brewka and Eiter articulated two general principles that should hold true for any prioritised default logic [5,6].

1. **Principle I - Preference:** Let T be a Reiter default theory[14] [17] with extensions E_1 and E_2 respectively generated by the defaults $R \cup \{d_1\}$ and $R \cup \{d_2\}$ where $d_1, d_2, \notin R \subseteq D$. Let $< \neq \varnothing$ be a strict partial order on D such that T is now a PDT. If $d_2 < d_1$ then E_2 cannot be a PDE of T.
1. **Principle II - Relevance:** Let T be a PDT with PDE E. Let $d = \frac{\theta:\phi}{\phi} \notin D$ such that $\theta \notin E$. Define a PDT $T' = \langle D \cup \{d\}, W, <' \rangle$. If $<' \cap D^2 =<$ then E is also a PDE of T'.

Principle I states that if E_1 and E_2 are Reiter extensions of the PDT that have almost the same generating defaults but one, such that d_1 generates E_1 and d_2 generates E_2, and if $d_2 < d_1$, then E_1 should be the PDE of the PDT. Principle II states that the addition of irrelevant defaults cannot change the PDEs unless the preference changes. Principle I is not satisfied by PDL (Eq. 11).

Corollary 1. *Equation 11 does not satisfy Principle I.*

Proof. (Based on [4, Section 5] and [6, Example 4]) Consider the PDT $T = \langle D, W, < \rangle$ with $W = \{a\}$, $D = \left\{ d_1 := \frac{b:c}{c}, d_2 := \frac{a:b}{b}, d_3 := \frac{a:\neg c}{\neg c} \right\}$ and $d_2 < d_3 < d_1$. Applying Eq. 11, we have $E_0 = Th(\{a\})$, $E_1 = E_0 + \neg c$, $E_2 = E_1 + b$ so $E := Th(\{a, b, \neg c\})$. The equivalent Reiter default theory gives E and also $E' := Th(\{a, b, c\})$ as extensions. E is generated by the defaults $\{d_2, d_3\}$ and E' is generated by the defaults $\{d_1, d_2\}$. However, the original preference states $d_3 < d_1$. Therefore, Principle I states that E cannot be an extension. However, by Eq. 11, E is an extension. ∎

Corollary 1 formalises Prakken's criticism of Eq. 11 (cited in [4]): in this example, there is nothing that could conflict with d_2 and it could always be applied as $W = \{a\}$. This means d_1 is always applicable. As d_1 is more $<$-preferred than d_3, d_1 should be applied first. Therefore, the correct extension should be $E' := Th(\{a, b, c\})$. This later lead Brewka and Eiter to conclude that PDL "does not take seriously what they believe" [5,6]. Brewka and Eiter's articulation of Principles I and II seeks to formalise the intuition that the preference should mean what it says.

We show here that Principle II is satisfied by Eq. 11 for general PDTs.

Theorem 2. *Let E be a PDE of $T = \langle D, W, < \rangle$ where $<$ does not have to be total. Let $d := \frac{\theta:\phi}{\phi} \notin D$ be a default such that $\theta \notin E$. Define a PDT $T' := \langle D \cup \{d\}, W, <' \rangle$ where $<' \cap D^2 =<$. E is also a PDE of T'.*

---

[14] Here, T is a PDT with no priority (partial order $<= \varnothing$), see [4, Proposition 6].

Proof. If E is a PDE of T then by Eq. 11 there exists some $LPDT$ $T^+ = \langle D, W, <^+ \rangle$ where $<^+$ is a linearisation of $<$ such that E is the unique PDE of T^+. As $<' \cap D^2 = <$, we can place d at any position along the chain $<^+$ to make a linearisation $<'^+$ of $<'$. As $\theta \notin E$, then d is not $<'^+$-least active at $E_0 = Th(W)$ (else $\theta \in E$). Similarly, the addition of any of the defaults in D to $E_i \subseteq E$ will not make d $<'^+$-least active either because $\theta \notin E$ implies $\theta \notin E_i$. Therefore, E is also a PDE of T'. ∎

Brewka and Eiter showed that their non-constructive version of PDL, modified to accommodate Prakken's intuition, does satisfy both these principles, unlike Eq. 11 [5,6]. We will prove that Eq. 11 can satisfy Principle I, provided that it reasons with the PDL analogue of $<_{SP}$ on its defaults. Consider the following example.

Example 7 (Continued from the Example in Corollary 1). We can transform T into its argumentation framework with $d_2 < d_3 < d_1 \mapsto r_2 <_D r_3 <_D r_1$.[15] It can be shown that $r_1 <_{SP} r_2 <_{SP} r_3$. Now consider the equivalent preference to $<_{SP}$ on the side of PDL, denoted as $d_1 <_{PDLSP} d_2 <_{PDLSP} d_3$. The Reiter extensions are $E = Th(\{a, b, \neg c\})$ and $E' = Th(\{a, b, c\})$, respectively generated by $\{d_2, d_3\}$ and $\{d_1, d_2\}$. Notice now that E is indeed the extension *and* $d_1 <_{PDLSP} d_3$, which means Principle I is satisfied.

Example 7 motivates the following definition, which formalises the idea of importing $<_{SP}$ from the argumentation semantics of PDL back into PDL.

Definition 6. *Let T be an LPDT. Let $< \cong <_D \mapsto <_{SP}$ be the corresponding SP order on the defeasible rules in its argumentation framework (Eqs. 7, 8 and 13). The **SP default priority**, $<_{PDLSP}$ on D is the total order that is order isomorphic to the SP order $<_{SP}$ on $\mathcal{R}_d$.*

Definition 6 transforms the prescriptive preference $<$ of a PDT to its corresponding descriptive preference $<_{PDLSP}$. This does not change the PDE, because Eq. 11 already selects the most active default at each stage, so explicitly incorporating "active" into $<$ to form $<_{PDLSP}$ means that selecting the $<_{PDLSP}$-greatest active rule is the same as selecting the $<$-greatest active rule.

We now prove that Principle I is always satisfied by Eq. 11 when using the SP default priority $<_{PDLSP}$. To do this, we first prove some properties of $Args(R)$ (Definition 1).

Lemma 1. *If $S = Args(R) \subseteq \mathcal{A}$ for some $R \subseteq \mathcal{R}_d$, then $DR(S) \subseteq R$, where $DR(S) := \bigcup_{A \in S} DR(A)$.*

Proof. For all $A \in S = Args(R)$, $DR(A) \subseteq R$ by definition. Let $r \in DR(S)$, then $(\exists A \in S) r \in DR(A) \subseteq R$, hence $r \in R$. ∎

Lemma 2. *If $S = Args(R) \subseteq \mathcal{A}$ for some $R \subseteq \mathcal{R}_d$ and $(\forall r \in R) Ante(r) \subseteq Conc(S)$, then $DR(S) = R$.*

---

[15] Where (e.g.) $d_1 := \frac{b:c}{c}$ means that $r_1 := (b \Rightarrow c)$, and similarly for d_i to r_i, $i \in \{2, 3\}$.

Proof. Let $r \in R$, then $Ante(r) \subseteq Conc(S)$ and $(\exists n \in \mathbb{N}) \, r = (\theta_1, \ldots, \theta_n \Rightarrow \phi)$. If $n = 0$ then $Ante(r) = \varnothing$, so r is always applicable and the argument $[\Rightarrow \phi] \in S$, so $r \in DR(S)$. If $n > 0$, as $Ante(r) \subseteq Conc(S)$, then for each θ_i for $1 \leq i \leq n$ there is some argument $A_i \in S$ such that $Conc(A_i) := \theta_i$. By Lemma 1, $DR(A_i) \subseteq R$. As $Args(R)$ contains all arguments freely constructed from all premises, strict rules and defeasible rules in R, we can construct the argument $B := [A_1, \ldots, A_n \Rightarrow \phi]$ such that $TopRule(B) = r$. Clearly, $DR(B) = \bigcup_{i=1}^{n} DR(A_i) \cup \{r\} \subseteq R$, hence $B \in S$ and hence $r \in DR(S)$. As r is arbitrary, we conclude $R \subseteq DR(S)$. By Lemma 1, it follows that $R = DR(S)$. ∎

Theorem 3. *Let $T := \langle D, W, < \rangle$ be an LPDT with corresponding Reiter default theory $T_\varnothing := \langle D, W \rangle$. For $i = 1, 2$ let E_i be an extension of $T_\varnothing$ generated by $R \cup \{d_i\} \subseteq D$, for $d_i \notin R$. Let $<_{PDLSP}$ be the SP default priority (Definition 6). If $d_2 <_{PDLSP} d_1$ then E_2 is not an extension of T.*

Proof. Let $i = 1, 2$. By [6, Section 2] and Definition 5, $NBD(E_i) = R \cup \{d_i\}$. Recall our order isomorphism $F : D \to \mathcal{R}_d$ implicit in Eq. 13. Then by the definition of the image of a function, Definition 5 and our notation from Footnote 15, $F(NBD(E_i)) = F(R) \cup \{r_i\} \subseteq \mathcal{R}_d$. Applying Definition 6, we have that $d_2 <_{PDLSP} d_1 \Leftrightarrow r_2 <_{SP} r_1$.

Now assume for contradiction that E_2 is the extension of T. By Theorem 1, the defeat graph of T, $DG(T)$, has its unique stable extension $\mathcal{E}_2$ such that $Conc(\mathcal{E}_2) = E_2$. In the paragraph after the statement of Theorem 1, we have mentioned $\mathcal{E}_2 = Args(F(NBD(E)))$. By Definition 5 and Theorem 1, and that E_2 is deductively closed, $(\forall r \in F(NBD(E_2))) \, Ante(r) \subseteq E_2 = Conc(\mathcal{E}_2)$. By Lemma 2, we conclude that $DR(\mathcal{E}_2) = F(NBD(E_2))$.

Now consider E_1. Define $\mathcal{E}_1 := Args(F(NBD(E_1))) \subseteq \mathcal{A}$, which is well-defined and *not* a stable extension of $DG(T)$, because $DG(T)$ has the unique stable extension $\mathcal{E}_2$. As E_1 is consistent by Eq. 11, it can be shown that $\mathcal{E}_1$ is conflict-free. Clearly, $r_1 \in DR(\mathcal{E}_1)$ because d_1 is a generating default of E_1. As there are no irrelevant rules,[16] there is some argument $A \in \mathcal{E}_1$ such that $r_1 := TopRule(A)$.

Either $A \in \mathcal{E}_2$ or $A \notin \mathcal{E}_2$. If $A \in \mathcal{E}_2$ then $DR(A) \subseteq DR(\mathcal{E}_2)$ and hence $r_1 \in DR(\mathcal{E}_2) = F(NBD(E_2)) = F(R) \cup \{r_2\}$, which is impossible. Therefore, $A \notin \mathcal{E}_2$. As $\mathcal{E}_2$ is the stable extension for $DG(T)$, we have that $(\exists B \in \mathcal{E}_2) \, B \hookrightarrow A$. Given this $B \in \mathcal{E}_2$ defeating A, we see that $DR(B) \subseteq DR(\mathcal{E}_2) = F(R) \cup \{r_2\}$.

Either $r_2 \in DR(B)$ or $r_2 \notin DR(B)$. If $r_2 \notin DR(B)$ then $DR(B) \subseteq F(R)$, and hence $B \in \mathcal{E}_1$ – contradiction as $\mathcal{E}_1$ is conflict free. If $r_2 \in DR(B)$, then $r_2 \in DR(B) - DR(A)$ and $r_1 \in DR(A) - DR(B)$, but by our hypothesis, $r_2 <_{SP} r_1$ and hence $B \prec_{SP} A$ (Eqs. 9 and 10), so $B \not\hookrightarrow A$ – contradiction. Therefore, $\mathcal{E}_2$ cannot be a stable extension of $DG(T)$. By Theorem 1, $E_2 = Conc(\mathcal{E}_2)$ cannot be a PDE of T. ∎

One way to see why Principle I failed for Eq. 11 is because if the priority relation $<$ of an LPDT $\langle D, W, < \rangle$ does not take the order of applicability of

---

[16] Recall after Definition 1, an irrelevant rule $r \in \mathcal{R}_d$ is one where $r \notin DR(\mathcal{A})$.

defeasible rules into account, then it is possible for a default of low priority to block a default of high priority because the former is applicable before the latter; this was what happened in the example of Corollary 1 (where $d_3 < d_1$ but $d_1 <_{PDLSP} d_3$). This is remedied by transforming the prescriptive preference $<$ into its corresponding descriptive preference $<_{PDLSP}$.

In summary, PDL is a way of using preferences to guide default reasoning. One inference mechanism is Eq. 11, which offers a constructive definition of extending facts with non-monotonic conclusions from the defaults. Brewka and Eiter articulated two principles (Principles I and II) that they argued any PDL should satisfy, and pointed out that Eq. 11 does not satisfy Principle I (Corollary 1). We apply the insights from the sound and complete argumentation semantics of PDL to show that Eq. 11, when reasoning with the PDL analogue of $<_{SP}$ (Definition 6), does satisfy Principle I. Further, we have shown that Principle II is already satisfied (Theorem 2).

5   Related Work

Preferences have been used to enhance the reasoning capabilities of NMLs and logic programs. A variety of approaches and attempts have been made to classify them [9]. For example, Schaub and Wang have uniformly characterised three different approaches to preferences in logic programming [18], where they clarified that an answer set (analogous to extensions in argumentation) is *preference-preserving* if the preference on the rules also reflects their order of applicability. Delgrande, Schaub and Tompits developed a transformation of arbitrary preferences on the rules of a logic program into a preference that is aligned with the applications of the rules such that the answer sets are preserved [8]. Our work here investigates analogues of such a transformation in structured argumentation theory inspired by ideas from how preferences are used in NMLs. It may be interesting to pursue some comparisons between how descriptive preferences similar to $<_{SP}$ are used in other NML systems, and how that relates to their ASPIC$^+$ argumentation semantics.

We are not the first to investigate descriptive preferences in argumentation. Dung has investigated the analogue of a preference-preserving answer set, called an *enumeration-based extension*, while articulating axioms suitable for the study of structured argumentation with preferences [11,12]. Dung defined an *ordinary attack relation* (a type of defeat relation) that satisfies all of his axioms as well as Brewka and Eiter's two principles. Dung then investigated soundness and completeness of enumeration-based extensions with respect to the ordinary attack relation. He discovered that enumeration-based extensions are stable with respect to this attack, but only exist when the underlying knowledge base is *well-ranked*. Intuitively, this means that the underlying preference is already descriptive. Our work in this paper has provided a way of transforming a prescriptive preference into a descriptive preference, such that a corresponding stable extension can always be shown to exist.

Finally, Liao et al. [13] have undertook a similar investigation using abstract normative systems to endow various NMLs with argumentation semantics, motivated by a deontic interpretation of defaults and their priorities. One result is an argumentation semantics for Brewka and Eiter's extended logic programming [5]. This work differs from ours as Liao et al. used abstract normative systems, which is a different structured argumentation theory to ASPIC$^+$. Although the version of abstract normative systems used by Liao et al. is analogous to ASPIC$^+$ where $\mathcal{R}_s = \varnothing$, it is still expressive enough to endow extended logic programming with argumentation semantics, and also elucidates the preferences used.

6   Conclusions and Future Work

We have defined the *structure-preference (SP) order* $<_{SP}$ on ASPIC$^+$ defeasible rules, which provides a descriptive account of the use of preferences in structured argumentation theory (Sect. 3). This argument preference is interesting because it can endow Brewka's PDL with sound and complete argumentation semantics (Sect. 4.1) and it makes the original PDL inference mechanism satisfy Brewka and Eiter's postulates (Sect. 4.2, Theorems 2 and 3).

Future work would incorporate $\mathcal{K}_p$ into $<_{SP}$. As there is no explicit structure,[17] intuitively $<_K$ should be unchanged but still somehow "prior" to $<_{SP}$ on $\mathcal{R}_d$, because rules cannot be applied without premises. By representing W as $\mathcal{K}_p$ instead of K_n, we can consider PDTs where W is inconsistent. Further, what would the SP order look like when $<_K$ and $<_D$ are *partial* orders instead of total orders? One approach could be to consider all possible linearisations $<_K^+$ and $<_D^+$, transform each into the appropriate total order $<_{SP}$ and then aggregate preferences using some appropriate social welfare function [19, Chap. 9]. Finally, it will be interesting to study the SP argument preference under other ASPIC$^+$ argument preference relations, or in other structured argumentation theories with preferences over the defeasible rules and a well-defined notion of argument construction.

Acknowledgements. We thank the anonymous referees for their constructive criticisms, which have improved the paper.

References

1. Besnard, P., Garcia, A., Hunter, A., Modgil, S., Prakken, H., Simari, G., Toni, F.: Introduction to structured argumentation. Argument Comput. **5**(1), 1–4 (2014)
2. Bondarenko, A., Dung, P.M., Kowalski, R.A., Toni, F.: An abstract, argumentation-theoretic approach to default reasoning. Artif. Intell. **93**(1), 63–101 (1997)
3. Brewka, G.: Nonmonotonic Reasoning: Logical Foundations of Commonsense. Cambridge Tracts in Theoretical Computer Science, vol. 12. Cambridge University Press (1991)

---

[17] This is unlike defeasible rules, which can be composed in series.

4. Brewka, G.: Adding priorities and specificity to default logic. In: MacNish, C., Pearce, D., Pereira, L.M. (eds.) JELIA 1994. LNCS, vol. 838, pp. 247–260. Springer, Heidelberg (1994). https://doi.org/10.1007/BFb0021977

5. Brewka, G., Eiter, T.: Preferred answer sets for extended logic programs. Artif. Intell. **109**(1), 297–356 (1999)

6. Brewka, G., Eiter, T.: Prioritizing default logic. In: Hölldobler, S. (ed.) Intellectics and Computational Logic, vol. 19, pp. 27–45. Springer, Heidelberg (2000). https://doi.org/10.1007/978-94-015-9383-0_3

7. Brewka, G., Truszczynski, M., Woltran, S.: Representing preferences among sets. In: AAAI (2010)

8. Delgrande, J.P., Schaub, T., Tompits, H.: A framework for compiling preferences in logic programs. Theory Practice Logic Program. **3**(2), 129–187 (2003)

9. Delgrande, J.P., Schaub, T., Tompits, H., Wang, K.: A classification and survey of preference handling approaches in nonmonotonic reasoning. Comput. Intell. **20**(2), 308–334 (2004)

10. Dung, P.M.: On the acceptability of arguments and its fundamental role in nonmonotonic reasoning, logic programming and n-person games. Artif. Intell. **77**, 321–357 (1995)

11. Dung, P.M.: An axiomatic analysis of structured argumentation for prioritised default reasoning. In: ECAI 2014, pp. 267–272. IOS Press (2014)

12. Dung, P.M.: An axiomatic analysis of structured argumentation with priorities. Artif. Intell. **231**, 107–150 (2016)

13. Liao, B., Oren, N., van der Torre, L., Villata, S.: Prioritized norms and defaults in formal argumentation. In: Deontic Logic and Normative Systems (2016)

14. Modgil, S., Prakken, H.: A general account of argumentation with preferences. Artif. Intell. **195**, 361–397 (2013)

15. Pigozzi, G., Tsoukiàs, A., Viappiani, P.: Preferences in artificial intelligence. In: Annals of Mathematics and Artificial Intelligence, pp. 1–41 (2014)

16. Prakken, H.: An abstract framework for argumentation with structured arguments. Argument Comput. **1**(2), 93–124 (2010)

17. Reiter, R.: A logic for default reasoning. Artif. Intell. **13**, 81–132 (1980)

18. Schaub, T., Wang, K.: A comparative study of logic programs with preference: preliminary report. In: Answer Set Programming (2001)

19. Shoham, Y., Leyton-Brown, K.: Multiagent Systems: Algorithmic, Game-Theoretic, and Logical Foundations. Cambridge University Press, Cambridge (2008)

20. Young, A.P., Modgil, S., Rodrigues, O.: Argumentation Semantics for Prioritised Default Logic. arXiv preprint arXiv:1506.08813v2 (2015). http://arxiv.org/abs/1506.08813. Accessed 22 May 2016

21. Young, A.P., Modgil, S., Rodrigues, O.: Prioritised default logic as rational argumentation. In: Proceedings of the 2016 International Conference on Autonomous Agents & Multiagent Systems, pp. 626–634. International Foundation for Autonomous Agents and Multiagent Systems (2016)

ASPIC-END: Structured Argumentation with Explanations and Natural Deduction

Jérémie Dauphin[(✉)] and Marcos Cramer

University of Luxembourg, Luxembourg City, Luxembourg
jeremie.dauphin@uni.lu

Abstract. We propose ASPIC-END, an adaptation of the structured argumentation framework ASPIC+ which can incorporate explanations and natural deduction style arguments. We discuss an instantiation of ASPIC-END that models argumentation about explanations of semantic paradoxes (e.g. the Liar paradox), and we show that ASPIC-END satisfies rationality postulates akin to those satisfied by ASPIC+.

1 Introduction

In order to develop tools that intelligently support scientists in their interpretation of data and evaluation of theories, it is important to develop formal models of the argumentation and reasoning about conflicting information found in many academic disciplines. One promising methodology for approaching this problem is *structured argumentation theory* [4], which allows for a fine-grained model of argumentation and argumentative reasoning based on a logical language and evaluated according to the principles developed in *abstract argumentation theory*.

One of the dominant formal frameworks for structured argumentation is the *ASPIC+ framework* [12]. In ASPIC+, arguments are built from axioms and premises as well as from strict and defeasible rules, in a similar manner as proofs are built from axioms and rules in a Hilbert-style proof system. Three kinds of attacks between arguments, *undermines*, *undercuts* and *rebuttals*, are defined between arguments, and finally an *argumentation semantics* from Dung-style abstract argumentation theory [1,8] is applied to determine which sets of arguments can be rationally accepted.

Scientific discourse is characterized not only by the exchange of arguments in favour and against various scientific hypotheses, but also by the attempt to scientifically *explain* observed phenomena. In the context of abstract argumentation, Šešelja and Straßer [16] have therefore proposed to incorporate the notion of *explanation* into argumentation theory, in order to model scientific debate more faithfully. So far, this incorporation of explanation into argumentation theory has not been extended to the case of structured argumentation. One goal of the

Jérémie Dauphin has received funding from the European Union's Horizon 2020 research and innovation programme under the Marie Skłodowska-Curie grant agreement No. 690974 for the project "MIREL: MIning and REasoning with Legal texts".

© Springer International Publishing AG, part of Springer Nature 2018
E. Black et al. (Eds.): TAFA 2017, LNAI 10757, pp. 51–66, 2018.
https://doi.org/10.1007/978-3-319-75553-3_4

current paper is to work towards filling this gap by presenting on the one hand a general framework for incorporating explanation into structured argumentation, and on the other hand a particular proposal for how to define explanations in instantiations of that framework within a specific domain.

Scientific arguments often involve hypothetical reasoning, which involves reasoning based on an assumption or hypothesis that is locally assumed to be true for the sake of the argument, but to which there is no commitment on the global level. Such hypothetical reasoning is captured well by natural deduction proof systems, whereas the Hilbert-style definition of arguments in ASPIC+ cannot account for such hypothetical reasoning.

We propose an adaptation of the ASPIC+ framework called *ASPIC-END* that allows for incorporating explanations and hypothetical reasoning. In order to illustrate the usage of ASPIC-END, we consider its application to argumentation about explanations of semantic paradoxes, a research topic within the field of philosophical logic, and present a specific instantiation of the framework that models a simple example from this domain.

In order to ensure that the ASPIC-END framework behaves as one would rationally expect, we have proved multiple rationality postulates about ASPIC-END, as was previously done for ASPIC+ [11].

The paper is structured as follows: In Sect. 2, we discuss related work and motivate ASPIC-END. In Sect. 3, we formally define the ASPIC-END framework, and in Sect. 4, we instantiate it for argumentation about explanation of semantic paradoxes. In Sect. 5, we present, motivate and prove six rationality postulates for ASPIC-END, and in Sect. 6 we conclude.

2   Related Work and Motivation for ASPIC-END

The work of Dung [8] introduced the theory of *abstract argumentation*, in which one models arguments by abstracting away from their internal structure to focus on the relations of conflict between them. In *structured argumentation*, one models also the internal structure of arguments through a formal language in which arguments and counterarguments can be constructed [4]. One important family of frameworks for structured argumentation is the family of ASPIC-like frameworks, consisting among others of the original ASPIC framework [13], the ASPIC+ framework [12], and the ASPIC- framework [7]. We briefly sketch ASPIC+, as it is the basis for our framework ASPIC-END.

In ASPIC+, one starts with a knowledge base and a set of rules which allow one to make inferences from given knowledge. There are two kinds of rules: *Strict rules* logically entail their conclusion, whereas *defeasible rules* only create a presumption in favour of their conclusion. Arguments are built either by introducing an element of the knowledge base into the framework, or by making an inference based on a rule and the conclusions of previous arguments. Attacks between arguments are constructed either by attacking a fallible premise of an argument (*undermining*), by attacking the conclusion of a defeasible inference made within an argument (*rebuttal*), or by questioning the applicability of such a rule (*undercutting*). Preferences between arguments can be derived from preferences between rules. An abstract argumentation framework has then been built and acceptable arguments can be selected using any abstract argumentation semantics.

Caminada and Amgoud [6] have introduced the notion of *rationality postulates* for structured argumentation frameworks. These are conditions that structured argumentation frameworks would rationally be expected to satisfy, such as closure under strict rules of the output and consistency of the conclusions given consistency of the strict rules. Caminada and Amgoud [6] showed that the original ASPIC system did not satisfy these postulates, but proposed minor changes that made it satisfy them. These changes have been incorporated into ASPIC+ [11].

ASPIC-END features three main differences from ASPIC+. The first is that it allows for arguments to introduce an assumption on which to reason hypothetically, just like in natural deduction. In natural deduction, hypothetical derivations are employed in the inference schemes called ¬-Introduction (or *proof by contradiction*), →-Introduction, and ∨-Elimination (or *reasoning by cases*). Allowing for the usage of defeasible rules within hypothetical reasoning leads to specific problems that have been studied for the inference scheme of reasoning by cases in a recent paper by Beirlaen et al. [3]. In the current paper we avoid these problems by not allowing defeasible rules within hypothetical reasoning. However, a conclusion made on the basis of an inference scheme involving hypothetical reasoning may still be incorporated into an argument that uses defeasible rules, so that there is some integration of defeasible and hypothetical reasoning. In order to keep the presentation simple, our formal definition of ASPIC-END will only cover the case of the inference scheme of proof by contradiction, but reasoning by cases and →-Introduction can be treated analogously. Our proof-by-contradiction arguments bear a vague similarity to Caminada's *S-arguments* [5], which can attack an argument by showing that its conclusion leads to an absurdity. But unlike S-arguments, proof-by-contradiction arguments can be embedded into more complex arguments which make use of the negated conclusion of the proof-by-contradiction argument to conclude something else.

The second difference is that ASPIC-END has a notion of *explanations* additionally to the notion of arguments. This feature is based on the work of Šešelja and Straßer [16], who have extended Dung-style abstract argumentation with *explananda* (phenomena that need to be explained) and an *explanatory relation*, which allows arguments to either explain these explananda or deepen another argument's explanation. In Sect. 3, we will need some definitions from [16]:

Definition 1. An explanatory argumentation framework (EAF) is a tuple ⟨$\mathcal{A}, \mathcal{X}, \rightarrow, \dashrightarrow$⟩, where $\mathcal{A}$ is a set of arguments, $\mathcal{X}$ is a set of explananda, $\rightarrow$ is an attack relation between arguments and $\dashrightarrow$ is an explanatory relation from arguments to either explananda or arguments.

Sets of admissible arguments are then selected:

Definition 2. Let $\Delta = \langle \mathcal{A}, \mathcal{X}, \rightarrow, --\rightarrow \rangle$ be an EAF, $A \in \mathcal{A}$ and $S \subseteq \mathcal{A}$. We say that S is *conflict-free* iff there are no arguments $B, C \in S$ such that $B \rightarrow C$. We say that S *defends* A iff for every $B \in \mathcal{A}$ such that $B \rightarrow A$, there exists $C \in S$ such that $C \rightarrow B$. We say that S is *admissible* iff S is conflict-free and for all $B \in S$, S defends B.

The most suitable admissible sets are then selected by also taking into account their explanatory power and depth. These are measured by first identifying the explanations present in each set of arguments.

Definition 3. Let $\Delta = \langle \mathcal{A}, \mathcal{X}, \rightarrow, --\rightarrow \rangle$ be an EAF, $S \subseteq \mathcal{A}$ and $E \in \mathcal{X}$. An *explanation* $X[E]$ for E offered by S is a set $S' \subseteq S$ such that there exists a unique argument $A \in S'$ such that $A --\rightarrow E$ and for all $A' \in S' \setminus \{A\}$, there exists a path in $--\rightarrow$ from A' to A.

In order to be able to compare sets of arguments on how many explananda they can explain and in how much detail, the two following measures are required:

Definition 4. Let $\Delta = \langle \mathcal{A}, \mathcal{X}, \rightarrow, --\rightarrow \rangle$ be an EAF and $S, S' \subseteq \mathcal{A}$. Let $\mathcal{E}$ be the set of explananda S offers an explanation for and $\mathcal{E}'$ the set of explananda S' offers an explanation for. We say that S is *explanatory more powerful than* S' ($S >_p S'$) if and only if $\mathcal{E} \supsetneq \mathcal{E}'$.

Definition 5. Let $\Delta = \langle \mathcal{A}, \mathcal{X}, \rightarrow, --\rightarrow \rangle$ be an EAF and $S, S' \subseteq \mathcal{A}$. We say that S is *explanatory deeper than* S' ($S >_d S'$) if and only if for each explanation X' offered by S', there is an explanation X offered by S such that $X' \subseteq X$ and for at least one such X and X' pair, $X' \subsetneq X$.

Šešelja and Straßer [16] define two procedures for selecting the most suitable sets of arguments. The first procedure (for the *argumentative core*) consists in selecting the most explanatory powerful conflict-free sets, from which the maximal most defended sets are then retained. The second procedure (for the *explanatory core*) selects the most explanatory powerful conflict-free sets, from which the most defended sets are taken, and then from those selects the minimal explanatory deepest sets. In our formalism, we will slightly alter and reformulate these procedures.

The third difference is that ASPIC-END allows for arguments about the correct rules of logical reasoning. In ASPIC+, such arguments cannot be modeled, as the rules of logical reasoning represented by strict rules, and arguments involving only strict rules can never be attacked. Argumentation about the correct rules of logical reasoning is quite common within the field of philosophical logic, and additionally occurs not only in other areas of philosophy, e.g. in philosophy of science, but also in the study of logic within fields other than philosophy, e.g. in relation to the applications of logic to linguistics, law and Artificial Intelligence. For example, our *prima facie* intuitions suggest that it is a law of logic that a sentence that is not true must be false. However, the Kripke-Feferman solution to the Liar paradox [9,15] suggests that some sentences, such as the Liar sentence, are neither true nor false, since giving them either one of the two truth values leads to a contradiction. This solution is not putting forward an argument against the falsehood of the sentence by rebutting it, nor is it undermining any of the argument's premises. It is undercutting the argument by attacking the inference made from the negation of truth to falsehood.

It is true that outside the academic disciplines of philosophy and logic, argumentation about the correct rules of logical reasoning is very rare. But the goal of structured argumentation frameworks like ASPIC+ and ASPIC-END is to be largely domain-independent, and to therefore incorporate domain-specific assumptions into instantiations of the framework rather than into the framework itself. Given that there are some domains in which arguments about the correct rules of logical reasoning are sometimes put forward, the restriction that disallows such arguments to be modeled in ASPIC+ should be moved from the definition of the framework to the definition of those instantiations of the framework in which such arguments should indeed be disallowed.

To allow such arguments about the correct laws of logic to be modeled in ASPIC-END, we replace strict rules by *intuitively strict rules* whose applicability can be questioned, as in the case of defeasible rules in ASPIC+, but which behave like strict rules when their applicability is accepted. This means that conclusions of intuitively strict rules cannot be rebutted, just as for strict rules in ASPIC+. Intuitively strict rules represent *prima facie laws of logic*, i.e. purportedly logical inference rules which make sense at first but are open to debate.

3   ASPIC-END

In this section, we define ASPIC-END and motivate the details of its definition.

Definition 6. An *argumentation theory* is a tuple $(\mathcal{L}, \mathcal{R}, n, \leq)$, where:

- $\mathcal{L}$ is a logical language closed under the two unary connectives negation ($\neg$) and assumability (*Assumable*) such that $\perp \in \mathcal{L}$.
- $\mathcal{R} = \mathcal{R}_{is} \cup \mathcal{R}_d$ is a set of intuitively strict ($\mathcal{R}_{is}$) and defeasible ($\mathcal{R}_d$) rules of the form $\varphi_1, \ldots, \varphi_n \rightsquigarrow \varphi$ and $\varphi_1, \ldots, \varphi_n \Rightarrow \varphi$ respectively, where $n \geq 0$ and $\varphi_i, \varphi \in \mathcal{L}$.
- $n : \mathcal{R} \rightarrow \mathcal{L}$ is a partial function.
- $\mathcal{R}_{ce} := \{(\perp \rightsquigarrow \alpha) \mid \alpha \in \mathcal{L}\} \subseteq \mathcal{R}_{is}$, $\forall r' \in \mathcal{R}_{is} \setminus \mathcal{R}_{ce}$, $n(r') \in \mathcal{L}$, and $\forall r \in \mathcal{R}_{ce}$, $n(r)$ is undefined.
- $\leq$ is a reflexive and transitive relation over R_d which represents preference, with $a < b$ iff $a \leq b$ and $b \not\leq a$.

Note that we interpret $\perp$ not just as any contradiction but as the conjunction of all formulas in the language.

We now inductively define how to construct arguments. At the same time, we define five functions on arguments that specify certain features of any given argument: $\mathsf{Conc}(A)$ denotes the conclusion of argument A. $\mathsf{As}(A)$ denotes the set of assumptions under which argument A is operating (so whenever $\mathsf{As}(A) \neq \emptyset$, A is a hypothetical argument). $\mathsf{Sub}(A)$ denotes the set of sub-arguments of A.

DefRules(A) denotes the set of all defeasible rules used in A. TopRule(A) denotes the last inference rule which has been used in the argument if such a rule exists, and is undefined otherwise.

Definition 7. An *argument* A on the basis of an argumentation theory $\Sigma = (\mathcal{L}, \mathcal{R}, n, \leq)$ has one of the following forms:

1. $A_1, \ldots, A_n \rightsquigarrow \psi$, where $A_1, \ldots, A_n$ are arguments such that there exists an intuitively strict rule $\mathsf{Conc}(A_1), \ldots, \mathsf{Conc}(A_n) \rightsquigarrow \psi$ in $\mathcal{R}_{is}$.
   $\mathsf{Conc}(A) := \psi$, $\mathsf{As}(A) := \mathsf{As}(A_1) \cup \cdots \cup \mathsf{As}(A_n)$,
   $\mathsf{Sub}(A) := \mathsf{Sub}(A_1) \cup \cdots \cup \mathsf{Sub}(A_n) \cup \{A\}$,
   $\mathsf{DefRules}(A) := \mathsf{DefRules}(A_1) \cup \cdots \cup \mathsf{DefRules}(A_n)$,
   $\mathsf{TopRule}(A) := \mathsf{Conc}(A_1), \ldots, \mathsf{Conc}(A_n) \rightsquigarrow \psi$.
2. $A_1, \ldots, A_n \Rightarrow \psi$, where $A_1, \ldots, A_n$ are arguments s.t. $\mathsf{As}(A_1) \cup \cdots \cup \mathsf{As}(A_n) = \emptyset$ and there exists a defeasible rule $\mathsf{Conc}(A_1), \ldots, \mathsf{Conc}(A_n) \Rightarrow \psi$ in $\mathcal{R}_d$.
   $\mathsf{Conc}(A) := \psi$, $\mathsf{As}(A) := \emptyset$,
   $\mathsf{Sub}(A) := \mathsf{Sub}(A_1) \cup \cdots \cup \mathsf{Sub}(A_n) \cup \{A\}$,
   $\mathsf{DefRules}(A) := \mathsf{DefRules}(A_1) \cup \cdots \cup \mathsf{DefRules}(A_n) \cup$
   $\{\mathsf{Conc}(A_1), \ldots, \mathsf{Conc}(A_n) \Rightarrow \psi\}$,
   $\mathsf{TopRule}(A) := \mathsf{Conc}(A_1), \ldots, \mathsf{Conc}(A_n) \Rightarrow \psi$.
3. $\mathsf{Assume}(\varphi)$, where $\varphi \in \mathcal{L}$. $\mathsf{Conc}(A) := \varphi$, $\mathsf{As}(A) := \{\varphi\}$, $\mathsf{Sub}(A) := \{\mathsf{Assume}(\varphi)\}$, $\mathsf{DefRules}(A) := \emptyset$, $\mathsf{TopRule}(A)$ is undefined.
4. $\mathsf{ProofByContrad}(\neg\varphi, A')$, where A' is an argument such that $\varphi \in \mathsf{As}(A')$ and $\mathsf{Conc}(A') = \bot$, with:
   $\mathsf{Conc}(A) = \neg\varphi$, $\mathsf{As}(A) = \mathsf{As}(A') \setminus \{\varphi\}$,
   $\mathsf{Sub}(A) = \mathsf{Sub}(A') \cup \{\mathsf{ProofByContrad}(\neg\varphi, A')\}$,
   $\mathsf{DefRules}(A) = \mathsf{DefRules}(A')$,
   $\mathsf{TopRule}(A)$ is undefined.

Notice that we do not allow for the use of defeasible rules within hypothetical arguments. We do however allow for the conclusions of defeasible arguments to be imported inside of a proof by contradiction. This is motivated by the fact that allowing for proofs by contradiction amounts to allowing for transpositions of any rule that can be used within a proof by contradiction, and transpositions are usually assumed only for strict rules in structured argumentation [6,11].

We now need to define the attack relation in our framework. Notice that in ASPIC-END, we also allow for an argument A to attack an argument B which makes an assumption φ if A concludes that φ is not assumable. For example, if one were to assume that the number 5 is yellow, since numbers do not have colors, it should be possible to attack the argument that introduces this assumption and any argument making an inference from this assumption.

Definition 8. Let $\Sigma = (\mathcal{L}, \mathcal{R}, n, \leq)$ be an argumentation theory and A, B two arguments on the basis of Σ. We say that A *attacks* B iff A *rebuts*, *undercuts* or *assumption-attacks* B, where:

- A *rebuts* argument B (on B') iff $\mathsf{Conc}(A) = \neg\varphi$ or $\neg\mathsf{Conc}(A) = \varphi$ for some $B' \in \mathsf{Sub}(B)$ of the form $B_1'', \ldots, B_n'' \Rightarrow \varphi$ and $\mathsf{As}(A) = \emptyset$.

- A *undercuts* argument B (on B') iff $\mathsf{Conc}(A) = \neg n(r)$ or $\neg\mathsf{Conc}(A) = n(r)$ for some $B' \in \mathsf{Sub}(B)$ such that $\mathsf{TopRule}(B') = r$, there is no $\varphi \in \mathsf{As}(B')$ such that $\neg\varphi = \mathsf{Conc}(A')$ or $\varphi = \neg\mathsf{Conc}(A')$ for some $A' \in \mathsf{Sub}(A)$, and there are arguments $B_1, ..., B_n$ such that $B_1 = B'$, $B_n = B$, $B_i \in \mathsf{Sub}(B_{i+1})$ for $1 \leq i < n$ and $\mathsf{As}(A) \subseteq \mathsf{As}(B_1) \cup \cdots \cup \mathsf{As}(B_n)$.
- A *assumption-attacks* B (on B') iff for some $B' \in \mathsf{Sub}(B)$ such that $B' = \mathsf{Assume}(\varphi)$, $\mathsf{Conc}(A) = \neg Assumable(\varphi)$ and $\mathsf{As}(A) = \emptyset$.

We require that any attacking argument A is making fewer assumptions than the B' it attacks, as to prevent arguments from attacking outside of their assumption scope. Note that in the case of rebuttal, since the attacked argument cannot have assumptions, we require that the attacking argument have none either.

In the case of undercutting, we also have the requirement that A does not use the contrary of any assumptions made by B' in any of its inferences, since the attack would not stand in the scope of B'. Additionally, we allow A to make use of any assumptions appearing in the chain of arguments leading B' to B, as these assumptions, even if they have been retracted, still constitute valid grounds on which to form an attack.

Similarly as in ASPIC+, one can also define a notion of successful attack by lifting the preference relation from rules to arguments as follows:

Definition 9. Let $\Sigma = (\mathcal{L}, \mathcal{R}, n, \leq)$ be an argumentation theory and A, B be two arguments on the basis of Σ. We define the *lifting of* $\leq$ *to arguments* $\preceq$ to be such that $A \preceq B$ iff there exists $r_a \in \mathsf{DefRules}(A)$, such that for all $r_b \in \mathsf{DefRules}(B)$, we have $r_a \leq r_b$. We also define $A \prec B$ by replacing $\leq$ with $<$ in the definition of $\preceq$.

Notice that this lifting corresponds to elitist weakest-link as described in [12]. We believe that this ordering is best suited for modeling philosophical and scientific arguments.

We now define what it means for an attack to be successful:

Definition 10. Let $\Sigma = (\mathcal{L}, \mathcal{R}, n, \leq)$ be an argumentation theory, A, B be two arguments on the basis of Σ. We say that A *successfully rebuts* B iff A rebuts B on B' for some argument B' and $A \not\prec B'$, and that A *defeats* B iff A assumption-attacks, undercuts or successfully rebuts B.

The aim of our system is to generate an EAF as defined in Sect. 2. For this three things need to be specified: A set $\mathcal{X}$ of explananda, a condition under which an argument explains an explanandum, and a condition under which an argument explains another argument. The first two of these three details are domain-specific, and are thus to be specified in an instantiation of the ASPIC-END framework. The third one, on the other hand, should be the same in all domains. The reason for this can be found in the informal clarification that Šešelja and Straßer [16] provided for what it means to say that an argument b explains an argument a: "argument b can be used to explain one of the premises of argument a [...] or the link between the premises and the conclusion."

In the context of structured argumentation, this informal clarification can be turned into a formal definition:

Definition 11. Let A, B be arguments. We say that B *explains* A (on A') iff $A' \in \mathsf{Sub}(A)$, $\mathsf{As}(B) \subseteq \mathsf{As}(A')$ and at least one of the following two cases holds:

- $A' \notin \mathsf{Sub}(B)$ and either $A' = (\rightsquigarrow \mathsf{Conc}(B))$ or $A' = (\Rightarrow \mathsf{Conc}(B))$.
- $\mathsf{Conc}(B) = n(\mathsf{TopRule}(A'))$ and $\nexists B' \in \mathsf{Sub}(B)$ such that $\mathsf{TopRule}(B') = \mathsf{TopRule}(A')$.

Intuitively, the idea behind this definition is that an argument B explains another argument A if B non-trivially concludes one of A's premises or one of the inference rules used by A.

We now have all the elements needed to build an EAF.

Definition 12. Let $\Sigma = (\mathcal{L}, \mathcal{R}, n, \leq)$ be an argumentation theory. Let $\mathcal{X}$ be a set of explananda, and let $\mathcal{C}$ be a criterion for determining whether an argument constructed from Σ explains a given explanandum $E \in \mathcal{X}$. The *explanatory argumentation framework* (EAF) *defined by* $(\Sigma, \mathcal{X}, C)$ is a tuple $\langle \mathcal{A}, \mathcal{X}, \rightarrow, \dashrightarrow \rangle$, where:

- $\mathcal{A}$ is the set of all arguments that can be constructed from Σ satisfying Definition 7;
- $(A, B) \in \rightarrow$ iff A defeats B, where $A, B \in \mathcal{A}$;
- $(A, E) \in \dashrightarrow$ iff criterion $\mathcal{C}$ is satisfied with respect to A and E, where $A \in \mathcal{A}$ and $E \in \mathcal{X}$;
- $(A, B) \in \dashrightarrow$ iff A explains B according to Definition 11, where $A, B \in \mathcal{A}$.

Once such a framework has been generated, we want to be able to extract the most interesting sets of arguments. Such a set should be able to explain as many explananda in as much detail as possible, while being self-consistent and plausible.

We define two kinds of extensions corresponding to the two selection procedures defined by Šešelja and Straßer [16]. As suggested in the informal discussion in their paper, we chose to give higher importance to the criterion of defense compared to the criterion of explanatory power. This prevents some absurd theories which manage to explain all explananda but cannot defend themselves against all attacks from beating plausible theories which fail to explain some of the explananda but are sound and fully defended.

Definition 13. Let $\Sigma = (\mathcal{L}, \mathcal{R}, n, \leq)$ be an argumentation theory, $\Delta = \langle \mathcal{A}, \mathcal{X}, \rightarrow, \dashrightarrow \rangle$ the EAF defined by Σ and $S \subseteq \mathcal{A}$ a set of arguments.

1. We say that S is *satisfactory* iff S is admissible and there is no $S' \subseteq \mathcal{A}$ such that $S' >_p S$ and S' is admissible.
2. We say that S is *insightful* iff S is satisfactory and there is no $S' \subseteq \mathcal{A}$ such that $S' >_d S$ and S' is satisfactory.
3. We say that S is an *argumentative core extension* (*AC-extension*) of Δ iff S is satisfactory and there is no $S' \supset S$ such that S' is satisfactory.

4. We say that S is an *explanatory core extension (EC-extension)* of Δ iff S is insightful and there is no $S' \subset S$ such that S' is insightful.

The AC-extensions are sets of arguments which represent the theories explaining the most explananda, together with all other compatible beliefs present in the framework. EC-extensions represent the core of those theories and only include the arguments which defend or provide details for them.

We define the conclusions of the arguments in a given extension as follows:

Definition 14. Let $\Sigma = (\mathcal{L}, \mathcal{R}, n, \leq)$ be an argumentation theory, $\Delta = \langle \mathcal{A}, \mathcal{X}, \rightarrow, --\rightarrow \rangle$ be the EAF defined by Σ and S be an extension of Δ. Then, we define the *conclusions of* S, denoted $\mathsf{Concs}(S)$, to be $\mathsf{Concs}(S) = \{\mathsf{Conc}(A) | A \in S \text{ s.t. } \mathsf{As}(A) = \emptyset\}$.

4 Modelling Explanations of Semantic Paradoxes in ASPIC-END

In this section, we discuss how ASPIC-END can be applied to modelling argumentation about explanations of semantic paradoxes, and illustrate this potential application with a simple example. We start by briefly motivating this application of structured argumentation theory.

Philosophy is an academic discipline in which good argumentative skills are a central part of every student's training. Philosophical texts are often much richer in explicit formulation of arguments than texts from other academic disciplines. For these reasons, we believe that modeling arguments from philosophical textbooks, monographs and papers can be an interesting test case for structured argumentation theory.

Different areas of philosophy vary with respect to how much logical rigor is commonly applied in the presentation of arguments. Even logically rigorous argumentation poses many interesting problems, as the rich literature on abstract and structured argumentation attests. In order to not confound these interesting problems with issues arising from the lack of logical rigor, it is a good idea to concentrate on the study of logically rigorous argumentation. Philosophical logic is an area of logic where logically rigorous arguments abound. One topic that has gained a lot of attention in philosophical logic is the study of semantic paradoxes such as the Liar paradox and Curry's paradox [2,10]. We therefore use the argumentation about the various explanations of the paradoxes that have been proposed in the philosophical literature as a test case for structured argumentation theory.

In our application of ASPIC-END to argumentation about explanations of semantic paradoxes, the explananda are the paradoxes (i.e. arguments that derive an absurdity under no assumption without using defeasible rules), which other arguments can explain by attacking the said derivation. So we instantiate the set $\mathcal{X}$ of explananda and criterion $\mathcal{C}$ for an explanation of an explanandum by an argument as specified in the following two definitions:

Definition 15. Let $\Sigma = (\mathcal{L}, \mathcal{R}, n, \leq)$ be an argumentation theory. For every argument A on the basis of Σ such that $\mathsf{DefRules}(A) = \emptyset, \mathsf{As}(A) = \emptyset$ and $\mathsf{Conc}(A) = \bot$, we stipulate an explanandum E_A, and say that $\mathsf{Source}(E_A) = A$. We define the set $\mathcal{X}$ of explananda based on Σ to be the set of all explananda E_A that we have thus stipulated.

Definition 16. Let $\Sigma = (\mathcal{L}, \mathcal{R}, n, \leq)$ be an argumentation theory, A, B arguments and E an explanandum based on Σ. We say that A *explains* E iff A defeats $\mathsf{Source}(E)$.

The following example illustrates an application of ASPIC-END to a version of the Liar paradox and two very simple explanations of it:[1]

Example: Define L to be the sentence "L is false". If L is true, i.e. "L is false" is true, then L is false, which is a contradiction. So L is not true, i.e. L is false. So "L is false" is true, i.e. L is true. So we have the contradiction that L is both true and false from no assumption.

A truth-value gap explanation: L is neither true nor false. When concluding that L is false because L is not true, we are making the assumption that any sentence is either true or false. This assumption does not hold for problematically self-referential sentences such as L.

A paracomplete explanation: The reasoning that led to the conclusion that L is not true is a proof by contradiction that derives a contradiction from the assumption that L is true. However, a proof by contradiction based on assumption ϕ can only be accepted once one accepts that the law of excluded middle holds for ϕ, i.e. that $\phi \vee \neg\phi$. However, the law of excluded middle should not be accepted for problematically self-referential statements like L, and thus also not to the statement "L is true". So "L is true" cannot be assumed for a proof by contradiction.

We now proceed to the ASPIC-END model of the reasoning and argumentation involved in the paradox and the two explananda. We use T, F and Psr to mean *true, false* and *problematically self-referential* respectively. The rules in our model are $\mathcal{R}_{is} = \{T(L) \rightsquigarrow T(F(L)); T(F(L)) \rightsquigarrow F(L); T(L), F(L) \rightsquigarrow \bot; \neg T(L) \rightsquigarrow F(L); F(L) \rightsquigarrow T(F(L)); T(F(L)) \rightsquigarrow T(L)\}$ with $n(\neg T(L) \rightsquigarrow F(L)) = r_1$ and $\mathcal{R}_d = \{ \Rightarrow \neg T(L) \wedge \neg F(L); \neg T(L) \wedge \neg F(L) \Rightarrow \neg r_1; \Rightarrow Psr(L); Psr(L) \Rightarrow \neg T(L) \wedge \neg F(L); \neg T(L) \wedge \neg F(L) \Rightarrow \neg Assumable(T(L))\}$. We also define the predicate $Expl$ to be: $Expl(A)$ iff $\mathsf{DefRules}(A) = \emptyset, \mathsf{As}(A) = \emptyset$ and $\mathsf{Conc}(A) = \bot$.

Infinitely many arguments can be constructed from this argumentation theory. However, the following set of arguments is the set of most relevant arguments, in the sense that other arguments will not defeat these arguments and will not add relevant new conclusions.

---

[1] See [10] for comprehensive presentations of truth-value gap and paracomplete explanations, besides many others.

$$A_1 = \mathsf{ProofByContrad}(\neg T(L), (\mathsf{Assume}(T(L)),$$
$$((\mathsf{Assume}(T(L)) \rightsquigarrow T(F(L))) \rightsquigarrow F(L)) \rightsquigarrow \bot)) \rightsquigarrow F(L)$$
$$A_2 = ((A_1 \rightsquigarrow T(F(L))) \rightsquigarrow T(L)), A_1 \rightsquigarrow \bot$$
$$B_1 = (\Rightarrow Psr(L)) \Rightarrow \neg T(L) \wedge \neg F(L)$$
$$B_2 = (\Rightarrow \neg T(L) \wedge \neg F(L)) \Rightarrow \neg r_1$$
$$C = ((\Rightarrow Psr(L)) \Rightarrow \neg T(L) \wedge \neg F(L)) \Rightarrow \neg Assumable(T(L))$$

We get the explanandum E with $\mathsf{Source}(E) = A_2$. B_2 defeats A_2 on A_1 and C defeats A_2 on $\mathsf{Assume}(T(L))$, thus they both explain E. B_1 explains B_2 by non-trivially concluding $\neg T(L) \wedge \neg F(L)$. The AC-extension is $\{B_1, B_2, C\}$ and the EC-extensions are $\{B_1, B_2\}$ and $\{C\}$ (Fig. 1).[2]

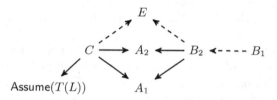

Fig. 1. The relevant arguments, explanandum, attacks and explanations from Example

5   Closure and Rationality Postulates

In this section, we show that ASPIC-END satisfies four rationality postulates analogous to the four postulates that Modgil and Prakken [11] have established for ASPIC+, as well as two new postulates motivated by the application of structured argumentation to the domain of philosophical logic.

The first postulate concerns the closure of the extensions under the sub-argument relation. The idea is that one cannot accept an argument while rejecting part of it.

Theorem 1. Let $\Sigma = (\mathcal{L}, \mathcal{R}, n, \leq)$ be an argumentation theory, $\Delta = \langle \mathcal{A}, \mathcal{X}, \rightarrow, -\rightarrow \rangle$ be the EAF defined by Σ and S be an AC-extension of Δ. Then, for all $A \in S$, $\mathsf{Sub}(A) \subseteq S$.

The proof of Theorem 1 rests on the following lemma, which can be proven in a straightforward way as in the case of ASPIC+ (see Lemma 35 of [11]):

---

[2] Notice that both solutions appear in the same AC-extension. This is only due to the brevity of our example. In a more comprehensive exposition of these explanations, arguments attacking other explanations would be included, and thus each AC-extension would contain no more than one solution.

Lemma 1. Let $\Sigma = (\mathcal{L}, \mathcal{R}, n, \leq)$ be an argumentation theory, $\Delta = \langle \mathcal{A}, \mathcal{X}, \rightarrow, \dashrightarrow \rangle$ be the EAF defined by Σ, $S \subseteq \mathcal{A}$ and $A, B \in \mathcal{A}$. We have that:

1. If S defends A and $S \subseteq S'$, then S' defends A.
2. If A defeats B' and $B' \in \mathsf{Sub}(B)$, then A defeats B.
3. If S defends A and $A' \in \mathsf{Sub}(A)$, then S defends A'.

Proof of Theorem 1: Let $A \in S$ and $A' \in \mathsf{Sub}(A)$. Suppose $S \cup \{A'\}$ is not conflict-free. Then, either some $B \in S$ defeats A', or A' defeats some $B' \in S$. Since S defends itself, if A' defeats $B' \in S$, then there exists B which defeats A'. So in both cases there exists $B \in S$ which defeats A'. But then by Lemma 1.2, B defeats A, so S is not conflict-free, which is a contradiction. So $S \cup \{A'\}$ is conflict-free. Also, since S defends A, by Lemma 1.3, S also defends A'. Hence, by maximality of the AC-extensions, $A' \in S$. □

Notice that this postulate does not hold for EC-extensions, as they are by definition minimal in their inclusion of arguments, and thus will often leave out low-level sub-arguments.

The second postulate concerns the closure of the conclusions under intuitively strict rules. In the case of ASPIC+, the corresponding postulate concerned the closure of the conclusions under all strict rules (see Theorem 13 in [11]). But since ASPIC-END allows for the rejection of intuitively strict rules, it is undesirable to consider the closure under all of them. Instead, we consider the closure under the accepted intuitively strict rule. The following two definitions define the set of *accepted* intuitively strict rules and the *closure* under a given set of intuitively strict rules:

Definition 17. Let $\Sigma = (\mathcal{L}, \mathcal{R}, n, \leq)$ be an argumentation theory, $\Delta = \langle \mathcal{A}, \mathcal{X}, \rightarrow, \dashrightarrow \rangle$ be the EAF defined by Σ and S be an extension of Δ. The *set of intuitively strict rules accepted by* S is $\mathcal{R}_{isa}(S) = \{r \in \mathcal{R}_{is} | \forall A \in \mathcal{A} \text{ s.t. } \mathsf{As}(A) = \emptyset \text{ and } \mathsf{Conc}(A) = \neg n(r) \text{ or } \neg\mathsf{Conc}(A) = n(r), \exists B \in S \text{ s.t. } B \text{ defeats } A\}$.

Definition 18. Let $\Sigma = (\mathcal{L}, \mathcal{R}, n, \leq)$ be an argumentation theory, $P \subseteq \mathcal{L}$ and $R' \subseteq \mathcal{R}_{is}$. We define the *closure of P under the set of rules R'*, denoted $Cl_{R'}(P)$, as the smallest set such that $P \subseteq Cl_{R'}(P)$, and when $(\varphi_1, ..., \varphi_n \rightsquigarrow \psi) \in R'$ and $\varphi_1, ..., \varphi_n \in Cl_{R'}(P)$, then $\psi \in Cl_{R'}(P)$.

Now the postulate on the closure under accepted intuitively strict rules can be formulated as follows:

Theorem 2. Let $\Sigma = (\mathcal{L}, \mathcal{R}, n, \leq)$ be an argumentation theory, $\Delta = \langle \mathcal{A}, \mathcal{X}, \rightarrow, \dashrightarrow \rangle$ be the EAF defined by Σ and S be an AC-extension of Δ. Then, $\mathsf{Conc}(S) = Cl_{\mathcal{R}_{isa}(S)}(\mathsf{Concs}(S))$.

Proof: We need to show that if $(\varphi_1, ..., \varphi_n \rightsquigarrow \psi) \in \mathcal{R}_{isa}(S)$ and $\varphi_1, ..., \varphi_n \in \mathsf{Concs}(S)$, then $\psi \in \mathsf{Concs}(S)$. Supposing these conditions are met, there exist arguments $A_1, ..., A_n$ with conclusions $\varphi_1, ..., \varphi_n$ respectively. We can then construct $A = A_1, ..., A_n \rightsquigarrow \psi$. Since $A_1, ..., A_n$ are defended by S and $\mathsf{TopRule}(A)$ is accepted by S, A is also defended by S, so $A \in S$. Hence, $\psi \in \mathsf{Concs}(S)$. □

The last two postulates presented in [11] are direct and indirect consistency, which state that when the set of strict rules is consistent, the set of conclusions and the closure of this set under strict rules are consistent.

We have three requirements for applying the consistency postulates. The first is that there cannot be non-defeasible arguments which contradict each other. The second requirement ensures that a formula and its negation are considered as contradictory and the third guarantees that no assumptions are prevented. The last two requirements are motivated by the consideration that in the applications of ASPIC-END not related to paradoxes, one would likely accept classical or intuitionistic logic, for both of which these requirements hold.

Definition 19. Let $\Sigma = (\mathcal{L}, \mathcal{R}, n, \leq)$ be an argumentation theory. We say that Σ is *consistency-inducing* iff:

1. there are no $A, B \in \mathcal{A}$ such that $\mathsf{DefRules}(A) = \mathsf{DefRules}(B) = \emptyset$ and $\mathsf{Conc}(A) = \neg\mathsf{Conc}(B)$,
2. for each $\varphi \in \mathcal{L}$ there is a rule r_φ of the form $\varphi, \neg\varphi \rightsquigarrow \bot \in \mathcal{R}_{is}$ such that $n(r_\varphi)$ is undefined,
3. there is no rule $r \in \mathcal{R}$ such that the conclusion of r is of the form $\neg Assumable(\varphi)$.

The following theorem establishes direct consistency for ASPIC-END:

Theorem 3. Let $\Sigma = (\mathcal{L}, \mathcal{R}, n, \leq)$ be a consistency-inducing argumentation theory, $\Delta = \langle \mathcal{A}, \mathcal{X}, \rightarrow, \dashrightarrow \rangle$ be the EAF defined by Σ and S be an AC or EC-extension of Δ. Then, there does not exist $\varphi \in \mathsf{Concs}(S)$ such that $\neg\varphi \in \mathsf{Concs}(S)$.

Proof: Suppose for a contradiction that there exists $\varphi \in \mathsf{Conc}(S)$ such that $\neg\varphi \in \mathsf{Conc}(S)$. Then, there exist two arguments $A, B \in S$ such that $\mathsf{Conc}(A) = \varphi$ and $\mathsf{Conc}(B) = \neg\varphi$. Since Σ is consistency-inducing, at least one of A and B has a defeasible sub-argument. For each maximal (w.r.t Sub) sub-argument C of A with a defeasible top rule, let A_C be the copy of A that has $\mathsf{Assume}(\mathsf{Conc}(C))$ instead of C (so $\mathsf{As}(A_C) = \{\mathsf{Conc}(C)\}$), and let D_C be $\mathsf{ProofByContrad}(\neg\mathsf{Conc}(C), A_C, B \rightsquigarrow \bot)$ (so D_C rebuts C). We can do this as well for every maximal sub-argument of B with a defeasible top rule. Then for at least one such sub-argument C of A or B, say of A, $A_C \not< C$ and $B \not< C$, hence $D_C \not< C$, and so D_C will defeat C. Then D_C defeats A on C. So some $F \in S$ defeats D_C. Since $B \in S$, F does not defeat B, so F defeats A_C. Since $\mathsf{Conc}(F) \neq \neg Assumable(\mathsf{Conc}(C))$ by item 3 of Definition 19 and F does not defeat A, $\mathsf{As}(F) = \{C\}$. By Theorem 1, $C \in S$. Let F' be hte copy of F that has C instead of $\mathsf{Assume}(C)$. Then F' defeats A. So some argument $G \in S$ defeats F'. but then G defeats F or C, which is a contradiction. ☐

Indirect consistency of AC-extensions follows from closure under accepted intuitively strict rules together with direct consistency:

Theorem 4. Let $\Sigma = (\mathcal{L}, \mathcal{R}, n, \leq)$ be a consistency-inducing argumentation theory, $\Delta = \langle \mathcal{A}, \mathcal{X}, \rightarrow, --\rightarrow \rangle$ be the EAF defined by Σ and S be an AC-extension of Δ. Then, there does not exist $\varphi \in Cl_{\mathcal{R}_{isa}(S)}(\mathsf{Concs}(S))$ such that $\neg \varphi \in Cl_{\mathcal{R}_{isa}(S)}(\mathsf{Concs}(S))$.

As explained in Sect. 2, we want ASPIC-END to be applicable to domains like philosophical logic, in which the correctness of logical rules can be up for debate. Among the proposals made by philosophers of how to handle the semantic paradoxes, there is paraconsistent dialetheism [14], which accepts some inconsistencies as true and uses a paraconsistent logic to avoid that everything can be derived. And in order to be able to show the internal structure of the paradox, we need to have an inconsistency arise from intuitively strict rules under no assumptions. For these reasons, the consistency postulates do not make sense for this kind of application of ASPIC-END.

However, there is a property similar to consistency that should still hold even when the intuitively strict rules lead to paradoxes and when the output extensions contain one that accepts paraconsistent dialetheism, namely that an extension should never be trivial, i.e. conclude everything.

For the non-triviality of the extensions, we require that rules are present in the framework which allow one to derive any formula from $\perp$.[3] We also require these rules of conjunction elimination from $\perp$ not to have a corresponding formula in $\mathcal{L}$ as a name, which prevents them from being attackable. Also, we require every other intuitively strict rule to have a name so that it can be attacked. We say that the argumentation theory is well-defined if it satisfies these requirements, and assume well-definedness in the non-triviality postulate stated in Theorem 5.

Theorem 5. Let $\Sigma = (\mathcal{L}, \mathcal{R}, n, \leq)$ be an argumentation theory, $\Delta = \langle \mathcal{A}, \mathcal{X}, \rightarrow, --\rightarrow \rangle$ be the EAF defined by Σ, and S be an AC or EC-extension of Δ. Then, $\perp \notin \mathsf{Concs}(S)$.

Proof: Suppose for a contradiction that $\perp \in \mathsf{Concs}(S)$. Then there exists a minimal (under sub-argument relation) argument $A \in S$ such that $\mathsf{Conc}(A) = \perp$ and $\mathsf{As}(A) = \emptyset$. Let $r = \mathsf{TopRule}(A)$. If $r \in \mathcal{R}_{is}$, then from Definition 6, $n(r) \in \mathcal{L}$ and so let $B = A \rightsquigarrow \neg n(r)$. Otherwise, let $B = A \rightsquigarrow \neg \perp$. By Definition 9, $B \not\prec A$. Then B undercuts or successfully rebuts A on A, so B defeats A. Since S is an AC- or EC-extension of Δ, it defends itself, so there exists $C \in S$ such that C defeats B. Suppose for a contradiction that C defeats B on $B' \neq B$. Since $\mathsf{Sub}(B) = \mathsf{Sub}(A) \cup \{B\}$, $B' \in \mathsf{Sub}(A)$. Then, by Lemma 1.2, C defeats A on B'. But S is conflict-free, so we have a contradiction. Hence, C defeats B on B. Since $B = A \rightsquigarrow \neg n(r)$, B cannot be rebutted nor assumption-attacked. Hence, C undercuts B on B. But from Definition 6 and since $\mathsf{TopRule}(B) \in \mathcal{R}_{ce}$, $n(\mathsf{TopRule}(B))$ is undefined, i.e. no argument undercuts B on B, a contradiction. Hence, $\perp \notin \mathsf{Concs}(S)$. $\qquad \square$

---

[3] As noted earlier, we interpret $\perp$ as the conjunction of all formulas in $\mathcal{L}$, so these rules are in effect conjunction elimination rules.

Indirect non-triviality of AC-extensions then follows from closure under accepted intuitively strict rules and direct non-triviality:

Theorem 6. Let $\Sigma = (\mathcal{L}, \mathcal{R}, n, \leq)$ be an argumentation theory, $\Delta = \langle \mathcal{A}, \mathcal{X}, \rightarrow , \dashrightarrow \rangle$ be the EAF defined by Σ and S be an AC-extension of Δ. Then, $\perp \notin Cl_{\mathcal{R}_{isa}(S)}(\mathsf{Concs}(S))$.

6   Conclusion and Future Work

We have proposed a modification of ASPIC+ called ASPIC-END, which incorporates a formal model of explanations, and features natural-deduction style arguments. We have shown how ASPIC-END can be instantiated for modelling argumentation about explanations of semantic paradoxes in ASPIC-END. Finally, we have shown that ASPIC-END satisfies rationality postulates analogous to those satisfied by ASPIC+, as well as non-triviality postulates that are relevant in the application to semantic paradoxes.

One topic of our future work on ASPIC-END is to study possible ways of instantiating explananda and explanations in other scientific domains. For explanations from the natural sciences, this might require an instantiation of ASPIC-END with a language covering causal notions. Furthermore, we will study the possibility of integrating the new results of Beirlaen et al. [3] on reasoning by cases in structured argumentation with our work on ASPIC-END.

References

1. Baroni, P., Caminada, M., Giacomin, M.: An introduction to argumentation semantics. Knowl. Eng. Rev. **26**(4), 365–410 (2011)
2. Beall, J.C., Glanzberg, M., Ripley, D.: Liar paradox. In: Zalta, E.N. (ed.) The Stanford Encyclopedia of Philosophy. Metaphysics Research Lab, Stanford University, Winter 2016 Edition (2016)
3. Beirlaen, M., Heyninck, J., Straßer, C.: Reasoning by cases in structured argumentation. In: Proceedings of SAC/KRR (2017, in press)
4. Besnard, P., Garcia, A., Hunter, A., Modgil, S., Prakken, H., Simari, G., Toni, F.: Introduction to structured argumentation. Argument Comput. **5**(1), 1–4 (2014)
5. Caminada, M.: A formal account of socratic-style argumentation. J. Appl. Logic **6**(1), 109–132 (2008)
6. Caminada, M., Amgoud, L.: On the evaluation of argumentation formalisms. Artif. Intell. **171**(5–6), 286–310 (2007)
7. Caminada, M., Modgil, S., Oren, N.: Preferences and unrestricted rebut. In: Computational Models of Argument - Proceedings of COMMA 2014, pp. 209–220 (2014)
8. Dung, P.M.: On the acceptability of arguments and its fundamental role in non-monotonic reasoning, logic programming and n-person games. Artif. Intell. **77**(2), 321–357 (1995)
9. Feferman, S.: Reflecting on incompleteness. J. Symbolic Logic **56**(01), 1–49 (1991)
10. Field, H.: Saving Truth from Paradox. Oxford University Press, Oxford (2008)
11. Modgil, S., Prakken, H.: A general account of argumentation with preferences. Artif. Intell. **195**, 361–397 (2013)

12. Modgil, S., Prakken, H.: The ASPIC+ framework for structured argumentation: a tutorial. Argument Comput. **5**(1), 31–62 (2014)
13. Prakken, H.: An abstract framework for argumentation with structured arguments. Argument Comput. **1**(2), 93–124 (2010)
14. Priest, G.: In Contradiction: A Study of the Transconsistent. Oxford University Press, Oxford (2006)
15. Reinhardt, W.N.: Some remarks on extending and interpreting theories with a partial predicate for truth. J. Philos. Logic **15**(2), 219–251 (1986)
16. Šešelja, D., Straßer, C.: Abstract argumentation and explanation applied to scientific debates. Synthese **190**(12), 2195–2217 (2013)

On the Links Between Argumentation-Based Reasoning and Nonmonotonic Reasoning

Zimi Li[1(✉)], Nir Oren[2(✉)], and Simon Parsons[3(✉)]

[1] Department of Computer Science, Graduate Center,
City University of New York, New York City, USA
zli2@gradcenter.cuny.edu
[2] Department of Computing Science, University of Aberdeen, Aberdeen, UK
n.oren@abdn.ac.uk
[3] Department of Informatics, King's College London, London, UK
simon.parsons@kcl.ac.uk

Abstract. In this paper we investigate the links between instantiated argumentation systems and the axioms for non-monotonic reasoning described in [15] with the aim of characterising the nature of argument based reasoning. In doing so, we consider two possible interpretations of the consequence relation, and describe which axioms are met by ASPIC$^+$ under each of these interpretations. We then consider the links between these axioms and the rationality postulates. Our results indicate that argument based reasoning as characterised by ASPIC$^+$ is—according to the axioms of [15]—non-cumulative and non-monotonic, and therefore weaker than the weakest non-monotonic reasoning systems considered in [15]. This weakness underpins ASPIC$^+$'s success in modelling other reasoning systems. We conclude by considering the relationship between ASPIC$^+$ and other weak logical systems.

1 Introduction

The rationality postulates proposed by Caminada and Amgoud [4] have been influential in the development of instantiated argumentation systems. These postulates identify desirable properties for the conclusions drawn from an argument based reasoning process, and focus on the effects of non-defeasible rules within an argumentation system. However, these postulates provide no desiderata with regards to the conclusions drawn from the defeasible rules found within an argumentation system. This latter type of rule is critical to argumentation, and identifying postulates for such rules is therefore important. At the same time, a large body of work exists which deals with non-monotonic reasoning (NMR). Such NMR systems (exemplified by approaches such as circumscription [18], default logic [23] and auto-epistemic logic [21]) introduce various approaches to handling defeasible reasoning, and axioms have been proposed to categorise such systems [15].

© Springer International Publishing AG, part of Springer Nature 2018
E. Black et al. (Eds.): TAFA 2017, LNAI 10757, pp. 67–85, 2018.
https://doi.org/10.1007/978-3-319-75553-3_5

In this paper we seek to combine the rich existing body of work on NMR with structured argumentation systems. We aim to identify what axioms structured argument systems, exemplified by ASPIC$^+$ [19] meet[1]. In doing so, we also wish to investigate the links between NMR axioms and the rationality postulates. This latter strand of work will, in the future, potentially allow us to identify additional rationality postulates which have not been considered to date.

2   The ASPIC$^+$ Argumentation Framework

ASPIC$^+$ [19] is a widely used formalism for structured argumentation, which satisfies the rationality postulates of [4][2]. Arguments within ASPIC$^+$ are constructed by chaining two types of inference rules, beginning with elements of a knowledge base. The first type of inference rule is referred to as a *strict* rule, and represents rules whose conclusion can be unconditionally drawn from a set of premises. This is in contrast to *defeasible* inference rules, which allow for a conclusion to be drawn from a set of premises as long as no exceptions or contrary conclusions exist.

Definition 1. *An* argumentation system *is a triple* $AS = \langle \mathcal{L}, \mathcal{R}, n \rangle$ *where:*

- $\mathcal{L}$ *is a logical language.*
- $^{-}$ *is a function from* $\mathcal{L}$ *to* $2^{\mathcal{L}}$*, such that:*
  - ϕ *is a contrary of* ψ *if* $\phi \in \overline{\psi}$*,* $\psi \notin \overline{\phi}$
  - ϕ *is a contradictory of* ψ *(denoted by* '$\phi = -\psi$'*), if* $\phi \in \overline{\psi}$*,* $\psi \in \overline{\phi}$
  - *each* $\phi \in \mathcal{L}$ *has at least one contradictory.*
- $\mathcal{R} = \mathcal{R}_s \cup \mathcal{R}_d$ *is a set of strict (*$\mathcal{R}_s$*) and defeasible (*$\mathcal{R}_d$*) inference rules of the form* $\phi_1, \ldots, \phi_n \rightarrow \phi$ *and* $\phi_1, \ldots, \phi_n \Rightarrow \phi$ *respectively (where* ϕ_i, ϕ *are meta-variables ranging over wff in* $\mathcal{L}$*), and* $\mathcal{R}_s \cap \mathcal{R}_d = \emptyset$*.*
- $n : \mathcal{R}_d \mapsto \mathcal{L}$ *is a naming convention for defeasible rules.*

We write $\phi_1, \ldots, \phi_n \rightsquigarrow \phi$ if $\mathcal{R}$ contains a strict rule $\phi_1, \ldots, \phi_n \rightarrow \phi$ or a defeasible rule $\phi_1, \ldots, \phi_n \Rightarrow \phi$.

Definition 2. *A* knowledge base *in an argumentation system* $\langle \mathcal{L}, \mathcal{R}, n \rangle$ *is a set* $\mathcal{K} \subseteq \mathcal{L}$ *consisting of two disjoint subsets* $\mathcal{K}_n$ *(the axioms) and* $\mathcal{K}_p$ *(the ordinary premises).*

An argumentation theory consists of an argumentation system and knowledge base.

Definition 3. *An* argumentation theory *AT is a pair* $\langle AS, \mathcal{K} \rangle$*, where AS is an argumentation system AS and* $\mathcal{K}$ *is a knowledge base.*

---

[1] ASPIC$^+$ was selected for this study due to its popularity, and its ability to model a variety of other structured systems [20].

[2] While additional rationality postulates have been proposed [24], we do not consider them in this paper.

An argumentation theory is *strict* iff $\mathcal{R}_d = \emptyset$ and $\mathcal{K}_p = \emptyset$, and is *defeasible* otherwise.

To ensure that reasoning meets norms for rational reasoning according to the *rationality postulates* of [4], an ASPIC$^+$ argumentation system's strict rules must be closed under transposition. That is, given a strict rule with premises $\varphi = \{\phi_1, \ldots, \phi_n\}$ and conclusion ϕ (written $\varphi \rightarrow \phi$), a set of n additional rules of the following form must be present in the system: $\{\overline{\phi}\} \cup \varphi \backslash \{\phi_i\} \rightarrow \overline{\phi_i}$ for all $1 \leq i \leq n$.

Arguments are defined recursively in terms of sub-arguments and through the use of several functions: $\mathtt{Prem}(A)$ returns all the premises of argument A; $\mathtt{Conc}(A)$ returns A's conclusion, and $\mathtt{TopRule}(A)$ returns the last rule used within the argument. $\mathtt{Sub}(A)$ returns all of A's sub-arguments. Given this, arguments are defined as follows.

Definition 4. *An argument A on the basis of an argumentation theory $AT = \langle \langle \mathcal{L}, \mathcal{R}, n \rangle, \mathcal{K} \rangle$ is:*

1. ϕ *if* $\phi \in \mathcal{K}$ *with:* $\mathtt{Prem}(A) = \{\phi\}$; $\mathtt{Conc}(A) = \{\phi\}$; $\mathtt{Sub}(A) = \{A\}$; $\mathtt{TopRule}(A) = undefined$.
2. $A_1, \ldots, A_n \rightarrow/\Rightarrow \phi$ *if A_i are arguments such that there respectively exists a strict/defeasible rule* $\mathtt{Conc}(A_1), \ldots, \mathtt{Conc}(A_n) \rightarrow/\Rightarrow \phi$ *in* $\mathcal{R}_s/\mathcal{R}_d$. $\mathtt{Prem}(A) = \mathtt{Prem}(A_1) \cup \ldots \cup \mathtt{Prem}(A_n)$; $\mathtt{Conc}(A) = \phi$; $\mathtt{Sub}(A) = \mathtt{Sub}(A_1) \cup \ldots \cup \mathtt{Sub}(A_n) \cup \{A\}$; $\mathtt{TopRule}(A) = \mathtt{Conc}(A_1), \ldots, \mathtt{Conc}(A_n) \rightarrow/\Rightarrow \phi$.

We write $\mathcal{A}(AT)$ to denote the set of arguments on the basis of the theory AT, and given a set of arguments $\mathbf{A}$, we write $\mathtt{Concs}(\mathbf{A})$ to denote the conclusions of those arguments, that is:

$$\mathtt{Concs}(\mathbf{A}) = \{\mathtt{Conc}(A) | A \in \mathbf{A}\}$$

Like other argumentation systems, ASPIC$^+$ utilises conflict between arguments—represented through attacks—to determine what conclusions are justified.

An argument can be attacked in three ways: on its ordinary premises, on its conclusion, or on its inference rules. These three kinds of attack are called undermining, rebutting and undercutting attacks, respectively.

Definition 5. *An argument A attacks an argument B iff A undermines, rebuts or undercuts B, where:*

- *A undermines B (on B') iff* $\mathtt{Conc}(A) = \overline{\phi}$ *for some $B' = \phi \in \mathtt{Prem}(B)$ and $\phi \in \mathcal{K}_p$.*
- *A rebuts B (on B') iff* $\mathtt{Conc}(A) = \overline{\phi}$ *for some $B' \in \mathtt{Sub}(B)$ of the form $B_1'', \ldots, B_2'' \Rightarrow \phi$.*
- *A undercuts B (on B') iff* $\mathtt{Conc}(A) = \overline{n(r)}$ *for some $B' \in \mathtt{Sub}(B)$ such that $\mathtt{TopRule}(B)$ is a defeasible rule r of the form $\phi_1, \ldots, \phi_n \Rightarrow \phi$.*

Note that, in ASPIC$^+$ rebutting is *restricted*: an argument with a strict $\mathtt{TopRule}$ can rebut an argument with a defeasible $\mathtt{TopRule}$, but not vice versa. ([5,16]

introduce the ASPIC- and ASPIC$_D^+$ systems which use unrestricted rebut). Finally, a set of arguments is said to be *consistent* iff there is no attack between any arguments in the set.

Attacks can be distinguished by whether they are preference-dependent (rebutting and undermining) or preference-independent (undercutting). The former succeed only when the attacker is preferred. The latter succeed whether or not the attacker is preferred. Within ASPIC$^+$ preferences over defeasible rules and ordinary premises are combined to obtain a preference ordering over arguments [19]. Here, we are not concerned about the means of combination, but, following [19], we only consider *reasonable* orderings. For our purposes, a reasonable ordering is one such that adding a strict rule or axiom to an argument will neither increase nor decrease its preference level.

Definition 6. *A preference ordering $\preceq$ is a binary relation over arguments, i.e., $\preceq \subseteq \mathcal{A} \times \mathcal{A}$, where $\mathcal{A}$ is the set of all arguments constructed from the knowledge base in an argumentation system.*

Combining these elements results in the following.

Definition 7. *A structured argumentation framework is a triple $\langle \mathcal{A}, att, \preceq \rangle$, where $\mathcal{A}$ is the set of all arguments constructed from the argumentation system, att is the attack relation, and $\preceq$ is a preference ordering on $\mathcal{A}$.*

Preferences over arguments interact with attacks such that *preference-dependent* attacks succeed when the attacking argument is preferred. In contrast *preference-independent* attacks always succeed. Attacks that succeed are called *defeats*. Using Definition 4 and the notion of defeat, we can instantiate an abstract argumentation framework from a structured argumentation framework.

Definition 8. *An (abstract) argumentation framework AF corresponding to a structured argumentation framework $SAF = \langle \mathcal{A}, att, \preceq \rangle$ is a pair $\langle \mathcal{A}, Defeats \rangle$ such that Defeats is the defeat relation on $\mathcal{A}$ determined by SAF.*

This abstract argumentation framework can be evaluated using standard argumentation semantics [8], defining the notion of an extension:

Definition 9. *Let $AF = \langle \mathcal{A}, Defeats \rangle$ be an argumentation framework, let $A \in \mathcal{A}$ and $E \subseteq \mathcal{A}$. E is said to be conflict-free iff there does not exist a $B, C \in E$ such that B defeats C. E is said to defend A iff for every $B \in \mathcal{A}$ such that B defeats A, there exists a $C \in E$ such that C defeats B. The characteristic function $F : 2^{\mathcal{A}} \to 2^{\mathcal{A}}$ is defined as $F(E) = \{A \in \mathcal{A} | E \text{ defends } A\}$. E is called (1) an admissible set iff E is conflict-free and $E \subseteq F(E)$; (2) a complete extension iff E is conflict-free and $E = F(E)$; (3) a grounded extension iff E is the minimal complete extension; (4) a preferred extension iff E is a maximal complete extension, where minimality and maximality are w.r.t. set inclusion; and (5) a stable extension iff E is a preferred extension which attacks all arguments in $\mathcal{A} - E$.*

We note in passing that other extensions have been defined and refer the reader to [1] for further details.

For a given semantics, if an argument is in an extension, it is said to be justified, given the information in the argumentation framework, and given the semantics that have been adopted. Dealing with structured arguments, we are not only interested in what arguments hold, but which propositions are the conclusions of arguments that hold, given some semantics. Thus we say that a proposition is a *justified conclusion* if it is the conclusion of an argument that is in an extension under some semantics. In fact, as [6] points out, the situation is more complex than that, since under some semantics there may be multiple extensions. Thus [6] defines the notions of sceptically, credulously and universally justified conclusions under a given semantics as follows.

Definition 10. *For* $T \in \{admissible, complete, preferred, grounded, stable\}$, *if* $AF = \langle A, Defeats \rangle$ *is an argumentation framework. we say that:*

- ϕ *is a* T *credulously justified conclusion of* AF *iff there exists an argument* A *and a* T *extension* E *such that* $A \in E$ *and* $\mathrm{Conc}(A) = \phi$.
- ϕ *is a* T *sceptically justified conclusion of* AF *iff for every* T *extension* E, *there exists an argument* $A \in E$ *such that* $\mathrm{Conc}(A) = \phi$.
- ϕ *is a* T *universally justified conclusion of* AF *iff there exists an argument* A *for every* T *extension* E, *such that* $A \in E$ *and* $\mathrm{Conc}(A) = \phi$.

3   Axiomatic Reasoning and ASPIC⁺

Kraus *et al.* [15], building on earlier work by Gabbay [11], identified a set of axioms which characterise non-monotonic inference in logical systems, and studied the relationships between sets of these axioms. Their goal was to characterise different kinds of reasoning; to pin down what it means for a logical system to be monotonic or non-monotonic; and—in particular—to be able to distinguish between the two. Table 1 presents the axioms of [15], which we will use to characterise reasoning in ASPIC⁺. The symbol $\mid\!\sim$ encodes a consequence relation, while $\models$ identifies the statements obtainable from the underlying theory. We have altered some of the symbols used in [15] to avoid confusion with the notation of ASPIC⁺. Equivalence is denoted $\equiv$ (rather than $\leftrightarrow$), and $\hookrightarrow$ (rather than $\rightarrow$) denotes the existence of a strict or defeasible rule.

Consequence relations that satisfy Ref, LLE, RW, Cut and CM are said to be *cumulative*, and [15] describes them as being the weakest interesting logical system. Cumulative consequence relations which also satisfy CP are *monotonic*, while consequence relations that are cumulative and satisfy M are called *cumulative monotonic*. Such relations are stronger than cumulative but not monotonic in the usual sense.

To determine which axioms ASPIC⁺ does or does not comply with, we must decide how different aspects of the axioms should be interpreted. We interpret the consequence relation $\mid\!\sim$ in two ways that are natural in the context of ASPIC⁺— describing these in detail later—and which fit with the high level meaning of "if α is in the knowledge base, then β follows", or "β is a consequence of α".

Table 1. The axioms from [15] that we will consider.

Abbr.	Axiom	Name
Ref	$\alpha \mid\!\sim \alpha$	Reflexivity
LLE	$\dfrac{\models \alpha \equiv \beta \qquad \alpha \mid\!\sim \gamma}{\beta \mid\!\sim \gamma}$	Left Logical Equivalence
RW	$\dfrac{\models \alpha \hookrightarrow \beta \qquad \gamma \mid\!\sim \alpha}{\gamma \mid\!\sim \beta}$	Right Weakening
Cut	$\dfrac{\alpha \wedge \beta \mid\!\sim \gamma \qquad \alpha \mid\!\sim \beta}{\alpha \mid\!\sim \gamma}$	Cut
CM	$\dfrac{\alpha \mid\!\sim \beta \qquad \alpha \mid\!\sim \gamma}{\alpha \wedge \beta \mid\!\sim \gamma}$	Cautious Monotonicity
M	$\dfrac{\models \alpha \hookrightarrow \beta \qquad \beta \mid\!\sim \gamma}{\alpha \mid\!\sim \gamma}$	Monotonicity
T	$\dfrac{\alpha \mid\!\sim \beta \qquad \beta \mid\!\sim \gamma}{\alpha \mid\!\sim \gamma}$	Transitivity
CP	$\dfrac{\alpha \mid\!\sim \beta}{\overline{\beta} \mid\!\sim \overline{\alpha}}$	Contraposition

Assuming such an interpretation of $\alpha \mid\!\sim \beta$ we can consider the meaning of the axioms. Some axioms are clear. For example, axiom T says that if β is a consequence of α, and γ is a consequence of β, then γ is a consequence of α. Other axioms are more ambiguous. Does $\alpha \wedge \beta \mid\!\sim \gamma$ in Cut mean that γ is a consequence of the conjunction $\alpha \wedge \beta$, or a consequence of α and β together? In other words is $\wedge$ a feature of the language underlying the reasoning system, or a feature of the meta-language in which the properties are written? Similarly, given the distinction between strict and defeasible rules, is $\alpha \hookrightarrow \beta$ a strict rule in ASPIC$^+$, a defeasible rule, or some statement in the property meta-language?

We interpret the symbols found in the axioms as follows:

- $\models \alpha$ means that α is an element of the relevant knowledge base.
- $\alpha \wedge \beta$ means both α and β, in particular in Cut and CM, $\wedge$ means that both α and β are in the knowledge base.
- $\alpha \equiv \beta$ is taken—as usual—to abbreviate the formula $(\alpha \hookrightarrow \beta) \wedge (\beta \hookrightarrow \alpha)$. We assume $\alpha \hookrightarrow \beta$ and $\beta \hookrightarrow \alpha$ have the same interpretation, i.e., both or neither are strict.
- $\alpha \hookrightarrow \beta$ has two interpretations. We have the *strict* interpretation in which $\alpha \hookrightarrow \beta$ denotes a strict rule $\alpha \to \beta$ in ASPIC$^+$, and the *defeasible* interpretation in which $\alpha \hookrightarrow \beta$ denotes either a strict or defeasible rule. We denote the latter interpretation by writing $\alpha \rightsquigarrow \beta$.

4   Axioms and Consequences in ASPIC+

In this section we examine which of the axioms ASPIC+ satisfies. Before doing so however, we must further pin down some aspects of ASPIC+ rules.

4.1   Preliminaries

To evaluate ASPIC+, we have to be a bit more precise about exactly what we are evaluating. We start by saying that we assume an arbitrary ASPIC+ argumentation theory $AT = \langle\langle\mathcal{L}, \mathcal{R}, n\rangle, \mathcal{K}\rangle$, in the sense that we say nothing about the contents of the knowledge base, or what domain-specific rules it contains. However, we distinguish between two classes of theory, with respect to the *base logic* that the theory contains.

The idea we capture by this is that in addition to domain specific rules—rules, for example, about birds and penguins flying—an ASPIC+ theory might also contain rules for reasoning in some logic. For example, we might equip an ASPIC+ theory with the axioms and inference rules of classical logic. Such a theory would be able to construct arguments using all the rules of classical logic, as well as all the domain-specific rules in the theory. The two base logics that we consider are classical logic, and what we call the "empty" base logic, where the ASPIC+ theory only contains domain-specific rules. (We make some observations about other base logics—intuitionistic logic and defeasible logic [2], but show no formal results for them.)

For each of the base logics, we consider the two different interpretations of the non-monotonic consequence relation $\vert\!\sim$ described above, identifying which axioms each interpretation satisfies. For our theory AT, we write AT_x to denote an extension of this augmentation theory also containing proposition x: $AT_x = \langle\langle\mathcal{L}, \mathcal{R}, n\rangle, \mathcal{K} \cup \{x\}\rangle$. An argument present in the latter, but not former, theory is denoted A^x.

4.2   Argument Construction

We begin by considering the consequence relation as representing argument construction. In other words, we interpret $\alpha \vert\!\sim \beta$ as meaning that if α is in the axioms or ordinary premises of a theory, we can construct an argument for β. More precisely:

Definition 11. *We write* $\alpha \vert\!\sim_{B,a} \beta$, *if for every* ASPIC+ *argumentation theory* $AT = \langle\langle\mathcal{L}, \mathcal{R}, n\rangle, \mathcal{K}\rangle$ *with base logic* B *such that* $\beta \notin \mathtt{Concs}(\mathcal{A}(AT))$, *it is the case that* $\beta \in \mathtt{Concs}(\mathcal{A}(AT_\alpha))$, *where* $B = \{\emptyset, c\}$, *representing the empty and classical base logics respectively.*

Proposition 1. *Ref, LLE, RW, Cut and CM hold for* $\vert\!\sim_{\emptyset,a}$ *in strict and defeasible theories.*

Proof. Consider an arbitrary theory $AT = \langle\langle \mathcal{L}, \mathcal{R}, n\rangle, \mathcal{K}\rangle$. *[Ref]* Given a theory AT_α, we have an argument $A^\alpha = [\alpha]$, so Ref holds for $\mathrel{|\!\sim}_{\emptyset,a}$. *[LLE]* Since $\alpha \mathrel{|\!\sim}_{\emptyset,a} \gamma$, AT_α contains a chain of arguments $A_1^\alpha, A_2^\alpha, \ldots, A_n^\alpha$ with $A_1^\alpha = [\alpha]$ and $\mathrm{Conc}(A_n^\alpha) = \gamma$. Given $\models \alpha \equiv \beta$, we have that both $\alpha \rightsquigarrow \beta$ and $\beta \rightsquigarrow \alpha$ are in the theory AT, so are in the theory AT_β. Within AT_β, we obtain a chain of arguments $B_0^\beta = [\beta], B_1^\beta = [B_0^\beta \rightsquigarrow \alpha], A_2^\beta, \ldots, A_n^\beta$. That is $\beta \mathrel{|\!\sim}_{\emptyset,a} \gamma$. Therefore, both strict and defeasible versions of LLE hold for $\mathrel{|\!\sim}_{\emptyset,a}$. *[RW]* Since $\gamma \mathrel{|\!\sim}_{\emptyset,a} \alpha$ in theory AT_γ, there is a chain of arguments $A_1^\gamma, A_2^\gamma, \ldots, A_n^\gamma$ with $A_1^\gamma = [\gamma]$ and $\mathrm{Conc}(A_n^\gamma) = \alpha$. Given $\models \alpha \hookrightarrow \beta$, theory AT must contain $\alpha \rightsquigarrow \beta$, as must AT_γ. In AT_γ, we have a chain of arguments $A_1^\gamma, \ldots, A_n^\gamma, A_{n+1}^\gamma = [A_n^\gamma \Rightarrow \beta]$. Thus, $\gamma \mathrel{|\!\sim}_{\emptyset,a} \beta$, and both strict and defeasible versions of RW hold for $\mathrel{|\!\sim}_{\emptyset,a}$. *[Cut]* Since $\alpha \wedge \beta \mathrel{|\!\sim}_{\emptyset,a} \gamma$, there is a chain of arguments $A_1^{\alpha,\beta}, A_2^{\alpha,\beta}, \ldots, A_n^{\alpha,\beta}$ with $A_1^{\alpha,\beta} = [\alpha]$, $A_2^{\alpha,\beta} = [\beta]$ in theory $AT_{\alpha,\beta}$, and $\mathrm{Conc}(A_n^{\alpha,\beta}) = \gamma$. In theory AT_α, since $\alpha \mathrel{|\!\sim}_{\emptyset,a} \beta$, there is a chain of arguments $B_1^\alpha, B_2^\alpha, \ldots, B_m^\alpha$ with $B_1^\alpha = [\alpha]$ and $\mathrm{Conc}(B_m^\alpha) = \beta$. There is also a chain of arguments $B_1^\alpha, B_2^\alpha, \ldots, B_m^\alpha, A_3^\alpha, \ldots, A_n^\alpha$. That is $\alpha \mathrel{|\!\sim}_{\emptyset,a} \gamma$. Therefore, cut holds for $\mathrel{|\!\sim}_{\emptyset,a}$. *[CM]* Since $\alpha \mathrel{|\!\sim}_{\emptyset,a} \gamma$ AT_α has a chain of arguments $A_1^\alpha, \ldots, A_n^\alpha$ with $A_1^\alpha = [\alpha]$ and $\mathrm{Conc}(A_n^\alpha) = \gamma$. $AT_{\alpha,\beta}$ has a similar chain of arguments $A_1^{\alpha,\beta}, \ldots, A_n^{\alpha,\beta}$, so $\alpha \wedge \beta \mathrel{|\!\sim}_{\emptyset,a} \gamma$. CM thus holds for $\mathrel{|\!\sim}_{\emptyset,a}$.

Since Ref, LLE, RW, Cut and CM hold, $\mathrel{|\!\sim}_{\emptyset,a}$ is cumulative for both strict and defeasible theories.

Proposition 2. *M and T hold for $\mathrel{|\!\sim}_{\emptyset,a}$ in strict and defeasible theories.*

Proof. Consider an arbitrary theory $AT = \langle\langle \mathcal{L}, \mathcal{R}, n\rangle, \mathcal{K}\rangle$. *[M]* Since $\beta \mathrel{|\!\sim}_{\emptyset,a} \gamma$, in the theory AT_β, there is a chain of arguments $A_1^\beta, A_2^\beta, \ldots, A_n^\beta$ with $A_1^\beta = [\beta]$ and $\mathrm{Conc}(A_n^\beta) = \gamma$. Given $\models \alpha \hookrightarrow \beta$, we have $\alpha \rightsquigarrow \beta$ in the theory AT, and also in the theory AT_α. In the latter, there is a chain of arguments $B_0^\alpha = [\alpha], B_1^\alpha = [B_0^\alpha \rightsquigarrow \beta], A_2^\alpha, \ldots, A_n^\alpha$. That is $\alpha \mathrel{|\!\sim}_{\emptyset,a} \gamma$. Therefore, both strict and defeasible versions of M hold for $\mathrel{|\!\sim}_{\emptyset,a}$. *[T]* Since $\beta \mathrel{|\!\sim}_{\emptyset,a} \gamma$, in AT_β, there is a chain of arguments $B_1^\beta, B_2^\beta, \ldots, B_m^\beta$ with $B_1^\beta = [\beta]$ and $\mathrm{Conc}(B_m^\beta) = \gamma$. Similarly, since $\alpha \mathrel{|\!\sim}_{\emptyset,a} \beta$, in AT_α, there is a chain of arguments $A_1^\alpha, A_2^\alpha, \ldots, A_n^\alpha$ with $A_1^\alpha = [\alpha]$ and $\mathrm{Conc}(A_n^\alpha) = \beta$. Combining this with $B_1^\alpha, B_2^\alpha, \ldots, B_m^\alpha$, we obtain the combined chain of arguments $A_1^\alpha, A_2^\alpha, \ldots, A_n^\alpha, B_2^\alpha, \ldots, B_m^\alpha$. That is $\alpha \mathrel{|\!\sim}_{\emptyset,a} \gamma$. Therefore, T holds for $\mathrel{|\!\sim}_{\emptyset,a}$.

Thus $\mathrel{|\!\sim}_{\emptyset,a}$ is cumulative monotonic for strict or defeasible theories. It is not, however, monotonic.

Proposition 3. *CP does not hold for $\mathrel{|\!\sim}_{\emptyset,a}$ in strict or defeasible theories.*

Proof. Consider an ASPIC$^+$ theory which contains: $\mathcal{K} = \{c\}$, $\mathcal{R}_s = \{\alpha, c \to d; \alpha, \overline{d} \to \overline{c}; c, \overline{d} \to \overline{\alpha}; \alpha \to e; \overline{e} \to \overline{\alpha}; d, e \to \beta; d, \overline{\beta} \to \overline{e}; \overline{\beta}, e \to \overline{d}\}$ We have $\alpha \mathrel{|\!\sim}_{\emptyset,a} \beta$ but not $\overline{\beta} \mathrel{|\!\sim}_{\emptyset,a} \overline{\alpha}$. Therefore, CP does not hold for $\mathrel{|\!\sim}_{\emptyset,a}$.

Having characterised $\mathrel{\vert\kern-0.3em\sim}_{\emptyset,a}$, we consider $\mathrel{\vert\kern-0.3em\sim}_{c,a}$. Clearly this will satisfy all the properties that are satisfied by $\mathrel{\vert\kern-0.3em\sim}_{\emptyset,a}$, since it includes all the inference rules of $\mathrel{\vert\kern-0.3em\sim}_{\emptyset,a}$. In addition, we have the following.

Proposition 4. *CP holds for $\mathrel{\vert\kern-0.3em\sim}_{c,a}$ in strict theories.*

Proof. Any strict ASPIC$^+$ theory with a classical base logic will generate the same set of consequences as classical logic. Furthermore, we know that CP is satisfied under classical logic. Therefore, the consequence relation $\mathrel{\vert\kern-0.3em\sim}_{c,a}$ satisfies CP for any strict theory.

Thus $\mathrel{\vert\kern-0.3em\sim}_{c,a}$ is monotonic for strict theories. However:

Proposition 5. *CP does not hold for $\mathrel{\vert\kern-0.3em\sim}_{c,a}$ in defeasible theories.*

Proof. Consider the counter-example from Proposition 3 where all rules are defeasible. Since the defeasible portion of the theory does not contain a rule of the form $\overline{\beta} \rightarrow \overline{d} \vee \overline{e}$, CP will not be satisfied.

4.3   Justified Conclusions

Next we interpret $\alpha \mathrel{\vert\kern-0.3em\sim} \beta$ as meaning that if α is in a theory, we can construct an argument for β such that β is in the set of justified conclusions (regardless of preferences). We will consider only the grounded and preferred semantics, but, as we will see, we have to bring in the ideas from Definition 10 since different kinds of justified conclusion lead to $\alpha \mathrel{\vert\kern-0.3em\sim} \beta$ satisfying different properties. We start with:

Definition 12. *Let $AF = \langle \mathcal{A}, Defeats \rangle$ be an abstract argumentation framework, we define*

$$\mathrm{Just}_g(\mathcal{A}(AT)) = \{\phi | \phi \text{ is a grounded justified conclusion}\}$$
$$\mathrm{Just}_p^c(\mathcal{A}(AT)) = \{\phi | \phi \text{ is a preferred credulously justified conclusion}\}$$
$$\mathrm{Just}_p^s(\mathcal{A}(AT)) = \{\phi | \phi \text{ is a preferred sceptically justified conclusion}\}$$
$$\mathrm{Just}_p^u(\mathcal{A}(AT)) = \{\phi | \phi \text{ is a preferred universally justified conclusion}\}$$

Note that we don't have to distinguish between different classes of grounded justified conclusion because, since there is exactly one grounded extension, the three different classes of grounded justified conclusion coincide. Then:

Definition 13. *We write $\alpha \mathrel{\vert\kern-0.3em\sim}_{B,j}^g \beta$, if for every ASPIC$^+$ argumentation theory $AT = \langle \langle \mathcal{L}, \mathcal{R}, n \rangle, \mathcal{K} \rangle$ with the B base logic such that $\beta \notin \mathrm{Just}_g(\mathcal{A}(AT))$, it is the case that $\beta \in \mathrm{Just}_g(\mathcal{A}(AT_\alpha))$, where $B = \{\emptyset, c\}$.*

Definition 14. *We write $\alpha \mathrel{\vert\kern-0.3em\sim}_{B,j}^{p,Sem} \beta$, if for every ASPIC$^+$ argumentation theory $AT = \langle \langle \mathcal{L}, \mathcal{R}, n \rangle, \mathcal{K} \rangle$ with the B base logic such that $\beta \notin \mathrm{Just}_p^{Sem}(\mathcal{A}(AT))$, it is the case that $\beta \in \mathrm{Just}_p^{Sem}(\mathcal{A}(AT_\alpha))$, where $B = \{\emptyset, c\}$ and $Sem \subseteq \{c, s, u\}$.*

We also write $\vdash^{p,*}_{\emptyset,j}$ to denote the union of $\vdash^{p,c}_{\emptyset,j}$, $\vdash^{p,s}_{\emptyset,j}$ and $\vdash^{p,u}_{\emptyset,j}$. Thus, $\alpha \vdash^{p,*}_{\emptyset,j} \beta$ means that a conclusion holds under (at least one of) the consequence relations. We write $\vdash^{p,\cap\{S\}}_{\emptyset,j}$ to denote that a conclusion holds under all the consequence relations in S. Thus, for example if we have that $\alpha \vdash^{p,\cap\{c,s\}}_{\emptyset,j} \beta$, then it is the case that $\alpha \vdash^{p,c}_{\emptyset,j} \beta$ and $\alpha \vdash^{p,s}_{\emptyset,j} \beta$. Similarly, when we say an axiom holds for $\vdash^{p,*}_{\emptyset,j}$, it means that the axiom holds for at least one of $\vdash^{p,s}_{\emptyset,j}$, $\vdash^{p,u}_{\emptyset,j}$, and $\vdash^{p}_{\emptyset,j}$. The same interpretation applies for axioms holding with respect to $\vdash^{p,\cap\{s\}}_{\emptyset,j}$.

It is worth noting the following result.

Proposition 6. *If* $\alpha \vdash^{g}_{B,j} \beta$ *or* $\alpha \vdash^{p,*}_{B,j} \beta$ *then* $\alpha \vdash_{B,a} \beta$.

Proof. Follows immediately from the definitions—for β to be a justified conclusion, there must first be an argument with β as a conclusion.

Since there are, in general, less justified conclusions of a theory than there are arguments, $\vdash^{g}_{\emptyset,j}$ and $\vdash^{p,*}_{\emptyset,j}$ are more restrictive notions of consequence than $\vdash_{\emptyset,a}$. It is therefore no surprise to find that fewer of the axioms from [15] hold. We have the following.

Proposition 7. *Ref, and the defeasible versions of LLE and RW, do not hold for* $\vdash^{g}_{\emptyset,j}$, $\vdash^{p,*}_{\emptyset,j}$ *in defeasible theories.*

*Proof. [**Ref**] Consider an ASPIC$^+$ theory that contains: $\mathcal{K}_n = \{\overline{\alpha}\}$ and $\mathcal{R} = \emptyset$. Here, we have an argument $A = [\overline{\alpha}]$. If a is in the knowledge base $\mathcal{K}_p$, we have another argument $B = [a]$. However, B is defeated by A, but not vice versa. So B is not in any extension. Thus, Ref does not hold for either $\vdash^{g}_{\emptyset,j}$ or $\vdash^{p,*}_{\emptyset,j}$. [**LLE (defeasible version)**] Consider an ASPIC$^+$ theory that contains $\mathcal{K}_n = \{c\}$ and $\mathcal{R} = \{\alpha \Rightarrow \beta; \beta \Rightarrow \alpha; \alpha \Rightarrow \gamma; c \to \overline{n_1}\}$ where $n(\beta \Rightarrow \alpha) = n_1$. Here, $\alpha \vdash^{g}_{\emptyset,j} \gamma$ and $\alpha \vdash^{p,*}_{\emptyset,j} \gamma$, but, $\beta \nvdash^{g}_{\emptyset,j} \gamma$ and $\beta \nvdash^{p,*}_{\emptyset,j} \gamma$. Therefore, the defeasible version of LLE does not hold for either $\vdash^{g}_{\emptyset,j}$ or $\vdash^{p,*}_{\emptyset,j}$. [**RW (defeasible version)**] Consider an ASPIC$^+$ theory that contains $\overline{\beta}$ in its axioms. For such a theory, β will not appear in any justified conclusions. Therefore, the defeasible version of RW does not hold for either $\vdash^{g}_{\emptyset,j}$ or $\vdash^{p,*}_{\emptyset,j}$.*

Proposition 8. *The strict version of LLE and RW hold for* $\vdash^{g}_{\emptyset,j}$ *and* $\vdash^{p,*}_{\emptyset,j}$ *in strict and defeasible theories.*

*Proof. Consider an arbitrary theory $AT = \langle\langle \mathcal{L}, \mathcal{R}, n\rangle, \mathcal{K}\rangle$. [**RW (strict version)**] Consider the extension E_γ in AT_γ containing an argument A^γ with $\mathrm{Conc}(A^\gamma) = \alpha$. Since $\models \alpha \rightsquigarrow \beta$, under the strict interpretation, we know that $\alpha \to \beta$ is in AT_γ. Therefore, we can construct an argument $B^\gamma = A^\gamma \to \beta$. Furthermore, the attackers of B are the attackers of A because $\mathrm{TopRule}(B)$ is a strict rule. Since A^γ is in the extension E_γ, B^γ is in the same extension E_γ. Therefore the strict version of RW holds for $\vdash^{g}_{\emptyset,j}$ and $\vdash^{p,*}_{\emptyset,j}$. [**LLE (strict version)**] Since $\models \alpha \equiv \beta$, under the strict interpretation, the rules $\beta \to \alpha$ and $\alpha \to \beta$ are in AT, AT_α, AT_β and $AT_{\alpha,\beta}$. Thus AT_α, AT_β, $AT_{\alpha,\beta}$ have the* *same extensions, just as for RW(strict version). If $\alpha \mathrel{\vert\!\sim}^{p,*}_{\emptyset,j} \gamma$, then $\beta \mathrel{\vert\!\sim}^{p,*}_{\emptyset,j} \gamma$. If $\alpha \mathrel{\vert\!\sim}^{g}_{\emptyset,j} \gamma$, then $\beta \mathrel{\vert\!\sim}^{g}_{\emptyset,j} \gamma$. Therefore, the strict version of LLE holds for $\mathrel{\vert\!\sim}^{g}_{\emptyset,j}$ and $\mathrel{\vert\!\sim}^{p,*}_{\emptyset,j}$.*

Proposition 9. *Cut holds for $\mathrel{\vert\!\sim}^{g}_{\emptyset,j}$ and $\mathrel{\vert\!\sim}^{p,s}_{\emptyset,j}$ in strict and defeasible theories.*

Proof. Since $\alpha \mathrel{\vert\!\sim}^{g}_{\emptyset,j} \beta$, the grounded justified conclusions of AT_α contain α and β. By adding β into the knowledge base, the grounded justified conclusions will not change – if the newly added β is not justified, then it has not effect; if the newly added β is justified, it will remain in the justified conclusions. The same argument applies for $\mathrel{\vert\!\sim}^{p,s}_{\emptyset,j}$.

Proposition 10. *Cut does not hold for either $\mathrel{\vert\!\sim}^{p,c}_{\emptyset,j}$ or $\mathrel{\vert\!\sim}^{p,u}_{\emptyset,j}$ in defeasible theories.*

Proof. We will give a counter-example. Consider the ASPIC$^+$ *theory that include $\mathcal{K} = \emptyset$ and $\mathcal{R} = \{a \Rightarrow c; c \Rightarrow b; b \Rightarrow \overline{c}; \overline{c} \Rightarrow r; \}$. The credulous or universal justified conclusions of AT_α are $\{a, b, c\}$. The credulous or universal justified conclusions of $AT_{\alpha,\beta}$ are $\{a, b, \overline{c}, r, c\}$. That is $a \wedge b \mathrel{\vert\!\sim}^{p}_{\emptyset,j} r$, $a \mathrel{\vert\!\sim}^{p}_{\emptyset,j} b$, but $a \mathrel{\not\vert\!\sim}^{p}_{\emptyset,j} r$. Therefore Cut does not hold for either $\mathrel{\vert\!\sim}^{p,c}_{\emptyset,j}$ or $\mathrel{\vert\!\sim}^{p,u}_{\emptyset,j}$.*

Proposition 11. *CM holds for $\mathrel{\vert\!\sim}^{g}_{\emptyset,j}$ in strict and defeasible theories.*

Proof. Since $\alpha \mathrel{\vert\!\sim}^{g}_{\emptyset,j} \gamma$, the grounded justified conclusions of AT_α contain α and γ. By adding β into the knowledge base, the grounded justified conclusions will not change. The justification is same as in the proof of Proposition 9.

Proposition 12. *CM does not hold for $\mathrel{\vert\!\sim}^{p,*}_{\emptyset,j}$ in defeasible theories.*

Proof. We will give counter-examples. Consider an ASPIC$^+$ *theory that include $\mathcal{K} = \emptyset$ and $\mathcal{R} = \{a \Rightarrow b; a \Rightarrow r; b \rightarrow \overline{n1}; r \rightarrow \overline{n2}; \}$, where $n(a \Rightarrow b) = n1$ and $n(a \Rightarrow r) = n2$. The credulous or universal justified conclusions of AT_α are $\{a, r, \overline{n1}, b, \overline{n2}\}$. And the credulous or universal justified conclusions of $AT_{\alpha,\beta}$ are $\{a, b, \overline{n2}\}$. That is $a \mathrel{\vert\!\sim}^{p}_{\emptyset,j} b$, $a \mathrel{\vert\!\sim}^{p}_{\emptyset,j} r$, but $a \wedge b \mathrel{\not\vert\!\sim}^{p}_{\emptyset,j} r$. Therefore CM does not hold for either $\mathrel{\vert\!\sim}^{p,c}_{\emptyset,j}$ or $\mathrel{\vert\!\sim}^{p,u}_{\emptyset,j}$. Now, consider an* ASPIC$^+$ *theory that include $\mathcal{K} = \emptyset$, $\mathcal{R} = \{a \Rightarrow r; r \Rightarrow b; b \Rightarrow \overline{r}\}$. The sceptical justified conclusions of AT_α are $\{a, b, r\}$. And the sceptical justified conclusions of $AT_{\alpha,\beta}$ are $\{a, b\}$. $a \mathrel{\vert\!\sim}^{p}_{\emptyset,j} b$, $a \mathrel{\vert\!\sim}^{p}_{\emptyset,j} r$, but $a \wedge b \mathrel{\not\vert\!\sim}^{p}_{\emptyset,j} r$. Therefore CM does not hold for $\mathrel{\vert\!\sim}^{p,s}_{\emptyset,j}$.*

Proposition 13. *M, T and CP do not hold for $\mathrel{\vert\!\sim}^{g}_{\emptyset,j}$ or $\mathrel{\vert\!\sim}^{p,*}_{\emptyset,j}$ in defeasible theories.*

Proof. We will give counter-examples. [M] Consider an ASPIC$^+$ *theory that contains $\mathcal{K}_n = \{\overline{\alpha}\}$ and $\mathcal{R} = \{\alpha \rightarrow \beta; \overline{\beta} \rightarrow \overline{\alpha}; \beta \Rightarrow \gamma\}$. Thus, $\beta \mathrel{\vert\!\sim}^{g}_{\emptyset,j} \gamma$ and $\beta \mathrel{\vert\!\sim}^{p,*}_{\emptyset,j} \gamma$, however, $\alpha \mathrel{\not\vert\!\sim}^{g}_{\emptyset,j} \gamma$ and $\alpha \mathrel{\not\vert\!\sim}^{p,*}_{\emptyset,j} \gamma$. Therefore, M does not hold for $\mathrel{\vert\!\sim}^{g}_{\emptyset,j}$ or $\mathrel{\vert\!\sim}^{p,*}_{\emptyset,j}$. [T] Consider an* ASPIC$^+$ *theory which includes $\mathcal{K} = \emptyset$ and*

$\mathcal{R} = \{\alpha \Rightarrow \beta; \beta \Rightarrow c; c \Rightarrow \gamma; \alpha \Rightarrow \overline{n_1}\}$ *where* $n(c \Rightarrow \gamma) = n_1$. *Thus,* $a \mathrel{\vdash\!\!\!\sim}^g_{\emptyset,j} b$, $b \mathrel{\vdash\!\!\!\sim}^g_{\emptyset,j} r$, $a \mathrel{\vdash\!\!\!\sim}^{p,*}_{\emptyset,j} b$ *and* $b \mathrel{\vdash\!\!\!\sim}^{p,*}_{\emptyset,j} r$, *but* $a \mathrel{\not\vdash\!\!\!\sim}^g_{\emptyset,j} r$ *and* $a \mathrel{\not\vdash\!\!\!\sim}^{p,*}_{\emptyset,j} r$. *Therefore,* T *does not hold for* $\mathrel{\vdash\!\!\!\sim}^g_{\emptyset,j}$ *or* $\mathrel{\vdash\!\!\!\sim}^{p,*}_{\emptyset,j}$. *[CP] Since contraposition does not hold for* $\mathrel{\vdash\!\!\!\sim}_{\emptyset,a}$, *by Proposition 3 it cannot hold for* $\mathrel{\vdash\!\!\!\sim}^g_{\emptyset,j}$ *or* $\mathrel{\vdash\!\!\!\sim}^{p,*}_{\emptyset,j}$.

If we consider only strict theories, the following holds.

Proposition 14. *Ref, CM, M and T hold for* $\mathrel{\vdash\!\!\!\sim}^g_{\emptyset,j}$ *and* $\mathrel{\vdash\!\!\!\sim}^{p,*}_{\emptyset,j}$ *in strict theories.*

Proof. If the theory is strict, then for any argumentation theory, all conclusions are justified. Therefore, for any strict theory, if $\alpha \mathrel{\vdash\!\!\!\sim}_{\emptyset,a} \beta$, *then* $\alpha \mathrel{\vdash\!\!\!\sim}^g_{\emptyset,j} \beta$ *and* $\alpha \mathrel{\vdash\!\!\!\sim}^{p,*}_{\emptyset,j} \beta$. *We know that* $\mathrel{\vdash\!\!\!\sim}_{\emptyset,a}$ *holds for Ref, CM, M and T, therefore,* $\mathrel{\vdash\!\!\!\sim}^g_{\emptyset,j}$ *and* $\mathrel{\vdash\!\!\!\sim}^{p,*}_{\emptyset,j}$ *holds for Ref, CM, M and T in strict theories.*

Proposition 15. *CP does not hold for* $\mathrel{\vdash\!\!\!\sim}^g_{\emptyset,j}$ *or* $\mathrel{\vdash\!\!\!\sim}^{p,*}_{\emptyset,j}$ *in strict theories.*

Proof. Since CP does not hold for $\mathrel{\vdash\!\!\!\sim}_{\emptyset,a}$ *under strict theories, CP can not hold for* $\mathrel{\vdash\!\!\!\sim}^g_{\emptyset,j}$ *or* $\mathrel{\vdash\!\!\!\sim}^{p,*}_{\emptyset,j}$.

This completes the characterisation of $\mathrel{\vdash\!\!\!\sim}^g_{\emptyset,j}$, $\mathrel{\vdash\!\!\!\sim}^{p,s}_{\emptyset,j}$, $\mathrel{\vdash\!\!\!\sim}^{p,c}_{\emptyset,j}$ and $\mathrel{\vdash\!\!\!\sim}^{p,u}_{\emptyset,j}$. As we argued above, adding classical logic as a base logic will create consequence relations that satisfy the same properties as each of these since they will includes all the same inference rules. In addition, we have the following:

Proposition 16. *CP holds for* $\mathrel{\vdash\!\!\!\sim}^g_{c,j}$ *and* $\mathrel{\vdash\!\!\!\sim}^{p,*}_{c,j}$ *in strict theories.*

Proof. As above, $\mathrel{\vdash\!\!\!\sim}_{c,a}$ *satisfies CP in strict theories. Since the strict part of the theory is always consistent, any conclusions from the argument construction are justified. Therefore, the consequence relation* $\mathrel{\vdash\!\!\!\sim}^g_{c,j}$ *and* $\mathrel{\vdash\!\!\!\sim}^{p,*}_{c,j}$ *satisfies CP for strict theories.*

4.4   Summary

The results for the two forms of consequence and the two base logics are summarized in Table 2. This shows, for example, that Ref is satisfied by $\mathrel{\vdash\!\!\!\sim}_{c,j}$ for strict theories whether the proposition in question is a premise or an axiom; that for defeasible theories, Ref is never satisfied by $\mathrel{\vdash\!\!\!\sim}_{c,j}$ for propositions that are premises, but is always satisfied for propositions that are axioms. Similarly, the table shows that CP does not hold for $\mathrel{\vdash\!\!\!\sim}_{\emptyset,a}$ for either strict or defeasible theories; that CP holds for $\mathrel{\vdash\!\!\!\sim}_{c,a}$ for strict theories, but not for defeasible theories.

Recall from Sect. 3 that a consequence relation which satisfies axioms Ref, LLE, RW, Cut and CM is said to be "cumulative", a cumulative consequence relation that also satisfies M is said to be "cumulative monotonic", and a consequence relation that satisfies CP is monotonic. Given this, it is clear that Table 2 is telling us that $\mathrel{\vdash\!\!\!\sim}_{\emptyset,a}$ is cumulative monotonic for both strict and defeasible theories, while $\mathrel{\vdash\!\!\!\sim}_{c,a}$ is monotonic for strict theories and cumulative monotonic for defeasible theories. Similarly, $\mathrel{\vdash\!\!\!\sim}^g_{\emptyset,j}$ is cumulative monotonic for strict theories, and cumulative for the strict portions of defeasible theories (if Ref is applied to axioms and LLE, RW and M are applied to strict rules only), but not even cumulative for the defeasible parts of defeasible theories (if Ref is applied to ordinary premises and LLE, RW or M are applied to defeasible rules, none of them hold). $\vdash^{p,s}_{\emptyset,j}$ is weaker than $\vdash^{g}_{\emptyset,j}$, since CM doesn't hold, and $\vdash^{p,\{c,u\}}_{\emptyset,j}$ is weaker still since Cut doesn't hold. Adding classical logic as a base logic means that CP holds, so $\vdash^{g}_{c,j}$ is monotonic for strict theories, and behaves exactly like $\vdash^{g}_{\emptyset,j}$ for defeasible theories. Again $\vdash^{p,s}_{c,j}$ is weaker than $\vdash^{g}_{c,j}$, since CM doesn't hold, and $\vdash^{p,\{c,u\}}_{c,j}$ is weaker still since Cut doesn't hold.

4.5 Discussion

What light do the results in Table 2 shine on ASPIC$^+$ and argumentation-based reasoning in general? We will answer that question by considering each of the consequence relations in turn.

Starting with $\vdash_{\emptyset,a}$, it is no surprise that the relation is cumulative monotonic and satisfies the axiom M which captures a form of monotonicity. It is clear from the detail of ASPIC$^+$, and indeed any argumentation system, that the number of arguments grows over time, and that once introduced, arguments do not disappear. However, the fact that $\vdash_{\emptyset,a}$ is not monotonic in the same strict sense as classical logic, and so is strictly weaker, as a result of not satisfying CP, is a bit more interesting. This is, of course, because arguments are not subject to the law of the excluded middle—it is perfectly possible for there to be arguments for α and $\overline{\alpha}$ from the same theory.

Turning to the various versions of consequence built around justified conclusions, they are perhaps more reasonable notions of consequence for ASPIC$^+$ than $\vdash_{\emptyset,a}$. If β is a justified conclusion of α, then there is an argument for β which holds despite any attacks (in the scenario we have considered, where all attacks may be defeats for some preference ordering—and therefore succeed—there can still be attacks on the argument for β, but the attacking arguments must themselves be defeated). This is quite a restrictive notion of consequence in a representation that allows for conflicting information, and as Table 2 makes clear, even $\vdash^{g}_{\emptyset,j}$, which is the strongest of the consequence relations based on justified conclusions, is a relatively weak notion of consequence and obeys less of the axioms than the non-monotonic logics analysed in [15], for example. For defeasible theories $\vdash^{g}_{\emptyset,j}$ is not cumulative, and only satisfies LLE and RW if the rules applied in those axioms are strict. As we pointed out above, at the time that [15] was published, cumulativity was considered the minimum requirement of a useful logic[3]. Whether or not one accepts this, it is clear that ASPIC$^+$ is weak. But is it too weak? To answer this, we should consider reason that $\vdash^{g}_{\emptyset,j}$ is not cumulative, which as Table 2 shows is due to LLE, RW and Ref.

---

[3] This position was doubtless a side-effect of the fact that at that time there were no logics that did not obey cumulativity. The subsequent discovery of logics of causality that are not cumulative suggests that this view should be revised.

Table 2. Summary of axioms satisfied by the argumentation-based consequence relations. Each row indicates which axiom is satisfied by a given consequence relation or not, and what conditions, if any, are required. See text for an explanation.

		Strict theories						Defeasible theories							
		$\mid\!\sim_{0,a}$	$\mid\!\sim_{c,a}$	$\mid\!\sim^{g}_{0,j}$	$\mid\!\sim^{p,*}_{0,j}$	$\mid\!\sim^{g}_{c,j}$	$\mid\!\sim^{p,*}_{c,j}$	$\mid\!\sim_{0,a}$	$\mid\!\sim_{c,a}$	$\mid\!\sim^{g}_{0,j}$	$\mid\!\sim^{p,s}_{0,j}$	$\mid\!\sim^{p,\{c,u\}}_{0,j}$	$\mid\!\sim^{g}_{c,j}$	$\mid\!\sim^{p,s}_{c,j}$	$\mid\!\sim^{p,\{c,u\}}_{c,j}$
Ref	Premise	Y	Y	Y	Y	Y	Y	Y	Y	N	N	N	N	N	N
	Axiom	Y	Y	Y	Y	Y	Y	Y	Y	Y	Y	Y	Y	Y	Y
LLE	Defeasible rule	Y	Y	Y	Y	Y	Y	Y	Y	N	N	N	N	N	N
	Strict rule	Y	Y	Y	Y	Y	Y	Y	Y	Y	Y	Y	Y	Y	Y
RW	Defeasible rule	Y	Y	Y	Y	Y	Y	Y	Y	N	N	N	N	N	N
	Strict rule	Y	Y	Y	Y	Y	Y	Y	Y	Y	Y	Y	Y	Y	Y
Cut		Y	Y	Y	Y	Y	Y	Y	Y	Y	Y	N	Y	Y	N
CM		Y	Y	Y	Y	Y	Y	Y	Y	Y	N	N	Y	N	N
M		Y	Y	Y	Y	Y	Y	Y	Y	N	N	N	N	N	N
T		Y	Y	Y	Y	Y	Y	Y	Y	N	N	N	N	N	N
CP		N	Y	N	N	Y	Y	N	N	N	N	N	N	N	N

LLE and RW only hold in the case of strict rules, either because the theory is strict, or because the case in question is of a strict rule in a defeasible theory. For both LLE and RW, the effect of the axiom is to extend an existing argument, either switching one premise for another (LLE), or adding a rule to the conclusion of an argument (RW). While having these axioms hold for defeasible rules would allow $\mid\!\sim^g_{\emptyset,j}$ to be cumulative for defeasible theories, this is not reasonable. Using LLE or RW to extend arguments with defeasible rules—by definition—means that the new arguments created by this extension can be defeated. Thus their conclusions may not be justified, and $\mid\!\sim^g_{\emptyset,j}$ must not be cumulative for defeasible rules. In other words $\mid\!\sim^g_{\emptyset,j}$ is not cumulative for defeasible rules exactly because it makes no sense for a system of defeasible rules to be cumulative.

This weakness raises the question of whether reasoning in ASPIC$^+$ can be strengthened. When we add classical logic as a base logic, we get a family of consequence relations that satisfy CP. Thus $\mid\!\sim^g_{c,j}$ is monotonic, but only if all elements are strict. For theories with defeasible elements, $\mid\!\sim^g_{c,j}$ cannot guarantee that CP will hold for arbitrary α and β, and, as above, LLE and RW will only hold for strict rules. Adding a base logic that is weaker than classical logic does not help in strengthening conclusions. If we add intuitionistic logic, for example, we don't get CP, because intuitionistic logic explicitly rejects this pattern of reasoning. A similar argument applies to Ref. Proposition 14 tells us that Ref holds for $\mid\!\sim^g_{\emptyset,j}$ and $\mid\!\sim^{p,*}_{\emptyset,j}$ for strict theories, meaning that α has to be an axiom[4]. If Ref were to hold for defeasible theories, α could be a premise. But premises can be defeated, again by definition, so it is not appropriate to directly conclude that any premise is a justified conclusion (it is necessary to go through the whole process of constructing arguments and establishing extensions to determine this).

From this we conclude that although $\mid\!\sim^g_{\emptyset,j}$ and $\mid\!\sim^g_{c,j}$ are not cumulative, and hence ASPIC$^+$ is, in some sense, weaker than non-monotonic logics like circumscription [18] and default logic [23], it is not clear that it is too weak. That is strengthening $\mid\!\sim^g_{\emptyset,j}$ or $\mid\!\sim^g_{c,j}$ so that they would be cumulative for defeasible theories would allow for conclusions that make no sense from the point of view of argumentation-based reasoning. Whether there are other ways to strengthen ASPIC$^+$ that do make sense is an open question, and one we intend to investigate in the future.

5   The Rationality Postulates

Finally, we consider the three postulates of [4] (which ASPIC$^+$ complies with), namely (1) closure under strict rules; and (2) direct and (3) indirect consistency. We ask whether the axioms discussed in this paper are equivalent to any of these postulates. In what follows, we assume that strict rules are consistent.

---

[4] This is exactly how defeasible logic [2] satisfies Ref.

5.1   Closure Under Strict Rules

Proposition 17. *An argumentation framework meets closure under strict rules if and only if the consequence relation for strict rules complies with right weakening (RW) with regards to justified conclusions.*

Proof. Given an argumentation framework AF, assume that α is in the justified conclusions. Therefore $\top \mathop{\vert\!\sim}_j \alpha$, and assume that there is a strict rule $\models \alpha \rightarrow \beta$. Using RW, we obtain $\top \mathop{\vert\!\sim}_j \beta$. Therefore RW implies closure under strict rules. Furthermore, having $\gamma \mathop{\vert\!\sim}_j \alpha$, as well as a strict rule $\alpha \rightarrow \beta$ results in $\gamma \mathop{\vert\!\sim}_j \beta$, i.e., the strict form of RW.

5.2   Direct Consistency

Direct consistency with regards to $\mathop{\vert\!\sim}_j$ requires that no extension contains inconsistent arguments (and therefore inconsistent conclusions). This is equivalent to the following axiom, unobtainable from the axioms discussed previously.

$$\frac{\alpha \mathop{\vert\!\sim}_j \beta}{\alpha \mathop{\not\vert\!\sim}_j \overline{\beta}}$$

5.3   Indirect Consistency

Proposition 18. *Assume we have direct consistency, and that strict rules are consistent. Any system which satisfies monotonicity under strict rules will satisfy indirect consistency, and vice-versa.*

Proof. From [4, Proposition 7], direct consistency and closure yield indirect consistency. We assume direct consistency, and monotonicity gives closure.

In this section we have shown that the rationality postulates described in [4] can be described using axioms from classical logic and non-monotonic reasoning. In future work, we intend to determine whether these axioms can help identify additional rationality postulates. In addition, we will investigate whether these axioms can represent the additional rationality postulates described in [24].

6   Related Work

There are several papers describing work that is similar in some respects to what we report here. Billington [2] describes Defeasible Logic, a logic that, as its name implies, differs from classical logic in that it deals with defeasible reasoning. In addition to introducing the logic, [2] shows that defeasible logic satisfies the axioms of reflexivity, cut and cautious monotonicity suggested in [11], thus satisfying what [11] describes as the basic requirements for a non-monotonic system (such a system is equivalent to a cumulative system in [15]). [13] subsequently established significant links between reasoning in defeasible logic and argumentation-based reasoning. To do this, [13] provides an argumentation system that makes use of defeasible logic as its underlying logic, and shows that the system is compatible with Dung's semantics [8]. Given Defeasible Logic's close relation to Prolog [22], this line of work is closely related to Defeasible Logic Programming (DeLP) [12], a formalism combining results of Logic Programming and Defeasible Argumentation. As a rule-based argumentation system, DeLP also has strict/defeasible rules and a set of facts. DeLP differs from ASPIC$^+$ in the types of attack relation it permits (no undermining) and in the way that it computes conclusions (it does not implement Dung's semantics).

[17] first introduce an argument system, containing two kinds of inference rules, namely, monotonic inference rules and non-monotonic inference rules. They show that most well-known non-monotonic systems, such as default logic, autoepistemic logic, negation as failure and circumscription, can be formulated as instances of their argument system. [3] continues this line of work, presenting an abstract framework for default reasoning which includes Theorist, default logic, logic programming, autoepistemic logic, non-monotonic modal logics, and certain instances of circumscription as special cases. [13] subsequently established significant links between reasoning in defeasible logic and argumentation-based reasoning. To do this, [13] provides an argumentation system that makes use of defeasible logic as its underlying logic, and shows that the system is compatible with Dung's semantics [8]. Similar to the current work, [14] investigates various consequence relations of deductive argumentation and their satisfaction of various properties. However, [14] focuses entirely on argument construction and says nothing about justified conclusions.

Also related are [9,10], which investigate cumulativity of ASPIC-like structured argumentation frameworks. Finally, [7] analyzes cautious monotonicity and cumulative transitivity with respect to Assumption-Based Argumentation.

7   Conclusions

In this paper we considered which of the axioms of [15] ASPIC$^+$ meets based on two different interpretations of the consequence relation. We demonstrated that, in terms of those axioms, the most natural forms of consequence in ASPIC$^+$ are rather weak. This is the case even when we assume ASPIC$^+$ theories contain all the inference rules of classical logic. However, as we discuss, strengthening the consequence relation (to, for example, be cumulative) neither makes sense in terms of argumentation-based reasoning, nor can easily be achieved by adding additional inference rules to ASPIC$^+$ theories. We also investigated the relationship between the axioms of [15] and the rationality postulates, and suggested an alternative, axiom based formulation of the latter.

As mentioned above, in the future we will investigate whether additional axioms can encode the rationality postulates described in [24]. We will also examine the properties of different interpretations of the logical symbols. For example, we assumed that $\equiv$ encodes the presence of two rules, but says nothing about their preferences or defeaters. Finally, we may consider other interpretations of the consequence relation. This paper therefore opens up several significant avenues of future investigation.

Acknowledgements. This work was partially supported by EPSRC grant EP/P010105/1.

References

1. Baroni, P., Caminada, M., Giacomin, M.: An introduction to argumentation semantics. Knowl. Eng. Rev. **26**(4), 365–410 (2011)
2. Billington, D.: Defeasible logic is stable. J. Log. Comput. **3**(4), 379–400 (1993)
3. Bondarenko, A., Dung, P.M., Kowalski, R.A., Toni, F.: An abstract, argumentation-theoretic approach to default reasoning. Artif. Intell. **93**(1), 63–101 (1997)
4. Caminada, M., Amgoud, L.: On the evaluation of argumentation formalisms. Artif. Intell. **171**(5), 286–310 (2007)
5. Caminada, M., Modgil, S., Oren, N.: Preferences and unrestricted rebut. In: Proceedings of 5th International Conference on Computational Models of Argument, pp. 209–220 (2014)
6. Croitoru, M., Vesic, S.: What can argumentation do for inconsistent ontology query answering? In: Liu, W., Subrahmanian, V.S., Wijsen, J. (eds.) SUM 2013. LNCS (LNAI), vol. 8078, pp. 15–29. Springer, Heidelberg (2013). https://doi.org/10.1007/978-3-642-40381-1_2
7. Čyras, K., Toni, F.: Non-monotonic inference properties for assumption-based argumentation. In: Black, E., Modgil, S., Oren, N. (eds.) TAFA 2015. LNCS (LNAI), vol. 9524, pp. 92–111. Springer, Cham (2015). https://doi.org/10.1007/978-3-319-28460-6_6
8. Dung, P.M.: On the acceptability of arguments and its fundamental role in non-monotonic reasoning, logic programming and n-persons games. Artif. Intell. **77**(2), 321–358 (1995)
9. Dung, P.M.: An axiomatic analysis of structured argumentation for prioritized default reasoning. In: European Conference on Artificial Intelligence, pp. 267–272 (2014)
10. Dung, P.M.: An axiomatic analysis of structured argumentation with priorities. Artif. Intell. **231**, 107–150 (2016)
11. Gabbay, D.M.: Theoretical foundations for non-monotonic reasoning in expert systems. In: Apt, K.R. (ed.) Proceedings of the NATO Advanced Study Institute on Logics and Models. NATO ASI Series, vol. 13, pp. 439–457. Springer, Heidelberg (1985). https://doi.org/10.1007/978-3-642-82453-1_15
12. García, A.J., Simari, G.R.: Defeasible logic programming: an argumentative approach. Theory Pract. Log. Prog. **4**(1+2), 95–138 (2004)
13. Governatori, G., Maher, M.J., Antoniou, G., Billington, D.: Argumentation semantics for defeasible logic. J. Log. Comput. **14**(5), 675–702 (2004)
14. Hunter, A.: Base logics in argumentation. In: COMMA, pp. 275–286 (2010)
15. Kraus, S., Lehmann, D., Magidor, M.: Nonmonotonic reasoning, preferential models and cumulative logics. Artif. Intell. **44**(1), 167–207 (1990)

16. Li, Z., Parsons, S.: On argumentation with purely defeasible rules. In: Beierle, C., Dekhtyar, A. (eds.) SUM 2015. LNCS (LNAI), vol. 9310, pp. 330–343. Springer, Cham (2015). https://doi.org/10.1007/978-3-319-23540-0_22
17. Lin, F., Shoham, Y.: Argument systems: a uniform basis for nonmonotonic reasoning. KR **89**, 245–255 (1989)
18. McCarthy, J.: Circumscription, a form of nonmonotonic reasoning. Artif. Intell. **13**, 27–39 (1980)
19. Modgil, S., Prakken, H.: A general account of argumentation with preferences. Artif. Intell. **195**, 361–397 (2012)
20. Modgil, S., Prakken, H.: The $ASPIC^+$ framework for structured argumentation: a tutorial. Argum. Comput. **5**(1), 31–62 (2014)
21. Moore, R.C.: Semantical considerations on nonmonotonic logic. Artif. Intell. **25**(1), 75–94 (1985)
22. Nute, D.: Defeasible logic. In: Bartenstein, O., Geske, U., Hannebauer, M., Yoshie, O. (eds.) INAP 2001. LNCS (LNAI), vol. 2543, pp. 151–169. Springer, Heidelberg (2003). https://doi.org/10.1007/3-540-36524-9_13
23. Reiter, R.: A logic for default reasoning. Artif. Intell. **13**(1), 81–132 (1980)
24. Wu, Y.: Between argument and conclusion - argument-based approaches to discussion, inference and uncertainty. Ph.D. thesis (2012)

Extended Explanatory Argumentation Frameworks

Jérémie Dauphin$^{(\boxtimes)}$ and Marcos Cramer

University of Luxembourg, Luxembourg City, Luxembourg
jeremie.dauphin@uni.lu

Abstract. Multiple extensions of Dung's argumentation frameworks (AFs) have been proposed in order to model features of argumentation that cannot be directly modeled in AFs. One technique that has already previously proven useful to study and combine such extensions is the meta-argumentation methodology involving the notion of a flattening. In order to faithfully model the interaction between explanation argumentation in scientific debates, Šešelja and Straßer have introduced *Explanatory Argumentation Frameworks* (EAFs). In this paper, we first prove that the flattening technique works as expected for recursive (higher-order) attacks. Then we apply this technique in order to combine EAFs with multiple other extensions that have been proposed to AFs, namely with recursive attacks, joint attacks and a support relation between arguments. This gives rise to *Extended Explanatory Argumentation Frameworks* (EEAFs). We illustrate the applicability of EEAFs by using them to model a piece of argumentation from a research-level philosophy book.

1 Introduction

Dung's argumentation frameworks (AFs) [7] are a powerful and flexible formal tool for formally modelling argumentative discourse. However, various researchers have felt the need to extend AFs in order to model features of argumentation that cannot be directly modeled in AFs, e.g. by enriching them with recursive (higher-order) attacks [2], joint attacks [9], a support relation between arguments [4,5], or explanatory features [10].

One technique that has already previously proven useful to study and combine such extensions is the meta-argumentation methodology involving the notion of a *flattening* [3]. A flattening is a function that maps some extended variant of argumentation frameworks into standard AFs. If the definition of the various argumentation semantics for that extended variant of AFs is independent from the definition of that flattening function, one wants the flattening to satisfy the property that it preserves these semantics, in the sense that applying

J. Dauphin has received funding from the European Union's Horizon 2020 research and innovation programme under the Marie Skłodowska-Curie grant agreement No. 690974 for the project "MIREL: MIning and REasoning with Legal texts".

© Springer International Publishing AG, part of Springer Nature 2018
E. Black et al. (Eds.): TAFA 2017, LNAI 10757, pp. 86–101, 2018.
https://doi.org/10.1007/978-3-319-75553-3_6

the flattening function, then calculating the extensions according to some argumentation semantics, and finally unflattening the extensions should yield the same result as directly calculating the extensions according to the corresponding argumentation semantics for the extended variant of argumentation frameworks. However, flattenings can also be used to define argumentation semantics for extended variants of AFs for which there is no definition of the semantics independent of flattenings. This approach has proven particularly useful for combining multiple extensions of AFs [3], because in this case, it is often much clearer what the "right" definition of a flattening is than what the "right" direct definition of the various argumentation semantics is.

Previous work on flattening argumentation frameworks with recursive attacks (AFRAs) was limited to second-order attacks [1,3], even though the original definition of recursive attacks was for arbitrarily deeply nested higher-order attacks [2]. This means that for the purpose of defining the flattening, attacking an attack between two arguments was allowed, but attacking such a second-order attack was already not allowed. In Sect. 3.1, we show how to define a flattening of arbitrary AFRAs, and prove that it conforms with the direct definition of the semantics of AFRAs.

The rest of the paper is devoted to applying the meta-argumentation methodology of flattening and unflattening in order to incorporate recursive attacks, joint attacks and a support relation between arguments into *Explanatory Argumentation Frameworks* (EAFs), which have been proposed by Šešelja and Straßer [10] in order to faithfully model the interaction between explanation and argumentation in scientific debate. EAFs feature *explananda* and an *explanatory relation* that can hold either between an argument and an explanandum, or between two arguments. We use the terms *Extended Explanatory Argumentation Frameworks* (EEAFs) for this enriched formalism that incorporates recursive attacks, joint attacks and a support relation into EAFs.

The explanatory relation from EAFs cannot be easily flattened. Therefore, for defining the semantics of EEAFs, we apply the meta-argumentation methodology by allowing the output of the flattening function to be an EAF rather than an AF. In other words, we flatten away recursive attacks, joint attacks and the support relation, but we do not flatten away explanations, instead making use of the semantics of EAFs instead of the semantics of standard AFs.

Finally, we illustrate the applicability of EEAFs by using them to model a piece of argumentation from the introduction to Hartry Field's book *Saving Truth from Paradox* [8], an important, relatively recent, monograph about semantic paradoxes, a major research topic within the field of philosophical logic.

The rest of the paper is structured as follows: In Sect. 2, we describe the various proposed extensions to AFs and outline the meta-argumentation methodology of flattening and unflattening. In Sect. 3, we first extend the meta-argumentation methodology to arbitrarily deeply nested AFRAs, and then use this methodology to formally define the semantics of EEAFs. In Sect. 4 we present an example that illustrates the applicability of EEAFs, before concluding the paper in Sect. 5.

2 Basics of Formal Argumentation

We use the standard notions of abstract argumentation frameworks as defined by Dung in 1995 [7].

2.1 Explanatory Argumentation Frameworks

In scientific debates, the discussions are usually centered around some phenomenons or evidence and the different parties propose theories to explain them. With this idea in mind, Šešelja and Straßer have extended abstract argumentation framework with explanatory features [10]. In these frameworks, there are not only arguments but also explananda. These are scientific phenomenons of which, unlike arguments, the acceptability is not being questioned.

Definition 1. An *explanatory argumentation framework* (EAF) is a tuple $\langle \mathcal{A}, \mathcal{X}, \rightarrow, \dashrightarrow, \sim \rangle$, where $\mathcal{A}$ is a set of arguments, $\mathcal{X}$ is a set of explananda, $\rightarrow \subseteq \mathcal{A} \times \mathcal{A}$ is an attack relation, $\dashrightarrow \subseteq \mathcal{A} \times (\mathcal{A} \cup \mathcal{X})$ is an explanation relation from arguments to either explananda or other arguments, and $\sim \subseteq \mathcal{A} \times \mathcal{A}$ is a symmetric incompatibility relation.

Note that the incompatibility relation's purpose is to differentiate between the opposing theories, as scientists usually do not accept multiple explanations of a given phenomenon at the same time.

Definition 2. Let $\langle \mathcal{A}, \mathcal{X}, \rightarrow, \dashrightarrow, \sim \rangle$ be an EAF. A set of arguments $S \subseteq \mathcal{A}$ is said to be *conflict-free* if and only if there are no arguments $a, b \in S$ such that $(a, b) \in \rightarrow \cup \sim$.

Note that the definition of admissible sets still stands but with the revised definition of conflict-freeness.

Definition 3. An *explanation* $X[e]$ for $e \in \mathcal{X}$ offered by a set of arguments S is a subset S' of S such that there exists a unique argument $a \in S'$ such that $a \dashrightarrow e$ and for all $a' \in S' \setminus a$, there exists a path in $\dashrightarrow$ from a' to a.

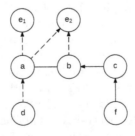

Fig. 1. Example EAF1

Example 1. Consider the EAF on Fig. 1. Note that the incompatibility relation has been represented by a straight line with no arrow between a and b.

Here we have two explananda, e_1 and e_2. a explains both e_1 and e_2 while b explains only e_2. Consider the conflict-free set $\{a, d, f\}$. It contains two explanations for e_1, namely $X_1[e_1] = \{a\}$ and $X_2[e_1] = \{a, d\}$. Similarly, it offers two explanations for e_2. The conflict-free set $\{b, f\}$ however offers an explanation only for e_2.

For our goal of selecting the best theory from our model, we need a way to compare how much and how well a given set of arguments is able to explain.

Definition 4. A set of arguments S_1 is *explanatory more powerful* than a set of arguments S_2 ($S_1 >_p S_2$) if and only if the set of explananda for which S_1 offers an explanation is a strict super-set of the set of explananda for which S_2 offers an explanation.

An explanation $X_1[e]$ is *explanatory deeper* than another explanation $X_2[e]$ ($X_1[e] >_d X_2[e]$) if and only if $X_2[e] \subset X_1[e]$.

In our previous example, we have that $\{a, d\} >_p \{b\}$ since $\{a, d\}$ offers an explanation for $\{e_1, e_2\}$ while $\{b\}$ only offers an explanation for $\{e_2\}$. Additionally, we have that $\{a, d\} >_d \{a\}$ and $\{a, d, f\} >_d \{a, f\}$.

Šešelja and Straßer [10] then propose two procedures for the selection of the best sets of arguments with respect to these notions. We have redefined them as extensions, in order to be more in line with abstract argumentation extensions, while preserving their concepts.

Definition 5. Let $\langle \mathcal{A}, \mathcal{X}, \rightarrow, \dashrightarrow, \sim \rangle$ be an EAF and $S \subseteq \mathcal{A}$ a set of arguments.

1. We say that S is *satisfactory* iff S is admissible and there is no $S' \subseteq \mathcal{A}$ such that $S' >_p S$ and S' is admissible.
2. We say that S is *insightful* iff S is satisfactory and there is no $S' \subseteq \mathcal{A}$ such that $S' >_d S$ and S' is satisfactory.
3. We say that S is an *argumentative core extension* (*AC-extension*) of Δ iff S is satisfactory and there is no $S' \supset S$ such that S' is satisfactory.
4. We say that S is an *explanatory core extension* (*EC-extension*) of Δ iff S is insightful and there is no $S' \subset S$ such that S' is insightful.

In our example, the AC-extension is $\{a, d, f\}$, while the EC-extension is $\{a, d\}$.

2.2 Argumentation Frameworks with Recursive Attacks

While EAFs add explanatory features to abstract argumentation frameworks, Baroni et al. [2] have developed an extension which enhances the expressive power of the attack relation. In their frameworks, they allow for attacks to target other attacks. This way, an argument may refute an attack relation between two other arguments without contesting the acceptability of any of them.

Definition 6. An *Argumentation Framework with Recursive Attacks* (AFRA) is a pair $\langle \mathcal{A}, \rightarrow \rangle$ where $\mathcal{A}$ is a set of arguments and $\rightarrow \subseteq \mathcal{A} \times (\mathcal{A} \cup \rightarrow)$ is an attack relation from arguments to either arguments or attacks.

For a given attack $\alpha = (A, X) \in \rightarrow$, we say that the source of α is $src(\alpha) = A$ and its target is $trg(\alpha) = X$.

Now that attacks can be targeted, we need to extend our notions of acceptance to also include them.

Definition 7. Let $F = \langle \mathcal{A}, \rightarrow \rangle$ be an AFRA, $\varphi \in \rightarrow, \psi \in (\mathcal{A} \cup \rightarrow)$ and $S \subseteq (\mathcal{A} \cup \rightarrow)$. We say that φ *defeats* ψ iff either $\psi = trg(\varphi)$ or $src(\psi) = trg(\varphi)$.

Additionally, we say that S is *conflict-free* iff there do not exist $\varphi, \psi \in S$ such that φ defeats ψ.

The notions of defense and admissibility then follows with a similar idea as in standard abstract argumentation frameworks.

Definition 8. Let $F = \langle \mathcal{A}, \rightarrow \rangle$ be an AFRA, $\varphi \in (\mathcal{A} \cup \rightarrow)$ and $S \subseteq (\mathcal{A} \cup \rightarrow)$. We say that S *defends* φ iff for every $\psi \in \rightarrow$ such that ψ defeats φ, there exists a $\delta \in S$ such that δ defeats ψ. We say that S is *admissible* iff S is conflict-free and defends its elements.

The complete semantics then follows with a similar definition as in classical abstract argumentation but using the adapted notions just defined.

Definition 9. Let $F = \langle \mathcal{A}, \rightarrow \rangle$ be an AFRA and $S \subseteq (\mathcal{A} \cup \rightarrow)$. We say that S is a *complete extension* of F iff S is admissible and contains every $\varphi \in (\mathcal{A} \cup \rightarrow)$ it defends.

2.3   Support in Abstract Argumentation

While classical abstract argumentation revolves around attacks, there has been research on extending it with a positive relation of support between arguments. We will first examine the formalism introduced by Cayrol and Lagasquie-Schiex called bipolar argumentation framework [5], as summarized by Boella et al. in [4].

Definition 10. A *bipolar argumentation framework* (BAF) is a triple $\langle \mathcal{A}, \rightarrow, \Rightarrow \rangle$ where $\mathcal{A}$ is a set of arguments, $\rightarrow \subseteq \mathcal{A} \times \mathcal{A}$ is an attack relation and $\Rightarrow \subseteq \mathcal{A} \times \mathcal{A}$ is a support relation.

Boella et al. [4] treat support in a deductive sense and thus introduce mediated attacks. The intuition behind these attacks is that if from a we can deduce b, then if we do not have b, we also cannot have a.

Definition 11. Let $\langle \mathcal{A}, \rightarrow, \Rightarrow \rangle$ be a bipolar argumentation framework. For $a, b \in \mathcal{A}$, there is a *mediated attack* from a to b if and only if there is a sequence $a_1 \Rightarrow a_2, ..., a_{n-1} \Rightarrow a_n$ such that $n \geq 2$, $a = a_1$ and $b \rightarrow a_n$.

They then define the semantics of bipolar argumentation frameworks with respect to their flattening. The flattened framework will consist of meta-arguments and an attack relation only, with the support relation from the BAFs being represented as a combination of auxiliary meta-arguments and attack relations.

Definition 12. Given a bipolar argumentation framework $\langle \mathcal{A}, \rightarrow, \Rightarrow \rangle$, the set of corresponding meta-arguments MA is $\{acc(a) \mid a \in \mathcal{A}\} \cup \{X_{a,b}, Y_{a,b} \mid a, b \in \mathcal{A}\} \cup \{Z_{a,b} \mid a, b \in \mathcal{A}\}$ and $\rightarrow_2 \subseteq MA \times MA$ is a binary relation on MA such that:

- For all $a, b \in \mathcal{A}$ such that $a \rightarrow b$, we have $acc(a) \rightarrow_2 X_{a,b}$, $X_{a,b} \rightarrow_2 Y_{a,b}$ and $Y_{a,b} \rightarrow_2 acc(b)$
- For all $a, b \in \mathcal{A}$ such that $a \Rightarrow b$, we have $acc(b) \rightarrow_2 Z_{a,b}$ and $Z_{a,b} \rightarrow_2 acc(a)$.

Example 2. The example represented in Fig. 2 is flattened in Fig. 3:

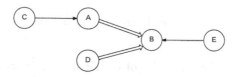

Fig. 2. Example bipolar argumentation framework

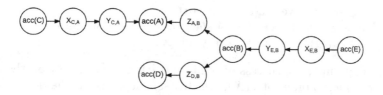

Fig. 3. Flattened BAF from Fig. 2

In the flattening, the mediated attacks are made apparent. By applying the semantics of classical abstract argumentation frameworks we can then retrieve the corresponding extensions of the BAF.

Note that Cohen et al. [6] have combined higher-order attacks and supports, with semantics defined directly on the higher-level frameworks. However, unlike with the flattening approach, it is unclear how to take further features into account in those direct semantics.

2.4 Joint Attacks

Another extension of AFs allows for joint attacks, where multiple arguments join forces to attack another argument.

Gabbay [9] calls this kind of relation a *joint attack*. He defines it as follows:

Definition 13. A *higher level argumentation framework* is a triple $(S, S^0, \rightarrow)$, where $S \neq \emptyset$ is a set of arguments, S^0 is the family of all finite non-empty subsets of S and $\rightarrow \subseteq S^0 \times S$ is an attack relation.

For simplicity of notation we will identify the singleton set $\{x\}$ with x.

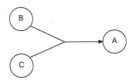

Fig. 4. Higher level argumentation framework

Similarly as before, the semantics of higher level networks will be defined in terms of their flattening. We define the flattening as follows:

Definition 14. Given a higher level argumentation framework $(S, S^0, \rightarrow)$, the set of corresponding meta-arguments MA is $\{acc(a), rej(a) \mid a \in \mathcal{A}\} \cup \{e(X) \mid X \in S^0\}$ and $\rightarrow_2 \subseteq MA \times MA$ is a binary relation on MA such that:

– For all $a \in \mathcal{A}$, we have $acc(a) \rightarrow_2 rej(a)$
– For all $X \in S^0$, and every $b \in \mathcal{A}$ such that $X \rightarrow b$, we have that $e(X) \rightarrow_2 acc(b)$ and $rej(a) \rightarrow_2 e(X)$ for every $a \in X$.

In the flattening, the success of a joint attack depends solely on the acceptance of the meta-argument $e(X)$, which itself depends on the acceptance of every argument in the coalition.

The flattening of the framework from Fig. 4 is depicted in Fig. 5.

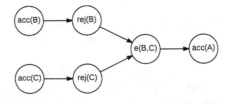

Fig. 5. Flattened version of the framework from Fig. 4

3   Aggregating Multiple Extensions of Abstract Argumentation Frameworks: EEAFs

In this section, we will introduce Extended Explanatory Argumentation Frameworks (EEAFs), an extension of EAFs from Sect. 2.1 with meta-argumentation features such as higher order attacks, support and joint attacks.

In order to motivate the semantics of EEAFs based on a flattening function, we will start by suggesting a flattening for AFRAs of any order. We will prove that this flattening leads to the same extensions as the AFRA semantics defined by Baroni et al. [2].

3.1   Flattening AFRAs

Boella et al. [3] define a flattening function for second-order AFRAs, which allows one to obtain for a given AFRA an equivalent abstract argumentation framework. We will now propose a flattening function for AFRAs of any order.

We will first define a function m which will associate each argument and each attack relation to the corresponding meta-argument. For an argument a, it will be the meta-argument $acc(a)$, while for an attack, it will be the Y auxiliary argument, since its acceptability is synonym of success for the attack.

Definition 15. Let $F = \langle \mathcal{A}, \rightarrow \rangle$ be an AFRA. The set of corresponding meta-arguments is $MA = \{acc(a) \mid a \in \mathcal{A}\} \cup \{X_{a,\psi}, Y_{a,\psi} \mid a \in \mathcal{A}, \ \psi \in (\mathcal{A} \cup \rightarrow)\}$. We define a partial function $m \colon (\mathcal{A} \cup \rightarrow) \mapsto MA$, such that:

- if $\varphi \in \mathcal{A}$, then $m(\varphi) = acc(\varphi)$.
- if $\varphi \in \rightarrow$ such that for some $\psi \in \mathcal{A}$ and some $\delta \in (\mathcal{A} \cup \rightarrow)$, $\varphi = (\psi, \delta)$, then $m(\varphi) = Y_{\psi, \delta}$.

We define the *flattening function* f to be $f(F) = \langle MA, \rightarrow_2 \rangle$, where $\rightarrow_2 \subseteq MA \times MA$ is a binary relations on MA such that

$$acc(a) \rightarrow_2 X_{a,\psi}, X_{a,\psi} \rightarrow_2 Y_{a,\psi} \text{ and } Y_{a,\psi} \rightarrow_2 m(\psi) \text{ for all } a \in \mathcal{A}, \psi \in (\mathcal{A} \cup \rightarrow)$$

One can then apply the classical abstract argumentation semantics such as complete, stable, preferred and grounded. We then need to define a function which can transform a meta-extension from the flattened AFRA to an extension for the original AFRA. A similar unflattening function has been introduced in [3], and has been slightly modified here to also unflatten attacks.

Definition 16. Given a set of meta-arguments $B \subseteq MA$, we define the *unflattening function* g as:

$$g(B) = \{a \mid acc(a) \in B\} \cup \{(a, \psi) \mid Y_{a,\psi} \in B\}$$

We also define a function $\bar{f}$ which provides a correspondence between a set of arguments and attacks from an AFRA and a set of meta-arguments from its flattened version.

Definition 17. Let $F = \langle \mathcal{A}, \rightarrow \rangle$ be an AFRA, $f(F) = \langle MA, \rightarrow_2 \rangle$ its flattening and $S \subseteq (\mathcal{A} \cup \rightarrow)$. We define the *correspondence function* $\bar{f} : \mathbb{P}(\mathcal{A} \cup \rightarrow) \mapsto \mathbb{P}(MA)$ as follows:

$$\bar{f}(S) = \{acc(a) \mid a \in S \cap \mathcal{A}\} \cup \{Y_{a,\psi} \mid (a, \psi) \in S \cap \rightarrow\} \cup$$
$$\{X_{b,\psi} \mid (a, b) \in S \cap \rightarrow, \; \psi \in \rightarrow\}$$

Notice that $g(\bar{f}(S)) = S$. We add the extra $X_{i,j}$ meta-arguments in order to represent the indirect attacks which the arguments in S might carry out, i.e. the attacks which are indirectly attacked by arguments in S due to them attacking the source of these attacks.

In [2], Baroni et al. define the semantics of AFRAs without having recourse to flattening. We will show that the process of flattening, applying complete semantics on the flattened frameworks and then unflattening it is equivalent to the directly applying the semantics they define for the complete semantics. We will show this gradually by first stating and proving three lemmas:

Lemma 1. Let $F = \langle \mathcal{A}, \rightarrow \rangle$ be an AFRA, $f(F) = \langle MA, \rightarrow_2 \rangle$ and $S \subseteq (\mathcal{A} \cup \rightarrow)$. S is conflict-free in F if and only if $\bar{f}(S)$ is conflict-free in $f(F)$.

Proof:

Let $F = \langle \mathcal{A}, \rightarrow \rangle$ be an AFRA, $f(F) = \langle MA, \rightarrow_2 \rangle$ and $S \subseteq (\mathcal{A} \cup \rightarrow)$.

1. $\Rightarrow$: Assume that S is conflict-free in F. Then, there is no $\varphi, \psi \in S$ such that $trg(\varphi) = \psi$ or $trg(\varphi) = src(\psi)$. Suppose for a contradiction that $\bar{f}(S)$ is not conflict-free in $f(F)$. This means that there exists two arguments $p, q \in \bar{f}(S)$ such that $p \rightarrow_2 q$. By the construction of $\rightarrow_2$ defined by the flattening function, there are only four possible cases, which all lead to the contradiction that S is not conflict-free. Therefore $\bar{f}(S)$ is conflict-free.

2. $\Leftarrow$: Suppose $\bar{f}(S)$ is conflict-free. Suppose for a contradiction that S is not conflict-free. Then, there exists $(a, \varphi), (b, \psi) \in S$ such that $\varphi = (b, \psi)$ or $\varphi = b$.

    In both cases we can reach the contradiction that $\bar{f}(S)$ is not conflict-free, therefore S is conflict-free. $\qquad\square$

Lemma 2. Let $F = \langle \mathcal{A}, \rightarrow \rangle$ be an AFRA, $f(F) = \langle MA, \rightarrow_2 \rangle$, $\varphi \in (\mathcal{A} \cup \rightarrow)$ and $S \subseteq (\mathcal{A} \cup \rightarrow)$. We have that:

φ is defended by S in F and if $\varphi = (a, \psi) \in \rightarrow$, we have $a \in S$, iff $m(\varphi)$ is defended by $\bar{f}(S)$ in $f(F)$ and if $\varphi \in \rightarrow$, then $acc(src(\varphi))$ is also defended by $\bar{f}(S)$.

Proof:

Let $F = \langle \mathcal{A}, \rightarrow \rangle$ be an AFRA, $f(F) = \langle MA, \rightarrow_2 \rangle$, $\varphi \in (\mathcal{A} \cup \rightarrow)$ and $S \subseteq (\mathcal{A} \cup \rightarrow)$.

1. $\Rightarrow$: Suppose that φ is defended by S in F and if $\varphi = (a, \psi) \in \rightarrow$, we have $a \in S$. Consider $m(\varphi)$ in $f(F)$. Suppose for some $p \in MA$, $p \rightarrow_2 m(\varphi)$. By

the construction of $\rightarrow_2$ defined by the flattening function, either $p = Y_{a,\varphi}$ for some $a \in \mathcal{A}$, or $p = X_{src(\varphi),trg(\varphi)}$.

In both cases, $m(\varphi)$ is defended by $\bar{f}(S)$. Hence, if φ is defended by S in F, then $m(\varphi)$ is defended by $\bar{f}(S)$ in $f(F)$. We now have to show that if $\varphi \in \rightarrow$, then $acc(src(\varphi))$ is also defended by $\bar{f}(S)$.

Suppose $\varphi \in \rightarrow$ and $p \in MA$ such that $p \rightarrow_2 acc(src(\varphi))$. Then, p must be of the form $Y_{a,src(\varphi)}$ for some $a \in \mathcal{A}$, and hence there exists $(a, src(\varphi)) \in \rightarrow$. Since $(a, src(\varphi))$ defeats φ, there exists some $\delta \in S$ such that δ defeats $(a, src(\varphi))$. We distinguish two cases:

Either $\delta = (b, a)$ or $\delta = (b, (a, src(\varphi)))$ for some $b \in \mathcal{A}$. In both cases, $acc(src(\varphi))$ is also defended by $\bar{f}(S)$.

Therefore, if φ is defended by S in F and if $\varphi = (a, \psi) \in \rightarrow$, we have $a \in S$, then $m(\varphi)$ is defended by $\bar{f}(S)$ in $f(F)$ and if $\varphi \in \rightarrow$, then $acc(src(\varphi))$ is also defended by $\bar{f}(S)$.

2. $\Leftarrow$: Suppose $m(\varphi)$ is defended by $\bar{f}(S)$ in $f(F)$ and if $\varphi \in \rightarrow$, then $acc(src(\varphi))$ is also defended by $\bar{f}(S)$. Consider φ in F. Suppose that for some $\psi \in \rightarrow$, ψ defeats φ. This means that either $\psi = (a, \varphi)$ or $\psi = (a, src(\varphi))$ for some $a \in \mathcal{A}$. In both cases, we can conclude that there exists a $\delta \in S$ such that δ defeats ψ by contradiction. Therefore, φ is defended by S.

We now have to show that if $\varphi = (a, \psi) \in \rightarrow$, we have $a \in S$, still under the assumption that $m(\varphi)$ is defended by $\bar{f}(S)$ in $f(F)$ and if $\varphi \in \rightarrow$, then $acc(src(\varphi))$ is also defended by $\bar{f}(S)$.

Suppose that $\varphi = (a, \psi) \in \rightarrow$. Then, by the construction of $\rightarrow_2$ defined by the flattening function, we have $X_{a,\psi} \rightarrow_2 Y_{a,\psi}$. Since $m(\varphi) = Y_{a,\psi}$ is defended by $\bar{f}(S)$, there exists $p \in \bar{f}(S)$ such that $p \rightarrow_2 X_{a,\psi}$. By the construction of $\rightarrow_2$, the only possibility is $p = acc(a)$. Hence, $acc(a) \in \bar{f}(S)$. Therefore, we have $a \in S$.

Thus, we can conclude that φ is defended by S in F and if $\varphi = (a, \psi) \in \rightarrow$, we have $a \in S$, if and only if $m(\varphi)$ is defended by $\bar{f}(S)$ in $f(F)$ and if $\varphi \in \rightarrow$, then $acc(src(\varphi))$ is also defended by $\bar{f}(S)$.   □

Lemma 3. Let $F = \langle \mathcal{A}, \rightarrow \rangle$ be an AFRA, $f(F) = \langle MA, \rightarrow_2 \rangle$ and $S \subseteq (\mathcal{A} \cup \rightarrow)$. We have that:

S is admissible in F and for every $(a, \psi) \in (S \cap \rightarrow)$, we have that $a \in S$
if and only if
$\bar{f}(S)$ is admissible in $f(F)$.

Proof: Let $F = \langle \mathcal{A}, \rightarrow \rangle$ be an AFRA, $f(F) = \langle MA, \rightarrow_2 \rangle$ and $S \subseteq (\mathcal{A} \cup \rightarrow)$.

1. $\Rightarrow$: Suppose $\bar{f}(S)$ is admissible in $f(F)$. Then, $\bar{f}(S)$ is conflict-free. Hence, according to Lemma 1, S is also conflict-free.

Let $\varphi \in S$. We need to show that φ is defended by S. We do this by applying Lemma 2, i.e. by establishing that $m(\varphi)$ is defended by $\bar{f}(S)$ in $f(F)$ and if $\varphi \in \rightarrow$, then $acc(src(\varphi))$ is also defended by $\bar{f}(S)$. We have $m(\varphi) \in \bar{f}(S)$ and $m(\varphi)$ is defended by $\bar{f}(S)$ since $\bar{f}(S)$ is admissible. By the definition of $\bar{f}$, for every $(a, \psi) \in (S \cap \to)$, we have $Y_{a,\psi} \in \bar{f}(S)$. Therefore, $acc(a) \in \bar{f}(S)$, since it is the only argument which can defend $Y_{a,\psi}$ from $X_{a,\psi}$'s attack and $\bar{f}(S)$ is admissible. This means that $acc(a)$ is defended by $\bar{f}(S)$. Thus, according to Lemma 2, every $\varphi \in S$ is defended by S, which means that S is admissible, and for every $(a, \psi) \in (S \cap \to)$, we have that $a \in S$.

2. $\Leftarrow$: Suppose S is admissible in F and for every $(a, \psi) \in (S \cap \to)$, we have that $a \in S$. Then, S is conflict-free and so, according to Lemma 1, $\bar{f}(S)$ is also conflict-free.

Let $p \in \bar{f}(S)$. p is either of the form $m(\varphi)$ for some $\varphi \in S$, or of the form $X_{a,b}$ for some $a, b \in MA$ and $(\psi, a) \in S$.

In both cases, p is defended by $\bar{f}(S)$. Hence, $\bar{f}(S)$ is admissible in $f(F)$.

Therefore, S is admissible in F and for every $(a, \psi) \in (S \cap \to)$, we have that $a \in S$, if and only if $\bar{f}(S)$ is admissible in $f(F)$. $\qquad\square$

Theorem 1. Let $F = \langle A, \to \rangle$ be an AFRA, $f(F) = \langle MA, \to_2 \rangle$ and $S \subseteq (A \cup \to)$. S is a complete extension of F if and only if $\bar{f}(S)$ is a complete extension of $f(F)$.

Proof:

Let $F = \langle A, \to \rangle$ be an AFRA, $f(F) = \langle MA, \to_2 \rangle$ and $S \subseteq (A \cup \to)$.

1. $\Rightarrow$: Suppose S is a complete extension of F. For every $(a, \psi) \in (S \cap \to)$, by the definition of defeat, a is defended by S, and thus $a \in S$. Therefore, by Lemma 3, $\bar{f}(S)$ is admissible.

Take some arbitrary $p \in MA$ and suppose that p is defended by $\bar{f}(S)$. Then, either $p = m(\varphi)$ for some $\varphi \in (A \cup \to)$, or $p = X_{a,b}$ for some $a, b \in A$.

   (a) Suppose that $p = m(\varphi)$ for some $\varphi \in (A \cup \to)$. Now assume that $\varphi \in \to$. Then, $m(\varphi) = Y_{src(\varphi),trg(\varphi)}$. By construction of $\to_2$, we have that $X_{src(\varphi),trg(\varphi)} \to_2 Y_{src(\varphi),trg(\varphi)}$. The only argument which can defend $Y_{src(\varphi),trg(\varphi)}$ from $X_{src(\varphi),trg(\varphi)}$ is $acc(src(\varphi))$. Since $\bar{f}(S)$ defends $Y_{src(\varphi),trg(\varphi)}$, we have that $acc(src(\varphi)) \in \bar{f}(S)$. As $\bar{f}(S)$ is admissible, $acc(src(\varphi))$ is defended by $\bar{f}(S)$. Hence, if $\varphi \in \to$, then $acc(src(\varphi))$ is defended by $\bar{f}(S)$.

   Therefore, by Lemma 2, φ is defended by S. Since S is a complete extension, this means that $\varphi \in S$. Therefore, $p = m(\varphi) \in \bar{f}(S)$.

   (b) Now suppose that $p = X_{a,b}$ for some $a, b \in A$. According to our assumptions, $\bar{f}(S)$ defends $X_{a,b}$. By construction of $\Rightarrow_2$, the only argument which attacks $X_{a,b}$ is $acc(a)$. Hence, there exists $Y_{c,a} \in \bar{f}(S)$ for some $c \in A$. So, by definition of $\bar{f}$, we have that $p = X_{a,b} \in \bar{f}(S)$.

   In either case, we have that $p \in \bar{f}(S)$. Hence, $\bar{f}(S)$ contains all arguments it defends. Since it is also admissible, $\bar{f}(S)$ is a complete extension of $f(F)$.

2. $\Leftarrow$: Suppose that $\bar{f}(S)$ is a complete extension of $f(F)$. Then, $\bar{f}(S)$ is admissible and contains all arguments it defends. According to Lemma 3, we have that S is admissible and for every $(a, \psi) \in (S \cap \to)$, we have that $a \in S$. Suppose that for some $\varphi \in (A \cup \to)$, φ is defended by S. Hence, by Lemma 2, $m(\varphi)$ is defended by $\bar{f}(S)$. Since $\bar{f}(S)$ is a complete extension of $f(F)$, $m(\varphi) \in \bar{f}(S)$.

Hence, by construction of $\bar{f}(S)$, we have that $\varphi \in S$. Therefore, for any $\varphi \in (\mathcal{A} \cup \rightarrow)$ such that φ is defended by S, we have $\varphi \in S$. Since S is also admissible, S is a complete extension of F.

Hence, S is a complete extension of F if and only if $\bar{f}(S)$ is a complete extension of $f(F)$. $\square$

3.2   Extended Explanatory Argumentation Frameworks

We will now extend EAFs, as seen in Sect. 2.1, by integrating them with the meta-argumentation techniques we have discussed so far.

Definition 18. An *extended explanatory argumentation framework* (EEAF) is a tuple $\langle \mathcal{A}, \mathcal{X}, \rightarrow, \dashrightarrow, \sim, \Rightarrow \rangle$, where $\mathcal{A}$ is a set of arguments, $\mathcal{X}$ is a set of explananda, $\dashrightarrow \subseteq (\mathcal{A} \times \mathcal{A}) \cup (\mathcal{A} \times \mathcal{X})$ is an explanatory relation, $\rightarrow \subseteq (\mathbb{P}(\mathcal{A}) \cup \dashrightarrow \cup \rightarrow) \times (\mathcal{A} \cup \dashrightarrow \cup \rightarrow \cup \Rightarrow)$ is a higher-order attack relation, $\sim \subseteq \mathcal{A} \times \mathcal{A}$ is an incompatibility relation and $\Rightarrow \subseteq \mathcal{A} \times \mathcal{A}$ is a support relation.

We then define the semantics of EEAFs in terms of their flattening.

Definition 19. Let $F = \langle \mathcal{A}, \mathcal{X}, \rightarrow, \dashrightarrow, \sim, \Rightarrow \rangle$ be an EEAF. The set of meta-arguments corresponding to F is $MA = \{acc(a), rej(a) \mid a \in \mathcal{A}\} \cup \{X_{m(\varphi),m(\psi)}, Y_{m(\varphi),m(\psi)} \mid \varphi \in (\mathcal{A} \cup \rightarrow \cup \dashrightarrow), \psi \in (\mathcal{A} \cup \rightarrow \cup \dashrightarrow \cup \Rightarrow)\} \cup \{e(S) \mid S \subseteq \mathcal{A}$ with at least two elements$\} \cup \{P_{a,\psi}, Q_{a,\psi} \mid a \in \mathcal{A}, \psi \in (\mathcal{A} \cup \mathcal{X})\} \cup \{Z_{a,b} \mid a, b \in \mathcal{A}\}$ and the set of meta-explananda is $MX = \mathcal{X}$. We define a partial function m which assigns for each element of the framework a corresponding meta-argument.

$$m : (\mathcal{A} \cup \rightarrow \cup \dashrightarrow \cup \Rightarrow) \mapsto MA.$$

such that:

- if $\varphi \in \mathcal{A}$, then $m(\varphi) = acc(\varphi)$;
- if $\varphi \in \mathcal{X}$, then $m(\varphi) = \varphi$;
- if $\varphi \in \Rightarrow$ such that for some $a, b \in \mathcal{A}$, $\varphi = (a \Rightarrow b)$, then $m(\varphi) = Z_{a,b}$;
- if $\varphi \in \rightarrow$ such that for some $S \subseteq \mathcal{A}$ with at least two elements and some $\psi \in (\mathcal{A} \cup \dashrightarrow \cup \rightarrow \cup \Rightarrow)$, $\varphi = (S \rightarrow \psi)$, then $m(\varphi) = e(S)$;
- if $\varphi \in \rightarrow$ such that for some $\psi \in (\mathcal{A} \cup \dashrightarrow \cup \rightarrow)$ and some $\delta \in (\mathcal{A} \cup \dashrightarrow \cup \rightarrow \cup \Rightarrow)$, $\varphi = (\psi \rightarrow \delta)$, then $m(\varphi) = Y_{\psi,\delta}$;
- if $\varphi \in \dashrightarrow$ such that for some $a \in \mathcal{A}$ and $\psi \in (\mathcal{A} \cup \mathcal{X})$, $\varphi = (a \dashrightarrow \psi)$, then $m(\varphi) = P_{a,\psi}$.

We define the *flattening function* f to be $f(F) = \langle MA, \mathcal{X}, \rightarrow_2, \dashrightarrow_2, \sim_2 \rangle$, where $\rightarrow_2, \sim_2 \subseteq MA \times MA$ and $\dashrightarrow_2 \subseteq MA \times (MA \cup \mathcal{X})$ are such that:

- $X_{m(\varphi),m(\psi)} \rightarrow_2 Y_{m(\varphi),m(\psi)}, Y_{m(\varphi),m(\psi)} \rightarrow_2 m(\psi)$ for all $\varphi, \psi \in (\mathcal{A} \cup \rightarrow \cup \dashrightarrow \cup \Rightarrow)$;
- $m(\varphi) \rightarrow_2 X_{m(\varphi),m(\psi)}$ if and only if $\varphi \rightarrow \psi$ and φ is not a set of arguments with at least two elements;

- $acc(a) \rightarrow_2 rej(a)$ for all $a \in \mathcal{A}$;
- $e(S) \rightarrow_2 m(\varphi)$ if and only if $S \rightarrow \varphi$ for $S \subseteq \mathcal{A}$ with at least 2 elements;
- $rej(a) \rightarrow_2 e(S)$ if and only if $a \in S$;
- $Z_{a,b} \rightarrow_2 acc(a)$ for all $a, b \in \mathcal{A}$;
- $acc(b) \rightarrow_2 Z_{a,b}$ if and only if $a \Rightarrow b$;
- $acc(a) \dashrightarrow_2 P_{a,\varphi}$, $P_{a,\varphi} \dashrightarrow_2 m(\varphi)$, $acc(a) \rightarrow_2 Q_{a,\varphi}$ and $Q_{a,\varphi} \rightarrow_2 P_{a,\varphi}$ if and only if $a \dashrightarrow \varphi$;
- $acc(a) \sim_2 acc(b)$ if and only if $a \sim b$.

Notice that the set of meta-arguments MA and the correspondence function m are defined through a simultaneous inductive definition, which is well-founded, because $\rightarrow$ is a well-founded relation (assuming that the set theory presupposed in Definition 6 is a standard set theory like ZFC that satisfies the Axiom of Foundation).

Note that we do not fully flatten the explanatory relation and flatten EEAFs into EAFs instead of AFs. This is due to the fact that the explanatory relation is not easily flattened, and extensions can still be extracted from explanatory argumentation frameworks via the two EAF extensions which are well-suited for our task. In order to do this, we need to define an unflattening function which will map a set of meta-arguments from a flattened EEAF to the corresponding set of arguments from the original EEAF.

Definition 20. Given an EEAF F and a set of meta-arguments $B \subseteq MA$ such that MA corresponds to F, we define the *unflattening function* g to be:

$$g(B) = \{a \mid acc(a) \in B\}$$

Notice that in the unflattening, we only care about the arguments and do not unflatten the meta-arguments which represent the other elements of EEAFs. This is due to the fact that we are only interested in selecting the arguments of the EEAF, which make up the argumentative and explanatory cores.

Definition 21. Let F be an EEAF and $G = f(F)$ its flattening. We say that $S \subseteq \mathcal{A}$ is an *AC-extension* of F iff $S = g(S')$, where S' is an AC-extension of G. Similarly, we say that $S \subseteq \mathcal{A}$ is an *EC-extension* of F iff $S = g(S')$, where S' is an EC-extension of G.

4   Applying EEAFs to the Liar Paradox

Let us now move on to an example, which focuses on two groups of solutions for the liar paradox. The arguments are extracted from *Saving Truth from Paradox* [8]. The first group is the solutions which weaken classical logic, namely the paracomplete, paraconsistent and semi-classical solutions. The second group is comprised of the underspill and overspill solutions.

We have the following arguments:

- E_p: This explanandum represents the paradox.

- *A*: The paracomplete, paraconsistent and semi-classical solutions which provide explanations for the paradox by weakening classical logic.
- *B*: The underspill and overspill solutions which provide their own explanation of the paradox by suggesting that for some predicates F, F is true of some objects that aren't F or vice-versa.
- *C*: We did not change logic to hide the defects in other flawed theories such as Ptolemaic astronomy, so why should we change the logic simply to hide these paradoxes?
- *D*: There is no known way of saving these flawed theories such as Ptolemaic astronomy and even if there was, there is little benefit to doing so.
- *F*: We have worked out the details of the new logics and they allow us to conserve the theory of truth.
- *G*: Changing the logic implies changing the meaning.
- *H*: Change of meaning is bad.
- *I*: The change is mere.
- *J*: This is no 'mere' relabelling.
- *K*: Change of truth schema is a change of the meaning of 'true'.
- *L*: The paradox forces a change of meaning.

The framework is represented in Fig. 6 and its flattening in Fig. 7. We have omitted less-relevant auxiliary arguments for the sake of visibility.

We get that the AC-extensions are $\{A, C, D, F, L, G, J, K\}$ and $\{B, C, D, F, L, G, J, K\}$. We can distinguish here the two rivaling solutions which are both selected. This is due to the fact that even though the author might have a preference for one or another, in the excerpt we have analyzed, he is merely defending the solutions represented in A from attacks and making no argument which attacks the solutions represented in B.

The EC-extensions are $\{A, D, F, L\}$, $\{A, D, F, J\}$ and $\{B, J\}$. Notice that there are two different EC-extensions which contain A, as there are two arguments which individually defend A from the coalition attack of $\{G, H\}$.

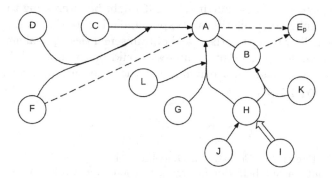

Fig. 6. EEAF representing the reasoning behind the excerpt

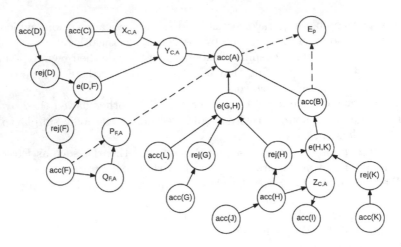

Fig. 7. Flattened EEAF representing the reasoning behind the excerpt

5   Conclusion and Future Work

We have examined several extensions of abstract argumentation frameworks that add explanatory features, recursive attacks, support and joint attacks. In the cases of recursive attacks, support and joint attacks, we have presented a flattening function, which allows us to instantiate these extended framework as standard AFs. We have shown that in the case of AFRAs, the complete semantics defined in terms of the flattening is equivalent to the complete semantics which has been defined directly on AFRAs. We have then aggregated these extensions into one framework, EEAFs, and defined the semantics in terms of its flattening to EAFs. Finally, we have explored an application of EEAFs to argumentation from a research-level philosophy book.

Concerning future work in the line of research of this paper, we plan to extend the result about the flattening of AFRAs to other argumentation semantics than the complete semantics. Furthermore, it might be interesting to investigate flattening the explanatory relation and explananda. Due to their intricate nature, it is not obvious how to flatten them and obtain semantics equivalent to the ones defined on EAFs. Another point of interest would be to apply EEAFs to other areas of scientific debates and examine whether the current features provide enough expressive power.

References

1. Baroni, P., Boella, G., Cerutti, F., Giacomin, M., van der Torre, L., Villata, S.: On the input/output behavior of argumentation frameworks. Artif. Intell. **217**, 144–197 (2014)
2. Baroni, P., Cerutti, F., Giacomin, M., Guida, G.: Encompassing attacks to attacks in abstract argumentation frameworks. In: Sossai, C., Chemello, G. (eds.) ECSQARU 2009. LNCS (LNAI), vol. 5590, pp. 83–94. Springer, Heidelberg (2009). https://doi.org/10.1007/978-3-642-02906-6_9

3. Boella, G., Gabbay, D.M., van der Torre, L., Villata, S.: Meta-argumentation modelling i: methodology and techniques. Stud. Logica **93**(2–3), 297–355 (2009)
4. Boella, G., Gabbay, D.M., van der Torre, L., Villata, S.: Support in abstract argumentation. COMMA **216**, 111–122 (2010)
5. Cayrol, C., Lagasquie-Schiex, M.C.: Bipolarity in argumentation graphs: towards a better understanding. IJAR **54**(7), 876–899 (2013)
6. Cohen, A., Gottifredi, S., García, A.J., Simari, G.R.: On the acceptability semantics of argumentation frameworks with recursive attack and support. In: Computational Models of Argument - Proceedings of COMMA 2016, pp. 231–242 (2016)
7. Dung, P.M.: On the acceptability of arguments and its fundamental role in nonmonotonic reasoning, logic programming and n-person games. Artif. Intell. **77**(2), 321–357 (1995)
8. Field, H.: Saving Truth from Paradox. Oxford University Press, Oxford (2008)
9. Gabbay, D.M.: Fibring argumentation frames. Stud. Logica **93**(2–3), 231–295 (2009)
10. Šešelja, D., Straßer, C.: Abstract argumentation and explanation applied to scientific debates. Synthese **190**(12), 2195–2217 (2013)

Probabilities on Extensions in Abstract Argumentation

Matthias Thimm[1](✉), Pietro Baroni[2], Massimiliano Giacomin[2], and Paolo Vicig[3]

[1] Institute for Web Science and Technologies (WeST), University of Koblenz-Landau, Koblenz, Germany
thimm@uni-koblenz.de
[2] Dip. Ingegneria dell'Informazione, University of Brescia, Brescia, Italy
[3] DEAMS, University of Trieste, Trieste, Italy

Abstract. Combining computational models of argumentation with probability theory has recently gained increasing attention, in particular with respect to abstract argumentation frameworks. Approaches following this idea can be categorised into the constellations and the epistemic approach. While the former considers probability functions on the subgraphs of abstract argumentation frameworks, the latter uses probability theory to represent degrees of belief in arguments, given a fixed framework. In this paper, we investigate the case where probability functions are given on the extensions of abstract argumentation frameworks. This generalises classical semantics in a straightforward fashion and we show that our approach also complies with many postulates for epistemic probabilistic argumentation.

1 Introduction

Computational models of argumentation are non-monotonic reasoning formalisms that focus on the role of arguments, i.e., defeasible reasons supporting a certain claim, and their relationships. In this context, the well-known formalism of abstract argumentation frameworks [11] abstracts from the inner structure of arguments and only models conflict between them, thus representing argumentation scenarios as directed graphs where arguments are vertices and an attack of one argument on another is modelled by a directed edge. Still, this approach is quite expressive, subsumes many other approaches to non-monotonic reasoning, and provides an active research field. Many research topics have been spawned around these frameworks including, among others, semantical issues [3], extensions on support [10], algorithms [9], and systems [30].

In their original form, abstract argumentation frameworks are a qualitative approach to non-monotonic reasoning as their semantics is set-based (it amounts to identifying sets of collectively acceptable arguments, called *extensions*) and inferences consist of statements regarding the acceptance status of arguments, which can be binary (an argument is simply "accepted" or "rejected") or three-valued (where a third option "undecided" is also possible). In recent years, many

E. Black et al. (Eds.): TAFA 2017, LNAI 10757, pp. 102–119, 2018.
https://doi.org/10.1007/978-3-319-75553-3_7

approaches have been developed that incorporate some quantitative aspects into abstract argumentation frameworks. These can be categorised into two families. In the first family, the syntactic representation of argumentation frameworks is extended with quantities, in order to incorporate more information explicitly. For example, in [25] arguments and attacks can be annotated with probabilities that model user-supplied information about the likelihood that these objects actually appear in the argumentation framework. This approach is also called the *constellations approach* to probabilistic argumentation [19]. The main aim of these works is then to generalise classical semantics and other notions to the extended approach. See also [12,32] for some other examples from this family based on weights and fuzzy logic, respectively. The other family is about bringing quantities into the semantics of vanilla argumentation frameworks themselves. Here, the syntactic representation is not extended and the aim is to derive quantitative information which is implicit in the topology of the graph. Concrete approaches within this family are, e.g., numerical ranking functions [1,7,18,27] and the equational approach [16,17]. The *epistemic approach* to probabilistic argumentation [5,21,22,28] considers the use of probability functions to capture the degrees of belief of an agent in (sets of) arguments (see [21] for a discussion). In this sense the epistemic approach shares some properties with both the families introduced above: on the one hand, the probability values are user-supplied, since they represent the belief of some agent, on the other hand, they can be put in relationship with the semantics of vanilla argumentation frameworks, since it is reasonable to assume that the beliefs of an agent take into account (and/or are constrained by) the topology of the graph.

In this paper, we contribute to the research trend on probabilistic argumentation by considering a further option, which consists in adding a probabilistic layer on top of classical semantics of abstract argumentation frameworks, i.e., we consider probability functions on extensions. This investigation is motivated by the fact that given an argumentation framework, capturing the attacks existing between arguments, each extension prescribed by an argumentation semantics can be regarded as an alternative answer to the question: "which arguments are able to survive the conflict together?". Thus the set of extensions can be regarded as a set of alternative reasonable options, each satisfying the "survival criterion" encoded by the argumentation semantics, which however does not provide any indication on which extension to select, in case the agents needs to finally choose one of them. This is required in particular in the case of practical reasoning where arguments concern reasons about what to do and alternative extensions may be put in correspondence with different available courses of action. In this context probabilities on extensions may encode additional information, external to the argumentation process, about which option is more likely to be selected by an agent. For instance suppose that in the context of some reasoning activity involving a health problem, two extensions emerge as reasonable, say one corresponding to undergoing surgery and the other to assuming a drug for a long time. The final choice is uncertain and is in the hands of the patient, whose (possibly non-rational) attitude towards the two options can be modeled by a probability assignment on the two extensions, e.g. you may assign a higher probability to the second extension if you know that the patient is particularly worried about the scars caused by surgery. These probability values could be acquired for instance using an approach to probabilistic user modeling, as proposed in [20].

Besides modelling the attitudes of a single agent, probabilities on extensions may be used to model collective attitudes too. Consider the case where two or more politicians argue about their government programmes and assume that their different positions are acceptable from an argumentative point of view. Then a probability assignment on the extensions corresponding to the positions of the candidates may reflect the outcomes of an opinion poll among the voters (note that the use of votes in the context of argumentation frameworks to support an initial numerical assessment, though not of probabilistic nature, has been considered in [14,24]).

Probability assignments on extensions provide then the basis for further inferential activities, for instance an argument can in general be included in different extensions and it is interesting to consider the probability that a specific argument (or sets of arguments) is selected. In the political example, different candidates, say all candidates, may share the argument that "we should cut taxes since this will promote economical growth", then the probability that this argument is accepted and that tax cuts are in the next government programme is 1, independently of the individual probabilities assigned to the various extensions/candidates (provided that you trust that politicians keep faith with their promises).

Altogether, the general idea is to provide a contribution to the investigation of integrated uncertain reasoning models encompassing both qualitative (in our case, based on abstract argumentation) and quantitative (in our case, probabilistic) evaluation aspects.

To provide a formal basis to this kind of modelling and reasoning activities, in this paper we investigate probability functions on extensions, and in particular,

1. we introduce our approach to probability functions over extensions and we draw some relationships with the maximum entropy principle and with imprecise probabilities (Sect. 3);
2. we investigate the properties of this extension, in particular wrt. rationality postulates usually considered for the epistemic approach (Sect. 4);
3. we investigate some computational issues of the approach (Sect. 5).

Necessary preliminaries are introduced in Sect. 2 and we conclude with a summary in Sect. 6.

2   Preliminaries

Abstract argumentation frameworks [11] take a very simple view on argumentation as they abstract away any detail about the internal structure of an argument, its origin and nature and so on. Abstract argumentation frameworks only capture the conflicts between arguments by means of a binary attack relation.

Definition 1. *An* abstract argumentation framework AF *is a tuple* AF $=$ (Arg, $\rightarrow$) *where* Arg *is a set of arguments and* $\rightarrow$ *is a relation* $\rightarrow\, \subseteq$ Arg $\times$ Arg.

For the sake of simplicity, in this paper we assume that the set Arg is finite. For two arguments $\mathcal{A}, \mathcal{B} \in$ Arg the relation $\mathcal{A} \rightarrow \mathcal{B}$ means that argument $\mathcal{A}$ attacks argument $\mathcal{B}$. We abbreviate $\mathsf{Att}_{\mathsf{AF}}(\mathcal{A}) = \{\mathcal{B} \mid \mathcal{B} \rightarrow \mathcal{A}\}$. Abstract argumentation frameworks can be concisely represented by directed graphs, where arguments are represented as nodes and edges model the attack relation.

Example 1. Consider the abstract argumentation framework $\mathsf{AF}_1 = (\mathsf{Arg}_1, \rightarrow_1)$ depicted in Fig. 1. Here it is $\mathsf{Arg}_1 = \{\mathcal{A}_1, \mathcal{A}_2, \mathcal{A}_3, \mathcal{A}_4, \mathcal{A}_5\}$ and $\rightarrow_1 = \{(\mathcal{A}_1, \mathcal{A}_2), (\mathcal{A}_2, \mathcal{A}_1), (\mathcal{A}_2, \mathcal{A}_3), (\mathcal{A}_3, \mathcal{A}_4), (\mathcal{A}_4, \mathcal{A}_5), (\mathcal{A}_5, \mathcal{A}_4), (\mathcal{A}_3, \mathcal{A}_5)\}$.

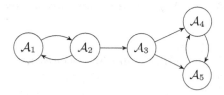

Fig. 1. The argumentation framework AF_1 from Example 1

An argumentation semantics is a formal criterion to determine the conflict outcomes. Two main approaches to semantics definition are available in the literature, namely the extension-based approach [11] and the labeling-based approach [33]. In this paper we focus on the extension-based approach, the reader is referred to [3] for a review and an analysis of the correspondence between the two approaches. An *extension* E of an argumentation framework $\mathsf{AF} = (\mathsf{Arg}, \rightarrow)$ is a set of arguments $E \subseteq \mathsf{Arg}$ that corresponds to a coherent and tenable view in the argumentation process underlying AF. Intuively an extension is a set of arguments which are "collectively acceptable" or "can survive the conflict together".

In the literature [3,8,11] a wide variety of different types of semantics has been proposed. The definition of a semantics typically builds on some basic properties that an extension should satisfy: arguably, conflict-freeness and admissibility are among the most important extension properties.

Definition 2. *An extension $E \subseteq$ Arg is* conflict-free *if for all $\mathcal{A}, \mathcal{B} \in E$ it is not the case that $\mathcal{A} \rightarrow \mathcal{B}$. An extension $E \subseteq$ Arg* defends *an argument $\mathcal{A} \in$ Arg if for all $\mathcal{C} \in$ Arg, if $\mathcal{C} \rightarrow \mathcal{A}$ then there is $\mathcal{B} \in E$ with $\mathcal{B} \rightarrow \mathcal{C}$. An extension $E \subseteq$ Arg is* admissible *if it is conflict-free and defends all its elements.*

We abbreviate by $cf(\mathsf{AF})$ the set of conflict-free extensions, by $mcf(\mathsf{AF})$ the maximal (wrt. set inclusion) conflict-free extensions, and by $adm(\mathsf{AF})$ the set of admissible extensions. Dung's traditional semantics are defined by imposing further constraints.

Definition 3. *Let* $\mathsf{AF} = (\mathsf{Arg}, \rightarrow)$ *be an abstract argumentation framework and E an admissible extension.*

- E is complete *if for all* $\mathcal{A} \in \mathsf{Arg}$, *if* E *defends* $\mathcal{A}$ *then* $\mathcal{A} \in E$.
- E is grounded *if and only if* E *is minimal among complete extensions*.
- E is preferred *if and only if* E *is maximal among complete extensions*.
- E is stable *if and only if* E *is complete and attacks all other arguments*.

All statements on minimality/maximality are meant to be with respect to set inclusion.

We denote by $comp(\mathsf{AF})$, $ground(\mathsf{AF})$, $pref(\mathsf{AF})$, and $st(\mathsf{AF})$ the sets of complete, grounded, preferred, and stable extensions of AF, respectively. Note that a grounded extension is uniquely determined and always exists [11], so we also abbreviate by $GE(\mathsf{AF})$ the unique grounded extension of AF, i.e., $ground(\mathsf{AF}) = \{GE(\mathsf{AF})\}$. Furthermore, we have the following relationships, cf. [3].

Proposition 1. *Let* $\mathsf{AF} = (\mathsf{Arg}, \rightarrow)$ *be an abstract argumentation framework. Then*

1. $st(\mathsf{AF}) \subseteq mcf(\mathsf{AF}) \subseteq cf(\mathsf{AF})$,
2. $st(\mathsf{AF}) \subseteq pref(\mathsf{AF}) \subseteq comp(\mathsf{AF}) \subseteq adm(\mathsf{AF}) \subseteq cf(\mathsf{AF})$, *and*
3. $ground(\mathsf{AF}) \subseteq comp(\mathsf{AF})$.

Besides the above mentioned four traditional semantics, a variety of further proposals have been considered in the literature such as *CF2 semantics* [2], which is not based on the admissibility property. However, in this paper we focus on complete, grounded, preferred, and stable semantics.

Example 2. We continue Example 1. There, the sets $E_1, \ldots, E_6$ given via

$$E_1 = \emptyset \qquad E_2 = \{\mathcal{A}_1\} \qquad E_3 = \{\mathcal{A}_2\}$$
$$E_4 = \{\mathcal{A}_1, \mathcal{A}_3\} \qquad E_5 = \{\mathcal{A}_2, \mathcal{A}_4\} \qquad E_6 = \{\mathcal{A}_2, \mathcal{A}_5\}$$

are admissible. Furthermore, $E_1, E_3, \ldots, E_6$ are complete, E_1 is grounded, and E_4, E_5, E_6 are both preferred and stable.

   As shown by the above example, in general argumentation semantics are *multi-extension* or *multiple-status* i.e. they may prescribe more than one extension for a given argumentation framework. When a semantics prescribes exactly one extension for every argumentation framework it is called *single-extension* or *single-status*. Among the semantics considered in this paper, only grounded semantics is single-status.

   The possible existence of multiple extensions gives rise to different notions of the justification status of an argument. Given a semantics $\mathcal{S}$, an argument $\mathcal{A}$ is *credulously justified* if there is an $\mathcal{S}$-extension E such that $\mathcal{A} \in E$; $\mathcal{A}$ is *skeptically justified* if for all $\mathcal{S}$-extensions E it holds that $\mathcal{A} \in E$. Note that, unless the set of extensions is empty, being skeptically justified implies being credulously justified and that the two notions coincide for single-extension semantics.

Example 3. We continue Example 2. Here, no argument is skeptically justified wrt. grounded, complete, preferred, and stable semantics. Furthermore, no argument is credulously justified wrt. grounded semantics and all arguments are credulously justified wrt. the other semantics.

3 Probabilities on Extensions

Let $\mathsf{AF} = (\mathsf{Arg}, \rightarrow)$ be fixed. As in the epistemic approach to probabilistic argumentation [5,21,22,28], we consider probability functions on sets of arguments, namely functions $P : 2^{\mathsf{Arg}} \rightarrow [0,1]$ with

$$\sum_{E \subseteq \mathsf{Arg}} P(E) = 1$$

The idea being that $P(E)$ indicates the probability that the extension E is selected as the final outcome of the semantics evaluation of AF. We denote as $\mathcal{P}_{\mathsf{AF}}$ the set of all such probability functions. For $P_1, P_2 \in \mathcal{P}_{\mathsf{AF}}$ we define $P_1 = P_2$ iff $P_1(E) = P_2(E)$ for all $E \subseteq \mathsf{Arg}$.

Central to our approach is the following definition.

Definition 4. *We say that $P \in \mathcal{P}_{\mathsf{AF}}$ is* semantically based *on a set $\mathcal{E} \subseteq 2^{\mathsf{Arg}}$, if $P(E) = 0$ for all $E \notin \mathcal{E}$.*

We denote as $\mathcal{P}_{\mathsf{AF}}^{\mathcal{E}} \subseteq \mathcal{P}_{\mathsf{AF}}$ the set of all probability functions that are semantically based on $\mathcal{E}$. For example, $\mathcal{P}_{\mathsf{AF}}^{mcf(\mathsf{AF})}$ is the set of all probability functions that are semantically based on the maximal conflict-free subsets of AF. Note that in many cases one can assume that the set $\mathcal{E}$ is known a priori, e.g. the set of extensions prescribed by a given semantics for a given argumentation framework can be computed using one of the available implemented systems for abstract argumentation [9,29,30]. In this case one can of course easily ensure that a probability function is semantically based on $\mathcal{E}$ by construction. The issue of studying computational procedures for indirectly enforcing that a probability function is semantically based on a set $\mathcal{E}$ and for transforming an arbitrary probability function into the "closest" one which is semantically based on a given set $\mathcal{E}$ are interesting issues of future work.

Example 4. We continue Example 3 and consider the probability functions $P_1, \ldots, P_7$ defined in Table 1. All these functions are semantically based on the admissible sets of AF_0, i.e., $P_1, \ldots, P_7 \in \mathcal{P}_{\mathsf{AF}_1}^{adm(\mathsf{AF}_1)}$. Furthermore, we have

- $P_4, \ldots, P_7 \in \mathcal{P}_{\mathsf{AF}_1}^{comp(\mathsf{AF}_1)}$,
- $P_4, P_5, P_6 \in \mathcal{P}_{\mathsf{AF}_1}^{st(\mathsf{AF}_1)} = \mathcal{P}_{\mathsf{AF}_1}^{pref(\mathsf{AF}_1)}$, and
- $P_7 \in \mathcal{P}_{\mathsf{AF}_1}^{ground(\mathsf{AF}_1)}$.

A first observation is that we obtain the same hierarchy of the probabilistic versions of semantics as in Proposition 1.

Proposition 2. *If $\mathcal{E} \subseteq \mathcal{E}'$ then $\mathcal{P}_{\mathsf{AF}}^{\mathcal{E}} \subseteq \mathcal{P}_{\mathsf{AF}}^{\mathcal{E}'}$. In particular*

1. $\mathcal{P}_{\mathsf{AF}}^{st(\mathsf{AF})} \subseteq \mathcal{P}_{\mathsf{AF}}^{mcf(\mathsf{AF})} \subseteq \mathcal{P}_{\mathsf{AF}}^{cf(\mathsf{AF})}$,
2. $\mathcal{P}_{\mathsf{AF}}^{st(\mathsf{AF})} \subseteq \mathcal{P}_{\mathsf{AF}}^{pref(\mathsf{AF})} \subseteq \mathcal{P}_{\mathsf{AF}}^{comp(\mathsf{AF})} \subseteq \mathcal{P}_{\mathsf{AF}}^{adm(\mathsf{AF})} \subseteq \mathcal{P}_{\mathsf{AF}}^{cf(\mathsf{AF})}$, and
3. $\mathcal{P}_{\mathsf{AF}}^{ground(\mathsf{AF})} \subseteq \mathcal{P}_{\mathsf{AF}}^{comp(\mathsf{AF})}$.

Table 1. Definition of probability functions from Example 4; $P_i(E) = 0$ for all remaining $E \notin \{E_1, \ldots, E_6\}$ for $i = 1, \ldots, 7$

	$E_1 = \emptyset$	$E_2 = \{A_1\}$	$E_3 = \{A_2\}$	$E_4 = \{A_1, A_3\}$	$E_5 = \{A_2, A_4\}$	$E_6 = \{A_2, A_5\}$
P_1	0.2	0.1	0.3	0.2	0.1	0.1
P_2	0	0.3	0.2	0.3	0.1	0.1
P_3	0	0.2	0.2	0.2	0.2	0.2
P_4	0	0	0	0.3	0.1	0.6
P_5	0	0	0	1/3	1/3	1/3
P_6	0	0	0	0.5	0.5	0.0
P_7	1	0	0	0	0	0

Proof. This follows directly from Definition 4 and Proposition 1. $\qquad\square$

Furthermore, as in the classical case we have that probabilistic reasoning wrt. grounded semantics is uniquely defined.

Proposition 3. $\left|\mathcal{P}_{\mathsf{AF}}^{ground(\mathsf{AF})}\right| = 1$.

Proof. As every AF *has a unique grounded extension* E, *any* P *semantically based on grounded semantics must have* $P(E) = 1$ *and* $P(E') = 0$ *for all other sets* E'. *Therefore,* P *is uniquely determined.* $\qquad\square$

Given a probability function $P \in \mathcal{P}_{\mathsf{AF}}$ representing uncertainty about which extension is selected, an agent may be focused on a single argument or, more generally on a set of arguments, and be interested in the probability that this argument or sets of arguments is included in the selected extension E. In other words the probability P can be extended to the events of the kind $(F \subseteq E)$ where F is a generic set of arguments and E is the selected extension. For a set of arguments F, this extended probability will be denoted as $P^\subseteq(E)$ and is derived from P as follows

$$P^\subseteq(F) = \sum_{E \in 2^{\mathsf{Arg}}, F \subseteq E} P(E) \tag{1}$$

For individual arguments $A \in \mathsf{Arg}$ we introduce a special notation

$$P^\in(A) \triangleq P^\subseteq(\{A\}) = \sum_{E \in 2^{\mathsf{Arg}}, A \in E} P(E) \tag{2}$$

Example 5. Continuing Example 4, we have, e.g.

$$P_2^\in(A_2) = P_2(E_3) + P_2(E_5) + P_2(E_6) = 0.4$$
$$P_4^\in(A_5) = P_4(E_6) = 0.6$$

The following propositions report some basic observations.

Proposition 4. *For $P \in \mathcal{P}_{AF}^{cf(AF)}$, $P^{\in}(\mathcal{A}) = 0$ for all self-attacking arguments $\mathcal{A}$.*

Proof. If $\mathcal{A}$ is self-attacking then $\mathcal{A}$ is not member of any conflict-free set E of AF. Therefore $P^{\in}(\mathcal{A}) = \sum_{\mathcal{A} \in E \in cf(AF)} P(E) = 0$. □

Proposition 5. *For $P \in \mathcal{P}_{AF}^{comp(AF)}$, $P^{\subseteq}(GE(AF)) = 1$ and $P^{\in}(\mathcal{A}) = 1$ for every argument $\mathcal{A} \in GE(AF)$.*

Proof. The statement follows from the fact that the grounded extension of AF is included in every complete extension of AF. □

While some basic results, as shown above, hold for every probability function P, provided that P is semantically based on a given set of extensions, more specific properties of the beliefs of an agent may depend on the actual probability function P adopted by the agent within $\mathcal{P}_{AF}^{\mathcal{E}}$. In case an agent has no information or criteria to adopt a specific P, the well-known *maximum entropy principle* [23,26] states that the uniform probability assignment is adopted. In our case, the assignment of uniform nonzero probability values is restricted to the prescribed set of extensions.

Definition 5. *Let $P \in \mathcal{P}_{AF}$. We say that P is semantically uniform on $\mathcal{E} \subseteq 2^{Arg}$, if $P \in \mathcal{P}_{AF}^{\mathcal{E}}$ and for all $E, E' \in \mathcal{E}$ we have $P(E) = P(E')$.*

Of course semantically uniform probability functions are uniquely determined, given AF and $\mathcal{E}$ and the value of $P^{\in}(\mathcal{A})$ for each argument $\mathcal{A}$ is easily characterised.

Proposition 6. *Let $\mathcal{E} \subseteq 2^{Arg}$.*

1. *If $P, P' \in \mathcal{P}_{AF}^{\mathcal{E}}$ are semantically uniform on $\mathcal{E}$, then $P = P'$, i.e. $\forall E \in \mathcal{E}$ $P(E) = P'(E)$.*
2. *If $P \in \mathcal{P}_{AF}^{\mathcal{E}}$ is semantically uniform on $\mathcal{E}$, then for all $\mathcal{A} \in Arg$*

$$P^{\in}(\mathcal{A}) = \frac{|\{E \in \mathcal{E} \mid \mathcal{A} \in E\}|}{|\mathcal{E}|}$$

Proof. This follows directly from Definition 5. □

Also we are interested to characterise the case where the set of possible extensions is restricted (e.g. from admissible extensions to complete extensions) while still applying the maximum entropy principle.

Definition 6. *$P \in \mathcal{P}_{AF}$ is a semantically uniform restriction of $P' \in \mathcal{P}_{AF}$, if P is semantically uniform on $\mathcal{E}$, P' is semantically uniform on $\mathcal{E}'$, and $\mathcal{E} \subseteq \mathcal{E}'$.*

Example 6. We continue Example 4. While both P_2 and P_3 are semantically based on $\mathcal{E} = \{E_2, \ldots, E_6\}$, only P_3 is semantically uniform wrt. $\mathcal{E}$. Furthermore, P_4, P_5, P_6 are semantically based on the stable/preferred extensions and P_5 is also semantically uniform on those. P_5 is also a semantically uniform restriction of P_3 and P_6 is a semantically uniform restriction of P_5.

The maximum entropy principle offers a simple criterion to select one representative element in the (usually uncountably large) set of probability functions that are semantically based on some set of extensions. By construction, the information content of this representative element is rather weak: in particular, as to individual arguments, it boils down to counting how often an argument appears in extensions, cf. item 2 of Proposition 6.

In general, given a set of probability functions, their *lower envelope* [31] can be regarded as another synthetic representative of the set itself.

Definition 7. *Given a set of probability functions $\mathcal{P}$ on a set $\mathcal{E}$ the lower envelope $\underline{P}$ of $\mathcal{P}$ is defined for each $E \in \mathcal{E}$ as $\underline{P}(E) = \inf_{P \in \mathcal{P}} P(E)$.*

The lower envelope of a set of probabilities has interesting formal properties since it belongs to the family of imprecise probabilities and in particular is a *coherent lower probability* [31] (see Theorem 1 below). In words, $\underline{P}(E)$ identifies the minimum degree of belief in E given the set $\mathcal{P}$. The function $\underline{P}$ can therefore be regarded as a sort of cautious representation of the information content of $\mathcal{P}$. Specialising this notion to our context we get the following definition.

Definition 8. *Given a set of probability functions $\mathcal{P} \subseteq \mathcal{P}_{\mathsf{AF}}$ we define*[1]

- $\underline{P}(E) = \inf_{P \in \mathcal{P}} P(E)$ *for every* $E \in 2^{\mathsf{Arg}}$
- $\underline{P}^{\subseteq}(E) = \inf_{P \in \mathcal{P}} P^{\subseteq}(E)$ *for every* $E \in 2^{\mathsf{Arg}}$
- $\underline{P}^{\in}(\mathcal{A}) = \inf_{P \in \mathcal{P}} P^{\in}(\mathcal{A})$ *for every* $\mathcal{A} \in \mathsf{Arg}$

It is worth noting that each coherent lower probability $\underline{P}$ function has a conjugate upper probability $\overline{P}$ which for each E is defined by the following conjugacy relation

$$\overline{P}(E) = 1 - \underline{P}(\neg E) \tag{3}$$

Thus for instance the upper probability that a given extension E is selected is equal to 1 minus the lower probability that E is not selected. Given the set of probability functions $\mathcal{P}$ of which $\underline{P}$ is the lower envelope, $\overline{P}$ can be equivalently characterized as the upper envelope of $\mathcal{P}$, replacing inf with sup and making other obvious adjustments in Definitions 7 and 8. In this sense, dually with respecty to $\underline{P}$, the function $\overline{P}$ can be regarded as a sort of optimistic representation of the information content of $\mathcal{P}$.

In general, for an event E, the interval $[\underline{P}(E), \overline{P}(E)]$ gives an account of the distance between a cautious and an optimistic reading of the set $\mathcal{P}$ with respect to E. In particular if $\underline{P}(E) = \overline{P}(E)$, the set $\mathcal{P}$ provides a precise information about the probability of E, while at the other extreme, if $\underline{P}(E) = 0$ and $\overline{P}(E) = 1$, the set $\mathcal{P}$ provides no information at all about the probability of E.

---

[1] Note that the definitional relation for $P^{\subseteq}(E)$ in (1) does not carry over to $\underline{P}^{\subseteq}(E)$, i.e. in general it does not hold that $\underline{P}^{\subseteq}(E) = \sum_{E' \in 2^{\mathsf{Arg}}, E \subseteq E'} \underline{P}(E')$. An analogous consideration applies to $\underline{P}^{\in}(\mathcal{A})$.

The reader is referred to [31] for an extensive treatment of these concepts. In particular in [31] the values $\underline{P}(E)$ and $\overline{P}(E)$ were given a behavioral interpretation in an idealized betting scheme on E.

To make this notion clearer, we recall that this interpretation is rooted in de Finetti's *subjective probability theory* [15], of which the theory of imprecise probabilities introduced in [31] is a generalisation.

In de Finetti's approach a (precise) probability assessment is a function $P : \mathcal{E} \rightarrow \mathbb{R}$, where $\mathcal{E}$ is an *arbitrary* (finite or infinite) set of events and $\mathbb{R}$ is the set of real numbers. For each event $E \in \mathcal{E}$, $P(E)$ is the "fair" price of a (unitary) bet on E, i.e. $P(E)$ is the amount of money that an agent is ready to pay to an opponent in order to receive the sum of 1 if E turns out to be **true** and 0 otherwise, and, indifferently, the sum that the agent is ready to receive from an opponent as a payment for the commitment to pay the sum of 1 if E turns out to be **true** and 0 otherwise. More formally, $P(E)$ is the price, according to the agent, of the *indicator* of E, denoted as $I(E)$, namely the random number which takes value 1 if E is **true**, and value 0 if E is **false**. It is assumed that the agent is indifferently ready to buy or sell $I(E)$ at price $P(E)$. In the case of buying, the random gain of the agent is $I(E) - P(E)$, while it is $P(E) - I(E)$ in the case of selling. A not necessarily unitary bet is characterized by a real coefficient (or stake) $s \in \mathbb{R}$, so that the gain of the agent is given by $s(I(E) - P(E))$. A positive (negative) value of s corresponds to a buying (selling) choice by the agent.

According to the betting interpretation, a probability assessment has to satisfy some conditions ensuring that the bet makes sense for both participants. In particular, de Finetti has established a property of coherence, called *dF-coherence* in the sequel.

Definition 9. *Given an arbitrary set of events $\mathcal{E}$, $P : \mathcal{E} \rightarrow \mathbb{R}$ is a dF-coherent probability if and only if $\forall n \in \mathbb{N}^+$, $\forall s_1, \dots, s_n \in \mathbb{R}$, $\forall E_1, \dots E_n \in \mathcal{E}$, it holds that*

$$max \left[\sum_{i=1}^{n} s_i (I(E_i) - P(E_i)) \right] \geq 0 \qquad (4)$$

where $\mathbb{N}^+$ is the set of positive integer numbers.

Intuitively dF-coherence states that for any finite combination of bets, the maximum value of the random gain of the agent is non-negative, hence the agent avoids a sure loss. It is well-known that dF-coherence implies several fundamental properties[2] of probability assessments, including in particular the fact that $0 \leq P(E) \leq 1$ for every event E and the following *self-conjugacy* relation:

$$P(E) = 1 - P(\neg E). \qquad (5)$$

Considering the same betting context, imprecise probabilities [31] can be introduced by lifting the assumption that the agent has a precise price estimation, used indifferently for buying or selling event indicators. Rather (as typical

---

[2] In fact, on finite algebras of events the notions of dF-coherent probabilities, finitely additive probabilities and σ-additive probabilities coincide.

in real markets) the agent considers, for each event E, two different prices, one for buying and one for selling $I(E)$, denoted respectively as $\underline{P}(E)$ and $\overline{P}(E)$. Clearly, $\underline{P}(E) \leq \overline{P}(E)$. Moreover, the agent is of course ready to buy also at any price lesser than $\underline{P}(E)$, which hence represents the *supremum buying price* for $I(E)$. Similarly, $\overline{P}(E)$ is the *infimum selling price* for $I(E)$. Given that, for any event E, $I(\neg E) = 1 - I(E)$, it turns out that buying an event is equivalent to selling its complement and vice versa. Hence, in the context of imprecise probabilities, the following *conjugacy relation* replaces condition (5):

$$\underline{P}(E) = 1 - \overline{P}(\neg E) \tag{6}$$

In virtue of the conjugacy relation, one can focus on lower or upper probabilities only.

Definition 10 provides the notion of coherence for lower probabilities [31].

Definition 10. *Given an arbitrary set of events $\mathcal{E}$, $\underline{P} : \mathcal{E} \to \mathbb{R}$ is a coherent lower probability if and only if $\forall n \in \mathbb{N} = \mathbb{N}^+ \cup \{0\}$, and for all real and non-negative $s_0, \ldots, s_n$, $\forall E_0, \ldots E_n \in \mathcal{E}$, it holds that*

$$max\left[\left[\sum_{i=1}^{n} s_i(I(E_i) - \underline{P}(E_i))\right] - s_0(I(E_0) - \underline{P}(E_0))\right] \geq 0 \tag{7}$$

The coherence condition requires that the maximum of the gain of the agent is non negative for every (including the empty) combination of buying bets with at most one selling bet of a single (arbitrarily selected) event E_0. In a sense Definition 10 allows the agent to use its supremum buying price for any buying transaction but also forces the agent to use the same price for (at most one) selling transaction. Intuitively, this ensures that the assessment $\underline{P}$ by the agent is not too unfair.

As already mentioned, the lower envelope theorem, one of the main results of the theory of imprecise probabilities developed in [31], provides a nice characterization of coherent lower probabilities by relating them to sets of precise probabilities.

Theorem 1 [31]. *Given a set $\mathcal{E}$, $\underline{P}$ is a coherent lower probability on $\mathcal{E}$ if and only if there is a set $\mathcal{P}$ of (precise) dF-coherent probabilities on $\mathcal{E}$ such that $\underline{P}(E) = \inf_{P \in \mathcal{P}} P(E)$ for every $E \in \mathcal{E}$.*

In words, a lower probability $\underline{P}$ is coherent if and only if it can be obtained as the lower envelope of a set ($\mathcal{P}$) of dF-coherent precise probabilities (P). This result provides both a constructive procedure for coherent lower probabilities and a motivation for their existence: when a set of different probability assessments is given, coherent lower probabilities arise by aggregating them in the least committed way.

Example 7. With reference to Table 1, let $\mathcal{P} = \{P_1, \ldots, P_6\}$, $\underline{P}$ be its lower envelope and $\overline{P}$ its conjugate upper envelope. We have $\underline{P}(E_1) = \underline{P}(E_2) =$

$\underline{P}(E_3) = \underline{P}(E_6) = 0$; $\underline{P}(E_4) = 0.2$; $\underline{P}(E_5) = 0.1$ and $\overline{P}(E_1) = 0.2$; $\overline{P}(E_2) = \overline{P}(E_3) = 0.3$; $\overline{P}(E_4) = \overline{P}(E_5) = 0.5$; $\overline{P}(E_6) = 0.6$. Also, for instance, $\underline{P}^{\in}(\mathcal{A}_2) = \inf_{P \in \mathcal{P}}\{P(E_3) + P(E_5) + P(E_6)\} = 0.4$ and dually $\overline{P}^{\in}(\mathcal{A}_2) = \sup_{P \in \mathcal{P}}\{P(E_3) + P(E_5) + P(E_6)\} = 0.7$. We have also $\underline{P}^{\in}(\mathcal{A}_1) = 0.3$; $\overline{P}^{\in}(\mathcal{A}_1) = 0.6$; $\underline{P}^{\in}(\mathcal{A}_3) = 0.2$; $\overline{P}^{\in}(\mathcal{A}_3) = 0.5$; $\underline{P}^{\in}(\mathcal{A}_4) = 0.1$; $\overline{P}^{\in}(\mathcal{A}_4) = 0.5$; $\underline{P}^{\in}(\mathcal{A}_5) = 0$; $\overline{P}^{\in}(\mathcal{A}_5) = 0.6$.

When the set $\mathcal{P}$ coincides with the set $\mathcal{P}^{\mathcal{E}}_{\mathsf{AF}}$ of *all* probability functions that are semantically based on $\mathcal{E}$, then for each argument $\mathcal{A}$ the possible values of $\underline{P}(\mathcal{A})$ and $\overline{P}(\mathcal{A})$ are limited, so that the provided information is either extremely precise (both values are either 0 or 1) or completely vague ($\underline{P}(\mathcal{A}) = 0$ and $\overline{P}(\mathcal{A}) = 1$).

Proposition 7. *Given the set of probability functions $\mathcal{P}^{\mathcal{E}}_{\mathsf{AF}}$ for some set of extensions $\mathcal{E}$, let $\underline{P}$ be its lower envelope and $\overline{P}$ its conjugate upper envelope. For each argument $\mathcal{A} \in \mathsf{Arg}$ it holds that:*

- $\underline{P}^{\in}(\mathcal{A}) = 1$ *iff* $\forall E \in \mathcal{E}$ $\mathcal{A} \in E$; $\underline{P}^{\in}(\mathcal{A}) = 0$ *otherwise;*
- $\overline{P}^{\in}(\mathcal{A}) = 1$ *iff* $\exists E \in \mathcal{E} : \mathcal{A} \in E$; $\overline{P}^{\in}(\mathcal{A}) = 0$ *otherwise.*

Proof. If $\forall E \in \mathcal{E}$ $\mathcal{A} \in E$ then $\forall P \in \mathcal{P}^{\mathcal{E}}_{\mathsf{AF}}$ it holds $P^{\in}(\mathcal{A}) = 1$ from which $\underline{P}^{\in}(\mathcal{A}) = \overline{P}^{\in}(\mathcal{A}) = 1$. Otherwise if $\exists E \in \mathcal{E} : \mathcal{A} \notin E$ then the probability function given by $P(E) = 1$ and $P(E') = 0$ for every $E' \neq E$ belongs to $\mathcal{P}^{\mathcal{E}}_{\mathsf{AF}}$ from which $P^{\in}(\mathcal{A}) = 0$ and $\underline{P}^{\in}(\mathcal{A}) = 0$. Analogously, if $\exists E \in \mathcal{E} : \mathcal{A} \in E$ the probability function given by $P(E) = 1$ and $P(E') = 0$ for every $E' \neq E$ belongs to $\mathcal{P}^{\mathcal{E}}_{\mathsf{AF}}$ from which $P^{\in}(\mathcal{A}) = 1$ and $\overline{P}^{\in}(\mathcal{A}) = 1$. Otherwise $\nexists E \in \mathcal{E} : \mathcal{A} \in E$ and then $\forall P \in \mathcal{P}^{\mathcal{E}}_{\mathsf{AF}}$ it holds $P^{\in}(\mathcal{A}) = 0$ from which $\overline{P}^{\in}(\mathcal{A}) = 0$.

In general, the lower (or upper) envelope and the upper envelope of a set of precise probabilities are not precise probabilities themselves. However in some special cases some interesting correspondences between lower (or upper) values and precise probability assignments can be obtained. This is in particular the case when considering the set $\mathcal{P}^{\mathcal{E}}_{\mathsf{AF}}$ of all probability functions that are semantically based on $\mathcal{E}$: it can be seen that for each argument $\mathcal{A}$ the lower probability value $\underline{P}^{\in}(\mathcal{A})$ induced by the lower envelope of $\mathcal{P}^{\mathcal{E}}_{\mathsf{AF}}$ coincides with the precise probability value $P^{\in}(\mathcal{A})$ induced by the precise probability $P \in \mathcal{P}_{\mathsf{AF}}$ which gives probability 1 to the intersection of the elements of $\mathcal{E}$.

Proposition 8. *Given the set of probability functions $\mathcal{P}^{\mathcal{E}}_{\mathsf{AF}}$ for some set of extensions $\mathcal{E}$, let $\underline{P}$ be its lower envelope and let $P \in \mathcal{P}_{\mathsf{AF}}$ be defined as $P(\bigcap_{E \in \mathcal{E}} E) = 1$, $P(E') = 0$ for every $E' \neq \bigcap_{E \in \mathcal{E}} E$. For each argument $\mathcal{A} \in \mathsf{Arg}$ it holds that $\underline{P}^{\in}(\mathcal{A}) = P^{\in}(\mathcal{A})$.*

Proof. By definition, $P^{\in}(\mathcal{A}) = 1$ if $\mathcal{A} \in \bigcap_{E \in \mathcal{E}} E$, $P^{\in}(\mathcal{A}) = 0$ otherwise. From Proposition 7 we have $\underline{P}^{\in}(\mathcal{A}) = 1$ if $\mathcal{A} \in \bigcap_{E \in \mathcal{E}} E$, $\underline{P}^{\in}(\mathcal{A}) = 0$ otherwise, which proves the statement.

A corollary of Proposition 8 concerns the set $\mathcal{P}_{\mathsf{AF}}^{comp(\mathsf{AF})}$ of probabilities semantically based on complete extensions. It follows from the fact that the grounded extension is the least complete extension and coincides with the intersection of all complete extensions and provides a nice counterpart of Proposition 5.

Corollary 1. *Given the set of probability functions $\mathcal{P}_{\mathsf{AF}}^{comp(\mathsf{AF})}$, let $\underline{P}$ be its lower envelope and let P be the unique member of $\mathcal{P}_{\mathsf{AF}}^{ground(\mathsf{AF})}$. For each argument $\mathcal{A} \in \mathsf{Arg}$ it holds that $\underline{P}^{\in}(\mathcal{A}) = P^{\in}(\mathcal{A})$.*

In general, similar considerations could be applied to strict subsets of $\mathcal{P}_{\mathsf{AF}}$ (e.g. satisfying some constraints induced by the beliefs of the considered agent(s)) in order to identify some representative and/or to analyse their information contents. This line of development is left to future work.

4   Comparison to Epistemic Probabilistic Argumentation

In this section we analyze our approach to semantically based probabilities with respect to some general properties considered in the literature for the epistemic approach [5,21,22,28].

First, unattacked arguments play a special role as they are, in a sense, unquestioned. The *Foundation* postulate from [21] requires that the probability of unattacked arguments is 1. In our context this is guaranteed if a probability function is based on a semantic notion at least as strong as completeness.

Proposition 9. *If $P \in \mathcal{P}_{\mathsf{AF}}^{comp(\mathsf{AF})}$ then $P^{\in}(\mathcal{A}) = 1$ for all unattacked arguments $\mathcal{A}$.*

Proof. If $\mathcal{A}$ is not attacked in AF then $\mathcal{A} \in E$ for every complete extension E of AF. Then $P^{\in}(\mathcal{A}) = \sum_{E \in comp(\mathsf{AF})} P(E) = 1$.   □

Furthermore, a central postulate in the above-mentioned approaches is *Coherence*, which states that the sum of the probabilities of two conflicting arguments must be at most one. In our context, conflict freeness is enough to guarantee this property.

Proposition 10. *If $P \in \mathcal{P}_{\mathsf{AF}}^{cf(\mathsf{AF})}$ then for every $\mathcal{A}, \mathcal{B} \in \mathsf{Arg}$ with $\mathcal{A} \to \mathcal{B}$, $P^{\in}(\mathcal{B}) \leq 1 - P^{\in}(\mathcal{A})$.*

Proof. Let $\mathcal{A}, \mathcal{B} \in \mathsf{Arg}$ with $\mathcal{A} \to \mathcal{B}$. Then for every $E \in cf(\mathsf{AF})$ it cannot be the case that both $\mathcal{A} \in E$ and $\mathcal{B} \in E$. Therefore

$$
\begin{aligned}
P^{\in}(\mathcal{A}) + P^{\in}(\mathcal{B}) &= \sum_{\mathcal{A} \in E \subseteq \mathsf{Arg}} P(E) + \sum_{\mathcal{B} \in E \subseteq \mathsf{Arg}} P(E) \\
&= \sum_{\mathcal{A} \in E \in cf(\mathsf{AF})} P(E) + \sum_{\mathcal{B} \in E \in cf(\mathsf{AF})} P(E) \\
&\leq \sum_{E \in cf(\mathsf{AF})} P(E) = 1
\end{aligned}
$$

□

The *Rationality* postulate [19] states that if an argument has a probability greater than 0.5 then any conflicting argument should have a probability lesser than 0.5. Since this property is implied by *Coherence*, we directly obtain the satisfaction of the Rationality postulate too.

Corollary 2. *If $P \in \mathcal{P}_{AF}^{cf(AF)}$ then for every $\mathcal{A}, \mathcal{B} \in$ Arg with $\mathcal{A} \rightarrow \mathcal{B}$, if $P^{\in}(\mathcal{A}) > 0.5$ then $P^{\in}(\mathcal{B}) \leq 0.5$.*

The postulate *Optimism* has been used [28] to establish a certain correspondence to traditional semantics. It states that the sum of the probability of an argument and the probabilities of its attackers should be at least 1. In our context this holds under stable semantics.

Proposition 11. *If $P \in \mathcal{P}_{AF}^{st(AF)}$ then for every $E \in st(AF)$, $\mathcal{A} \in E$, $P^{\in}(\mathcal{A}) \geq 1 - \sum_{\mathcal{B} \rightarrow \mathcal{A}} P^{\in}(\mathcal{B})$.*

Proof. We have that $P^{\in}(\mathcal{A}) = 1 - \sum_{\mathcal{A} \notin E} P(E)$. By definition every stable extension S attacks all arguments not included in S. Then in particular every stable extension not including $\mathcal{A}$ includes an attacker of $\mathcal{A}$ from which it follows that $\sum_{\mathcal{B} \rightarrow \mathcal{A}} P^{\in}(\mathcal{B}) \geq \sum_{\mathcal{A} \notin E} P(E)$ from which $P^{\in}(\mathcal{A}) \geq 1 - \sum_{\mathcal{B} \rightarrow \mathcal{A}} P^{\in}(\mathcal{B})$.

Moreover two extreme cases have been considered in [21], namely *maximal* (respectively, *minimal*) epistemic probabilities where the probability of every argument is 1 (respectively 0). In our context they can be put in direct correspondence with special topological cases. Assuming probabilities which are semantically based on conflict-free sets, a maximal probability can be obtained only for argumentation frameworks with an empty attack relation.

Proposition 12. *If $P \in \mathcal{P}_{AF}^{cf(AF)}$ then $P^{\in}(\mathcal{A}) = 1$ for every argument $\mathcal{A} \in$ Arg only if $\rightarrow = \emptyset$.*

Proof. From the fact that $P^{\in}(\mathcal{A}) = 1$ for every argument $\mathcal{A} \in$ Arg it follows that it must be the case that $P(\text{Arg}) = 1$ and $P(E) = 0$ for every E such that $E \subsetneq$ Arg. For such a probability P to belong to $\mathcal{P}_{AF}^{cf(AF)}$ it must be the case that Arg is conflict-free, i.e. $\rightarrow = \emptyset$.

By the way when $\rightarrow = \emptyset$, the whole set of arguments Arg is the unique extension prescribed by all semantics considered in this paper but the conflict-free and the admissible semantics. Thus the maximal probability is also the unique probability compatible with those semantics when no attacks are present.

Conversely, it is clear thar a minimal probability is achieved only when the empty set has probability 1.

Proposition 13. *For $P \in \mathcal{P}_{AF}$ then $P^{\in}(\mathcal{A}) = 0$ for every argument $\mathcal{A} \in$ Arg if and only if $P(\emptyset) = 1$ and $P(E) = 0$ for every E such that $\emptyset \subsetneq E \subseteq$ Arg.*

Then, the minimal probability can be semantically based only if the empty set belongs to the set of extensions. In particular, the following proposition is directly derived from basic properties of the grounded and complete semantics (and is related with Proposition 5).

Proposition 14. *Let $P_\emptyset \in \mathcal{P}_{AF}$ be defined as $P_\emptyset(\emptyset) = 1$ and $P_\emptyset(E) = 0$ for every E such that $\emptyset \neq E \subseteq \mathsf{Arg}$. $P_\emptyset \in \mathcal{P}_{AF}^{comp(AF)}$ iff $GE(AF) = \emptyset$ iff $\forall \mathcal{A} \in \mathsf{Arg}$ $\exists \mathcal{B} \in \mathsf{Arg} : \mathcal{B} \to \mathcal{A}$.*

Proof. $P_\emptyset \in \mathcal{P}_{AF}^{comp(AF)}$ holds if and only if the empty set is a complete extension, which in turn holds if and only if $GE(AF) = \emptyset$, given that the grounded extension $GE(AF)$ is the minimal complete extension. By well-known properties of the grounded semantics [11] $GE(AF) = \emptyset$ holds if and only if every argument has at least an attacker (since every unattacked argument belongs to $GE(AF)$).

5   Computational Issues

We now discuss some computational issues of our approach, in particular, we make some straightforward comments on computational complexity.

Our approach is about probabilistic reasoning [26] with abstract argumentation frameworks. In general, bringing quantities into a qualitative reasoning problem also adds computational complexity. When reasoning with infinite sets such as $\mathcal{P}_{AF}^{\mathcal{E}}$ several properties of this set ensure that this can be done effectively. The next result shows that the set $\mathcal{P}_{AF}^{\mathcal{E}}$ is well-behaved wrt. important properties.

Proposition 15. *For every $\mathcal{E} \subseteq 2^{\mathsf{Arg}}$, $\mathcal{P}_{AF}^{\mathcal{E}}$ is a connected, closed, and convex set.*

Proof. Let $P_1, P_2 \in \mathcal{P}_{AF}^{\mathcal{E}}$, $\delta \in (0,1)$, and define the δ-convex combination $P_3 \in \mathcal{P}_{AF}$ of P_1 and P_2 via

$$P_3(E) = \delta P_1(E) + (1 - \delta) P_2(E)$$

for all $E \subseteq 2^{\mathsf{Arg}}$. Then for $E' \notin \mathcal{E}$ we have

$$P_3(E') = \delta P_1(E') + (1 - \delta) P_2(E') = 0$$

and therefore $P_3 \in \mathcal{P}_{AF}^{\mathcal{E}}$ showing that $\mathcal{P}_{AF}^{\mathcal{E}}$ is convex. Every convex set is also connected.

To show closure, let $P_1, P_2, \ldots$ be a sequence of probability functions in $\mathcal{P}_{AF}^{\mathcal{E}}$ such that $\lim_{i \to \infty} P_i(E)$ exists for all $E \subseteq 2^{\mathsf{Arg}}$ and define $P \in \mathcal{P}_{AF}$ via

$$P(E) = \lim_{i \to \infty} P_i(E)$$

Note that it is straightforward to see that indeed $P \in \mathcal{P}_{AF}$. Then for $E' \notin \mathcal{E}$ we have

$$P(E') = \lim_{i \to \infty} P_i(E') = \lim_{i \to \infty} 0 = 0$$

and therefore $P \in \mathcal{P}_{AF}^{\mathcal{E}}$ showing that $\mathcal{P}_{AF}^{\mathcal{E}}$ is closed.   □

Note that due to the above result pertaining the closure of sets $\mathcal{P}_{\mathsf{AF}}^{\mathcal{E}}$, we can substitute "infimum" by "minimum" in Definition 8. Due to connectedness and convexity, minima and maxima can be effectively computed by convex optimisation techniques[3]. We are currently investigating how to exploit this for algorithmic issues.

Regarding computational complexity, the following result immediately follows from well-known complexity results for abstract argumentation, see e.g. [13].

Proposition 16. *Let* AF *be an abstract argumentation framework and* $P \in \mathcal{P}_{\mathsf{AF}}^{\mathcal{E}}$ *semantically uniform.*

1. *Deciding whether* $P(\mathcal{A}) > 0$ *for some* $\mathcal{A} \in$ Arg *is*
   (a) NP-complete for $\mathcal{E} = comp(\mathsf{AF})$,
   (b) NP-complete for $\mathcal{E} = pref(\mathsf{AF})$,
   (c) NP-complete for $\mathcal{E} = st(\mathsf{AF})$, and
   (d) in P *for* $\mathcal{E} = ground(\mathsf{AF})$.
2. *Deciding whether* $P(\mathcal{A}) = 1$ *for some* $\mathcal{A} \in$ Arg *is*
   (a) in P *for* $\mathcal{E} = comp(\mathsf{AF})$,
   (b) Π_2^P-complete for $\mathcal{E} = pref(\mathsf{AF})$,
   (c) coNP-complete for $\mathcal{E} = st(\mathsf{AF})$, and
   (d) in P *for* $\mathcal{E} = ground(\mathsf{AF})$.

Proof. Observe that $P(\mathcal{A}) > 0$ *is equivalent to asking whether* $\mathcal{A}$ *is credulously inferred. Correspondingly,* $P(\mathcal{A}) = 1$ *is equivalent to asking whether* $\mathcal{A}$ *is skeptically inferred. For the complexity of these problems see e.g. [13].* $\square$

6   Summary

We proposed a novel perspective to combine probability theory with abstract argumentation. In our approach, we combine classical extension-based semantics with quantitative uncertainty by considering probability functions on extensions and analysing some relevant reasoning tasks. We did some preliminary investigation and showed that our proposal faithfully generalises classical semantics and is compatible with some postulates considered in the epistemic approach to probabilistic argumentation. Some relationships with imprecise probability theory were also pointed out and finally, we made some observations regarding computational complexity.

The work reported in this paper is preliminary and a deeper investigation of the proposed formalism and of its potential applications is called for. In particular, the development of algorithmic approaches for using our framework is part of ongoing work. Finally, concerning the issue of where do the probability values come from, we suggest that an interesting direction of investigation is learning or estimating the probabilities of extensions or of arguments from the past choices of an agent or of a community of agents (e.g. an electoral body) in similar decision contexts.

---

[3] The size of the optimization problem depends of course on the size of the set $\mathcal{E}$ which might be large in some cases. The reader may refer to [4,6] for studies on the size of the set of extensions prescribed by a given semantics.

Acknowledgments. The authors are grateful to the anonymous referees for their helpful comments.

References

1. Amgoud, L., Ben-Naim, J.: Argumentation-based ranking logics. In: Proceedings of the 14th International Conference on Autonomous Agents and Multiagent Systems (AAMAS 2015), pp. 1511–1519 (2015)
2. Baroni, P., Giacomin, M., Guida, G.: SCC-recursiveness: a general schema for argumentation semantics. Artif. Intell. **168**(1–2), 162–210 (2005)
3. Baroni, P., Caminada, M., Giacomin, M.: An introduction to argumentation semantics. Knowl. Eng. Rev. **26**(4), 365–410 (2011)
4. Baroni, P., Dunne, P.E., Giacomin, M.: On extension counting problems in argumentation frameworks. In: Proceedings of the 3rd International Conference on Computational Models of Argument (COMMA 2010), pp. 63–74 (2010)
5. Baroni, P., Giacomin, M., Vicig, P.: On rationality conditions for epistemic probabilities in abstract argumentation. In: Proceedings of the Fifth International Conference on Computational Models of Argumentation (COMMA 2014), pp. 121–132 (2014)
6. Baumann, R., Strass, H.: On the maximal and average numbers of stable extensions. In: Black, E., Modgil, S., Oren, N. (eds.) TAFA 2013. LNCS (LNAI), vol. 8306, pp. 111–126. Springer, Heidelberg (2014). https://doi.org/10.1007/978-3-642-54373-9_8
7. Bonzon, E., Delobelle, J., Konieczny, S., Maudet, N.: A comparative study of ranking-based semantics for abstract argumentation. In: Proceedings of the 30th AAAI Conference on Artificial Intelligence (AAAI 2016), pp. 914–920 (2016)
8. Caminada, M.: Semi-stable semantics. In: Proceedings of the First International Conference on Computational Models of Argument (COMMA 2006), pp. 121–130 (2006)
9. Charwat, G., Dvorak, W., Gaggl, S.A., Wallner, J.P., Woltran, S.: Methods for solving reasoning problems in abstract argumentation - a survey. Artif. Intell. **220**, 28–63 (2015)
10. Cohen, A., Gottifredi, S., Garcia, A.J., Simari, G.R.: A survey of different approaches to support in argumentation systems. Knowl. Eng. Rev. **29**(5), 513–550 (2014)
11. Dung, P.M.: On the acceptability of arguments and its fundamental role in non-monotonic reasoning, logic programming and n-person games. Artif. Intell. **77**(2), 321–358 (1995)
12. Dunne, P.E., Hunter, A., McBurney, P., Parsons, S., Wooldridge, M.: Weighted argument systems: basic definitions, algorithms, and complexity results. Artif. Intell. **175**(2), 457–486 (2011)
13. Dvořák, W.: Computational aspects of abstract argumentation. Ph.D. thesis, Technische Universität Wien (2012)
14. Eğilmez, S., Martins, J., Leite, J.: Extending social abstract argumentation with votes on attacks. In: Black, E., Modgil, S., Oren, N. (eds.) TAFA 2013. LNCS (LNAI), vol. 8306, pp. 16–31. Springer, Heidelberg (2014). https://doi.org/10.1007/978-3-642-54373-9_2
15. de Finetti, B.: Theory of Probability, vol. I. Wiley, Hoboken (1974)
16. Gabbay, D.: Equational approach to argumentation networks. Argum. Comput. **3**(2–3), 87–142 (2012)

17. Gabbay, D., Rodrigues, O.: Probabilistic argumentation: an equational approach. Logica Universalis **9**(3), 345–382 (2015)
18. Grossi, D., Modgil, S.: On the graded acceptability of arguments. In: Proceedings of the 24th International Joint Conference on Artificial Intelligence (IJCAI 2015), pp. 868–874 (2015)
19. Hunter, A.: A probabilistic approach to modelling uncertain logical arguments. Int. J. Approx. Reason. **54**(1), 47–81 (2013)
20. Hunter, A.: Modelling the persuadee in asymmetric argumentation dialogues for persuasion. In: Proceedings of the Twenty-Fourth International Joint Conference on Artificial Intelligence (IJCAI 2015), pp. 3055–3061 (2015)
21. Hunter, A., Thimm, M.: Probabilistic argumentation with incomplete information. In: Proceedings of the 21st European Conference on Artificial Intelligence (ECAI 2014), pp. 1033–1034, August 2014
22. Hunter, A., Thimm, M.: On partial information and contradictions in probabilistic abstract argumentation. In: Proceedings of the 15th International Conference on Principles of Knowledge Representation and Reasoning (KR 2016), pp. 53–62, April 2016
23. Kern-Isberner, G.: Characterizing the principle of minimum cross-entropy within a conditional-logical framework. Artif. Intell. **98**(1–2), 169–208 (1998)
24. Leite, J., Martins, J.: Social abstract argumentation. In: Proceedings of the 22nd International Joint Conference on Artificial Intelligence (IJCAI 2011), pp. 2287–2292 (2011)
25. Li, H., Oren, N., Norman, T.J.: Probabilistic argumentation frameworks. In: Modgil, S., Oren, N., Toni, F. (eds.) TAFA 2011. LNCS (LNAI), vol. 7132, pp. 1–16. Springer, Heidelberg (2012). https://doi.org/10.1007/978-3-642-29184-5_1
26. Paris, J.B.: The Uncertain Reasoner's Companion - A Mathematical Perspective. Cambridge University Press, Cambridge (2006)
27. Santini, F.: Graded justification of arguments via internal and external endogenous features. In: Schockaert, S., Senellart, P. (eds.) SUM 2016. LNCS (LNAI), vol. 9858, pp. 352–359. Springer, Cham (2016). https://doi.org/10.1007/978-3-319-45856-4_26
28. Thimm, M.: A probabilistic semantics for abstract argumentation. In: Proceedings of the 20th European Conference on Artificial Intelligence (ECAI 2012), pp. 750–755, August 2012
29. Thimm, M., Villata, S.: The first international competition on computational models of argumentation: results and analysis. Artif. Intell. **252**, 267–294 (2017)
30. Thimm, M., Villata, S., Cerutti, F., Oren, N., Strass, H., Vallati, M.: Summary report of the first international competition on computational models of argumentation. AI Mag. **37**(1), 102–104 (2016)
31. Walley, P.: Statistical Reasoning with Imprecise Probabilities. Chapman and Hall, London (1991)
32. Wu, J., Li, H., Oren, N., Norman, T.J.: Gödel fuzzy argumentation frameworks. In: Proceedings of the 6th International Conference on Computational Models of Argument (COMMA 2016), pp. 447–458 (2016)
33. Wu, Y., Caminada, M.: A labelling-based justification status of arguments. Stud. Logic **3**(4), 12–29 (2010)

A Forward Propagation Algorithm
for the Computation of the Semantics
of Argumentation Frameworks

Odinaldo Rodrigues[✉]

Department of Informatics, King's College London, London, UK
odinaldo.rodrigues@kcl.ac.uk

Abstract. In this paper we propose a novel algorithm for the computation of the semantics of argumentation frameworks. The algorithm can generate all complete extensions and thus can be used in problems involving the grounded, complete, preferred and stable semantics. The algorithm takes advantage of the constraints imposed on legal labelling functions to prune the search space of possible solutions.

1 Introduction

This paper describes a new algorithm for the computation of the semantics of argumentation frameworks based on the idea of *forward propagation* of **in** labels of accepted argument. The basic mechanism is very simple: the construction of complete extensions is done by attempting to re-label **in** all undecided (**und**) arguments that could *potentially* be labelled **in** by a labelling function and checking whether the resulting function can be made "legal".

The algorithm works on the strongly connected components (SCCs) of an argumentation framework which are arranged into layers following the direction of attacks. Because of the dependencies between the valid assignments of labels of attacking and attacked arguments, a solution for one layer may impose constraints on the possible solutions for SCCs of subsequent layers. In such cases, we say that the solution of one layer *conditions* the possible legal label assignments of the attacked SCC. So we take this idea further by looking at the consequences of legally labelling an argument **in** in an SCC: we search for labelling assignments of the SCC satisfying an increasing set of constraints. All solutions thus found are combined in the way described in [11].

We start with the undecided arguments of an SCC that could potentially be labelled **in** in *some* solution. By labelling one of these arguments **in**, we are forced to label all of its attackers **out** (i.e., reject them). If all attackers of an argument are re-labelled **out**, then the argument must be re-labelled **in**, imposing new constraints on the labels of the arguments that it attacks, and so forth. Forcing the attacker of an argument to be labelled **out** is done analogously by requiring that at least one of the attacker's attacker is labelled **in**, so the whole process can be done through a series of recursive forward propagation operations of **in** labels each of smaller complexity than the original one.

© Springer International Publishing AG, part of Springer Nature 2018
E. Black et al. (Eds.): TAFA 2017, LNAI 10757, pp. 120–136, 2018.
https://doi.org/10.1007/978-3-319-75553-3_8

Searching for extensions in this way has several advantages. The constraints can *prune* the search space considerably by ruling out assignments that violate the admissibility conditions. Thus, the algorithm is designed to incrementally "fill in" the gaps of an admissible but partially uncommitted labelling function by successively swapping labels from **und** to **in** or **out**, and as a result generating all complete extensions along the way. As a by-product, we can pick an argument of interest and attempt to construct a legal labelling assignment that labels the argument in a particular way (e.g., **in**), without necessarily having to look at all solutions of the SCC or the argumentation framework as a whole.

The rest of the paper is structured as follows. In Sect. 2, we provide some background material for the paper. This is followed by the presentation of the algorithm itself in Sect. 3.[1] In Sect. 4, we compare our algorithm with others in the literature. Section 5 provides some empirical evaluation of the algorithm and we conclude in Sect. 6 with a discussion and some future work.

2   Background

An abstract argumentation framework is a system for reasoning about arguments proposed by Dung [9] and defined in terms of a directed graph $\langle \mathcal{A}, \mathcal{R} \rangle$, where $\mathcal{A}$ is a *finite* non-empty set of arguments and $\mathcal{R}$ is a binary relation on $\mathcal{A}$, called the *attack relation*. If $(X, Y) \in \mathcal{R}$, we say that X attacks Y and denote it in the graph with an edge from X to Y. In what follows, $X^- = \{Y \in \mathcal{A} | (Y,X) \in \mathcal{R}\}$; and $X^+ = \{Y \in \mathcal{A} | (X,Y) \in \mathcal{R}\}$. For sets $E \subseteq \mathcal{A}$, E^- and E^+ are defined in an obvious way via set union. We write $E \rightarrow X$ as a shorthand for $X \in E^+$. The *path-equivalence relation* $\sim_{\mathcal{R}} \subseteq \mathcal{A}^{\in}$ is defined as $X \sim_{\mathcal{R}} Y$ iff $X = Y$ or there is a path from X to Y and a path from Y to X in $\mathcal{R}$. A *strongly connected component* (SCC) is an equivalence class of arguments under $\sim_{\mathcal{R}}$.

One of the main purposes of an argumentation framework is to provide a way of reasoning about the *status* of its arguments, i.e., whether an argument is accepted or is defeated by other arguments. Arguments that have no attacks are always accepted. However, an attack from X to Y may not be sufficient to defeat Y, because X may itself be defeated, and thus the statuses of arguments need to be determined systematically. In Dung's original formulation, this is usually done through *acceptability* conditions for the arguments. A semantics can then be defined in terms of *extensions*—subsets of $\mathcal{A}$ with special properties. A set $E \subseteq \mathcal{A}$ is said to be *conflict-free* if for all elements $X, Y \in E$, we have that $(X, Y) \notin \mathcal{R}$. Although a conflict-free set only contains elements that do not attack each other, this does not necessarily mean that all arguments in the set are properly supported. Well-supported sets satisfy special *admissibility* criteria. An argument $X \in \mathcal{A}$ is *acceptable with respect to E*, if for all $Y \in X^-$, $E \cap Y^- \neq \varnothing$. A set E is *admissible* if it is conflict-free and all of its elements are acceptable with respect to itself. An admissible set E is a *complete extension* iff E contains all arguments which are acceptable with respect to itself; E is called

---

[1] For easier understanding the algorithm is broken into functional sub-components.

a *preferred extension* iff E is a $\subseteq$-maximal complete extension; and E is stable if E is preferred and $E \cup E^+ = \mathcal{A}$.

Dung's semantics can also be presented in terms of a Caminada *labelling function* of the form $\lambda : \mathcal{A} \longrightarrow \{in,out,und\}$ satisfying certain conditions [4, 5,15]. Let dom denote the domain of a function and λ a labelling function, we define $in(\lambda) = \{X \in$ dom $\lambda | \lambda(X) = $ **in**$\}$; $und(\lambda) = \{X \in$ dom $\lambda | \lambda(X) = $ **und**$\}$; and $out(\lambda) = \{X \in$ dom $\lambda | \lambda(X) = $ **out**$\}$. The notion of extension is recovered from the set $in(\lambda)$ for some labelling function λ. Furthermore, we say that an argument X is *illegally labelled* **in** by λ, if $X^- \not\subseteq out(\lambda)$; X is *illegally labelled* **out** by λ, if $X^- \cap in(\lambda) = \varnothing$; and X is *illegally labelled* **und** by λ, if either $X^- \subseteq out(\lambda)$ or $X^- \cap in(\lambda) \neq \varnothing$. Finally, X is *super-illegally labelled* **in** if it is attacked by an argument that is legally labelled **in** or labelled **und** [12]. A labelling function is legal it does not illegally label any arguments.

2.1   Computing Extensions via Decomposition into SCCs

Baroni et al. proposed a general recursive schema for argumentation semantics in [1]. The schema employs the decomposition of an argumentation framework into SCCs and can be used to obtain Dung's admissibility-based semantics. Based on that, many researchers showed how to compute the extensions of argumentation frameworks under several semantics. Baumann adapted the Modgil-Caminada's algorithms [12] to compute extensions under the grounded, preferred and stable semantics in what he called "split" frameworks [2]. Preliminary experimental results of the advantages of these techniques were then shown in [3]. Liao described the use of the decomposition idea for computation of argumentation semantics in a more general way [11].

The overall process can be summarised as follows. Firstly, the SCCs of an argumentation framework are arranged into layers following the direction of attack. Then the solutions for each layer are computed using an appropriate algorithm for the semantics at hand and the solutions of the previous layers. Finally, the solutions of subsequent layers are combined in a systematic way. To illustrate this idea, consider the argumentation framework $\mathcal{N}$ in Fig. 1 with SCCs $S_1 = \{X\}$, $S_2 = \{W, Y\}$ and $S_3 = \{A, B, C, D, E\}$. Following the attack relation, these SCCs can be arranged into two layers, the first containing S_1 and S_2 and the second containing S_3. The solutions of the SCCs in a given layer are all independent from each other, but the attacks between arguments of different layers create dependencies of the solutions of an SCC on the solutions of the SCCs attacking it. For example, the computation of the solutions of S_3 depends on the labels assigned to X and W, and thus on the solutions of S_1 and S_2. As S_1 and S_2 have no external attackers, their solutions can be computed completely independently of the rest of the framework. S_2 has three legal assignments: one in which both W and Y are labelled **und** and the other two in which one of them is labelled **in** and the other is labelled **out**. $X = $ **in** is the only solution to S_1, so each of the partial solutions to S_2 must be augmented with the assignment

$X = $ **in**, giving all partial solutions to layer 0: $f_1 : X = Y = $ **in**, $W = $ **out**, $f_2 : X = W = $ **in**, $Y = $ **out**, and $f_3 : X = $ **in**, $W = Y = $ **und**.[2] [11].

Now consider the computation of the solutions for S_3. We say that S_3's solutions are *conditioned* by the labels of the external attackers X and W in the partial solutions f_1, f_2 and f_3. In any such solution, $X = $ **in**, but the label of W could be either **out**, **in** or **und**. In order to generate all complete extensions for $\mathcal{N}$, each partial solution f_1, f_2 and f_3 needs to be *expanded* with the solutions for S_3 under the constraints that they impose.

Definition 1 (Initial Conditioned Solution for an SCC). *Let f be a conditioning solution for an SCC S. The initial solution for S conditioned by f $\lambda_S^f : S \longmapsto \{out,und,in\}$ is a legal labelling function whose set $in(\lambda_S^f)$ is $\subseteq$-minimal with respect to all legal labelling functions conditioned by f.*

λ_S^f is the "minimal" (grounded) solution for S under f. It is a special case of forward propagation from external attackers starting with the all undecided labelling (all-**und**). The Discrete Gabbay-Rodrigues Iteration Schema [10] is an example of a method that can perform this propagation very efficiently.

Since $f_1(X) = $ **in** and $f_1(W) = $ **out**, the search for the solutions for S_3 conditioned by f_1 consists of the search for all solutions to S_3 with the constraint $A = $ **out** or the search of all possible ways to "expand" $\lambda_S^{f_1}$ by swapping labels from **und** to **in** or **out**. Similarly, since the $f_2(X) = f_2(W) = $ **in**, under f_2 we need to satisfy the constraint $A = B = $ **out**. A similar reasoning applies to solution f_3 in which we have the "implicit" constraint $\lambda_S^{f_3}(B) \neq $ **in** (since $\lambda_S^{f_3}(W) = $ **und**). More generally speaking, the whole process can be thought of as follows: given a SCC S, a conditioning solution f, and a partial labelling function λ_S^f, compute the set Λ of all expansions of λ_S^f satisfying some constraints.

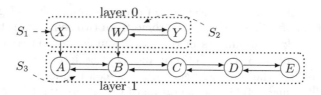

Fig. 1. A complex argumentation framework and its decomposition into layers.

Decomposition breaks the argumentation problem into smaller sub-problems, but an algorithm is still needed to find the solutions for each SCC. Modgil-Caminada's algorithm for preferred extensions is one algorithm that can be adapted for this [12].

---

[2] This is called the *horizontal combination* of solutions of the layer.

2.2 Modgil-Caminada's Algorithm for Preferred Extensions

For space limitations we cannot present Modgil-Caminada's algorithm in full, but we will describe it in general terms. This should suffice for our discussion.

Since preferred extensions are associated with maximal sets of arguments that are labelled **in**, Modgil-Caminada's algorithm starts with the labelling function that labels all arguments **in** (all-**in**) and then successively "corrects" illegally labelled arguments via a so-called *transition step*. Eventually, all illegal labels get corrected, and the set of arguments labelled **in** will correspond to an extension – those that are maximal will correspond to the preferred extensions.[3] A transition step consists of the following. If the argument X is illegally labelled **in**, then it is re-labelled **out**, if it can be legally re-labelled so. Otherwise it is re-labelled **und**. Afterwards, the labels of all arguments in X^+ that become illegally labelled **out** by the fact that X has been re-labelled from **in** to **out** or **und**, are then also changed to **und**. The algorithm applies transition steps as follows. If there is any argument X in λ that is super illegally labelled **in**, then the algorithm performs a single transition step on X generating a new labelling function λ' and then calls itself recursively from λ'. If there is no such argument, the algorithm will instead iterate through *all* arguments that are illegally labelled **in**; apply a transition step on each; and call itself recursively from the new labelling functions thus generated. Eventually, all labels will become legal and the algorithm will simply return the labelling functions with maximal sets of arguments labelled **in**.

In Sects. 4 and 5, we will see that the strategy used by Modgil-Caminada's algorithm may result in a very high number of operations.

3   A New Algorithm for Enumeration and Decision Problems of Argumentation Semantics

Our algorithm's strategy takes advantage of the constraints that a legal labelling function must satisfy. These constraints come from two sources: *(i)* the labels of the external attacking arguments in the conditioning solutions (which already partially determine the SCC's solution); and *(ii)* the internal constraints arising from re-labelling the seed argument **in**. The constraints *help* to reduce the search space. The successful implementation of this strategy relies on an efficient propagation mechanism (see Sect. 6) and a bottom-up method for constructing all extensions.

This way of looking into the problem has two major implications. By generating all complete extensions, the method can be used in problems involving the grounded, complete, stable and preferred semantics. For the grounded semantics, all we need to do is to propagate the (unique) conditioning solution; for the preferred semantics, we generate alternative solutions but only keep those that maximise the set of nodes labelled **in**; and for the stable semantics we exclude preferred solutions with undecided nodes. Secondly, because we only work on an

---

[3] Unlike ours, Modgil-Caminada's algorithm does not guarantee the generation of all complete extensions.

individual argument at a time, we can define decision procedures for argument acceptability that do not need to necessarily generate all extensions.

In order to lighten the notation, we will drop the subscript and superscript in λ_S^f when the context makes the SCC S and the conditioning solution f clear. Given a partial solution λ conditioned by a solution f, an argument X of an SCC S can potentially be re-labelled from **und** to **in** if it satisfies the following conditions: (I1) $\lambda(X) = \mathbf{und}$; (I2) $X \notin X^-$ (it does not attack itself); and (I3) $\{Y \in X^- | f(Y) = \mathbf{und}\} = \varnothing$. [4] The set $possIns_S \subseteq S$ is the set of nodes satisfying conditions (I1)–(I3). Thus the starting point for Algorithm 1 is an SCC S; the set $possIns_S$; a conditioning solution f for previous layers; and a partial solution λ for S conditioned by f. The algorithm will compute the set Λ of all complete (or preferred)[5] labelling functions that "expand" λ by successively searching for complete/preferred labelling functions that label an element of $possIns_S$ **in**. Each search is done via Algorithm 4, which we now explain.

Algorithm 1 Finding extensions from a given set of arguments

Input: $possIns_S$, a SCC S, a conditioning labelling function f, a conditioned legal labelling function λ for S, and a set of candidate labelling functions Λ
Output: true (success) or false (failure) and an updated set Λ

```
1  Function findExtsFromArgs(S,possIns_S,f,λ,Λ)
2      while possIns_S ≠ ∅ do
3          Pick X ∈ possIns_S
4          possIns_S ← possIns_S \{X}
5          findExtsFromArg(X, S, f, λ, Λ)
6      end while
7  end
```

In Sect. 2, we saw that an argument X is legally labelled **in** in a solution λ if all arguments that it attacks are labelled **out** and that if all arguments that attack X are labelled **out** then X must be labelled **in** in λ. Thus, to re-label X **in** we must re-label **out** all arguments that it attacks. By re-labelling some arguments **out**, we may also be forced to re-label **in** some other arguments, and so forth. We call this process the *forward propagation of the* **in** *label*. All attackers of X must also be labelled **out** for X to be legally labelled **in** in λ. Thus, all external attackers of X must be labelled **out** (by f) and all internal attackers that are still labelled **und** must be re-labelled **out**. This can be done by ensuring that every internal attacker Y that is labelled **und**, gets an attacker Z to be legally re-labelled **in**. We call the process of ensuring that all attackers of X are legally labelled **out** *the backward propagation of the* **in** *label*. Backward propagations can be done in terms of one or more forward propagations and this is the motivation for the title of the algorithm.

Although we start with "good enough" candidates, i.e., arguments satisfying (I1)–(I3), both types of propagations may fail, since we have no control over the assignments of conditioning solutions and the propagations may result in inconsistent label requirements. A failed propagation simply means that we

---

[4] We know that $\lambda(X) = \mathbf{und}$ by (I1), but we still want to make sure that X can be re-labelled **in** which is not the case if an external attacker $Y \in X^-$ has $f(Y) = \mathbf{und}$.

[5] The set Λ is updated according to the desired semantics (see Sect. 3.3).

cannot construct a legal labelling function meeting the required constraints, so
we backtrack to any available alternatives. We now explain the details.

3.1 Propagating Forwards

A forward propagation essentially requires changing the label of a *seed node* in
a partial solution λ from **und** to **in** and then following the direction of attacks
to re-label any nodes that may have thus have been rendered illegally labelled.
One important aspect of a forward propagation is that (if successful) it will
generate a single solution λ' from a partial solution λ which, by construction, has
less undecided nodes than λ itself. A forward propagation is carried out by the
function `propagateIN` in Algorithm 2. Figure 2 illustrates the labelling function
λ' obtained as the result of a successful forward propagation from $X = $ **in** and
$f = \varnothing$ and a failed forward propagation from $W_2 = $ **in** and $f = \varnothing$. The latter fails
because by labelling W_2 **in**, we must label U **out**, which then requires T to be
labelled **in**, which in turn requires W_2 to be labelled **out**, which is not possible.

Algorithm 2 Forward propagation of an IN label

Input: argument X to label **in**, its SCC S, a partial legal labelling function λ, and a
$\qquad$ conditioning labelling function f

Output: false if failure; or true if successful, with the new partial labelling function λ'

1 **Function** propagateIN(X,S,f,λ,λ')
2 $\quad$ **if** $\{Z \in X^+ \mid \lambda(Z) = \mathbf{in}\} \neq \varnothing$ **then**
3 $\quad\quad$ **return** *false*
4 $\quad$ **else**
5 $\quad\quad$ $\lambda' \leftarrow \lambda;\ \lambda'(X) = \mathbf{in}$
6 $\quad\quad$ **forall** $Y \in \{Z \in X^+ \mid \lambda'(Z) = \mathbf{und}\}$ **do**
7 $\quad\quad\quad$ $\lambda'(Y) \leftarrow \mathbf{out}$
8 $\quad\quad$ **end forall**
9 $\quad\quad$ $newIns \leftarrow \{Z \in S \mid \lambda'(Z) = \mathbf{und}$ and for all $Y \in Z^-,\ \lambda'(Y) = \mathbf{out}\}$
10 $\quad\quad$ **while** $newIns \neq \varnothing$ **do**
11 $\quad\quad\quad$ Pick $W \in newIns$
12 $\quad\quad\quad$ **if** propagateIN$(W,S,f,\lambda',\lambda'')$ **then**
13 $\quad\quad\quad\quad$ $\lambda' \leftarrow \lambda''$
14 $\quad\quad\quad\quad$ $newIns \leftarrow newIns \backslash \{W\}$
15 $\quad\quad\quad$ **else**
16 $\quad\quad\quad\quad$ **return** *false*
17 $\quad\quad\quad$ **end if**
18 $\quad\quad$ **end while**
19 $\quad\quad$ **return** *true*
20 $\quad$ **end if**
21 **end**

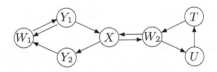

SCC S, $f = \varnothing$, λ is all-**und**

`propagateIN`$(X, S, f, \lambda, \lambda')$ succeeds with
$in(\lambda') = \{X, U\}$, $out(\lambda') = \{T, W_2, Y_2\}$,
$und(\lambda') = \{W_1, Y_1\}$

`propagateIN`$(W_2, S, f, \lambda, \lambda')$ fails since we
cannot label W_2 both **in** and **out**

Fig. 2. Results of forward propagations from $X = $ **in** and $W_2 = $ **in**.

If the forward propagation from a node X is successful, we must then ensure that all of X's attackers are legally labelled **out** in order to guarantee that the solution is legal. This is done by a backward propagation.

3.2 Propagating Backwards

In the example in Fig. 2, it is easy to see that λ' is not legal, since $\lambda'(X) = \mathbf{in}$, $Y_1 \rightarrow X$, but $\lambda'(Y_1) = \mathbf{und}$. We can perform a backward propagation from X by performing one or more forward propagations using any of the attackers of each attacker of X as the seed. $X^- = \{Y_1, W_2\}$, so we want a labelling function that labels at least one of the arguments in Y_1^- *and* in W_2^- in (X itself already satisfies the latter). It is easy to see that the labelling function $\lambda'' = \{X = U = W_1 = \mathbf{in}, Y_2 = W_2 = T = Y_1 = \mathbf{out}\}$ satisfies these requirements.

Naturally, a backward propagation may also fail. Consider the network in Fig. 3. After a successful forward propagation from $X = \mathbf{in}$, $f = \varnothing$, and $\lambda = \text{all-}\mathbf{und}$, we get the labelling function $\lambda' = \{X = \mathbf{in}, Y = \mathbf{out}, W_1 = W_2 = W_3 = \mathbf{und}\}$, which is not legal, since $W_3 \rightarrow X$ and $\lambda'(W_3) \neq \mathbf{out}$. So we attempt to backward propagate from X, $f = \varnothing$ and λ'. We need to label W_3 **out**, which requires labelling $W_2 = \mathbf{in}$, which is not possible since it attacks itself, and thus the backward propagation fails. What this means in practice is that X cannot be part of any extension (this reasoning can be used in decision problems).

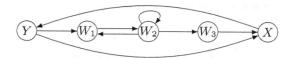

Fig. 3. Backward propagation.

Unlike a forward propagation, a backward propagation can generate multiple labelling functions. Consider the SCC S in the network in Fig. 4(L). A call to `propagateIN(X, S, f, λ, λ')` will succeed with $\lambda' = \{X = \mathbf{in}, Y = \mathbf{out}, Z_1 = Z_2 = Z_3 = Z_4 = W_1 = W_2 = \mathbf{und}\}$. We must now legally label both W_1 and W_2 **out**. But here we have a choice between labelling Z_1 or Z_2 **in**. So both $\lambda'_C = \{X = \mathbf{in}, Y = \mathbf{out}, Z_1 = \mathbf{in}, W_1 = W_2 = Z_3 = \mathbf{out}, Z_2 = Z_4 = \mathbf{und}\}$ (Fig. 4(C)) and $\lambda'_R = \{X = \mathbf{in}, Y = \mathbf{out}, Z_2 = \mathbf{in}, W_1 = W_2 = Z_4 = \mathbf{out}, Z_1 = Z_3 = \mathbf{und}\}$ (Fig. 4(R)) are returned in Λ from an invocation to `propagateOUT(X, S, f, λ', Λ)`.

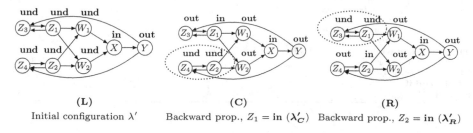

(L)	(C)	(R)
Initial configuration λ'	Backward prop., $Z_1 = \mathbf{in}$ (λ'_C)	Backward prop., $Z_2 = \mathbf{in}$ (λ'_R)

Fig. 4. A sample network and two successful backward propagations from $X = \mathbf{in}$.

There are two more important considerations to make. First, all of the attackers of the seed node must be labelled **out**. Therefore any solution returned by `propagateOUT` must satisfy this requirement. Our implementation approach in Algorithm 3 was to work with two lists. *makeOuts* contains the nodes that still need to be labelled **out** and starts with all undecided attackers of the seed node (Algorithm 3, line 5). At least one solution must be found labelling all of these nodes **out**. If this is not possible, `propagateOUT` simply fails (Algorithm 3, line 27). This essentially complements `propagateIN` to guarantee the *correctness* of the algorithm. The solutions are stored in the list *sols*, which is initialised with the result of the forward propagation of the seed node (Algorithm 3, line 12). For each node in *makeOuts*, *sols* is replaced with a new set of satisfying solutions. Each successive node is then checked against all new solutions thus generated which, by construction, label **out** all of the previously removed nodes in *makeOuts*. If we successfully exhaust all of the nodes in *makeOuts*, then `propagateOUT` succeeds and returns all corresponding solutions (line 30). Otherwise, it fails and Λ is not updated.

Algorithm 3 Backward propagation of an IN label

Input: argument X labelled **in**, its SCC S, a conditioning labelling function f, and a labelling function λ' obtained from propagating $X = $ **in** forward

Output: false or true with a set of new partial labelling functions Λ

```
 1  Function propagateOUT(X,S,f,λ',Λ)
 2      if there exists W ∈ X⁻ such that W is in a previous layer and f(W) ≠ out or there exists
        W ∈ X⁻ such that W ∈ S and f(W) = in then
 3          return false
 4      else
 5          makeOuts(X)← {W ∈ X⁻ | W ∈ S and λ'(W) = und}
 6          forall W ∈ makeOuts(X) do
 7              makeIns(W)← {Z ∈ W⁻ | Z ∈ S and λ'(Z) = und}
 8              if makeIns(W) = ∅ then
 9                  return false
10              end if
11          end forall
12          sols← {λ'}
13          while makeOuts(X) ≠ ∅ do
14              Pick W ∈ makeOuts(X) such that |makeIns(W)| is minimal
15              makeOuts(X)← makeOuts(X)\{W}
16              newSols← ∅
17              forall λ' ∈ sols do
18                  forall Y ∈ makeIns(W) do
19                      if findExtsFromArg(Y,f,λ',newSols) then
20                          success← true
21                      end if
22                  end forall
23              end forall
24              if success then
25                  sols← newSols
26              else
27                  return false
28              end if
29          end while
30          Λ← newSols
31          return true
32      end if
33  end
```

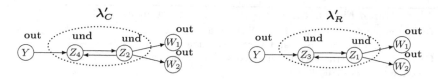

Fig. 5. Undecided sub-cycles within solutions.

The story does not end here though, and this takes us to the second important consideration which has to do with *completeness*. The result of a successful backward propagation may still leave some nodes of an SCC in what are effectively *induced sub-SCCs ring-fenced by* **out**-*labelled nodes*. Consider the network of Fig. 4(L) again. In order to legally label X **in** we need to label W_1 and W_2 **out**. We have seen that this can be done by labelling either Z_1 or Z_2 **in**, giving us the solutions λ'_C (extension $\{X, Z_1\}$) and λ'_R ($\{X, Z_2\}$) of Fig. 4(C) and (R), respectively. However, λ'_C leaves Z_2 and Z_4 undecided, whereas λ'_R leaves Z_1 and Z_3 undecided. In order to break these cycles (and hence guarantee completeness w.r.t. all complete extensions), all we have to do is to simply treat Z_4–Z_2 and Z_3–Z_1 as "sub-SCCs" and restart the whole process from the same original conditioning solution but now with initial conditioned solutions λ'_C and λ'_R (see Fig. 5). This is implemented in lines 8 and 10 of Algorithm 4. In our example, λ'_C will generate sub-solutions $Z_4 = $ **in**, $Z_2 = $ **out** and $Z_4 = $ **out**, $Z_2 = $ **in**; whereas λ'_R will generate sub-solutions $Z_3 = $ **in**, $Z_1 = $ **out** and $Z_3 = $ **out**, $Z_1 = $ **in**. The search will eventually terminate because recursive calls are only made with initial solutions containing less **und** labels than their parents' and the fact that the argumentation graph is finite.

3.3 Combining All Steps

Algorithm 1 will attempt to label **in** all candidate arguments that can be possibly labelled **in**. We then generate all possible solutions starting from each of these arguments with Algorithm 4. This requires to attempt to propagate forward from $X = $ **in** (line 3). If this is successful, it will generate a new labelling function λ'' with at least two less undecided arguments than λ'. We then attempt to propagate backwards from λ'' (line 5), to guarantee that all attackers of X are legally labelled **out**. If this is successful, it will generate a number of possible solutions Λ', which we add to the current set of solutions (line 7). These solutions may still leave some undecided nodes, so we restart the process from each solution σ in Λ' and the remaining candidate undecided nodes (lines 8 and 10), adding again the results to the set of solutions (line 12). At this point, we can filter out the solutions that do not yield preferred extensions if needed (see Algorithm 5).

Algorithm 4 Finding extensions from a given argument

Input: argument X to label **in**, its SCC S, a conditioning labelling function f, a legal
     labelling function λ, and a set of candidate labelling functions Λ

Output: false or true with an updated set of candidate labelling functions Λ

```
 1 Function findExtsFromArg(X,S,f,λ,Λ)
 2 |   λ' ← λ; λ'(X) = in
 3 |   if propagateIN(X,S, f,λ',λ'') then
 4 |   |   Λ' ← ∅
 5 |   |   if propagateOUT(X,f,λ'',Λ') then
 6 |   |   |   for σ ∈ Λ' do
 7 |   |   |   |   updateExts(σ,Λ)
 8 |   |   |   |   possIns_S ← {Y ∈ S|σ(Y) = und, Y ∉ Y⁺, {X ∈ Y⁻\S|f(X) = und} = ∅}
 9 |   |   |   |   if possIns ≠ ∅ then
10 |   |   |   |   |   Λ'' ← ∅; findExtsFromArgs(S,possIns_S,f,σ,Λ'')
11 |   |   |   |   |   for σ' ∈ Λ'' do
12 |   |   |   |   |   |   updateExts(σ',Λ)
13 |   |   |   |   |   end for
14 |   |   |   |   end if
15 |   |   |   end for
16 |   |   else
17 |   |   |   return false
18 |   |   end if
19 |   else
20 |   |   return false
21 |   end if
22 end
```

Algorithm 5 Updating the set of candidate solutions

Input: a solution λ and a set of candidate solutions Λ

Output: an updated set of candidate solutions Λ, according to the semantics

```
 1 Function updateExts(λ,Λ)
 2 |   if preferred semantics then
 3 |   |   Remove all solutions γ in Λ whose set of in-nodes is contained in the set of in-nodes of λ
 4 |   end if
 5 |   Λ ← Λ ∪ {λ}
 6 end
```

Proposition 1 (Soundness and Completeness). *Let S be an SCC, f an
admissible conditioning labelling function, and λ a labelling function for S conditioned by f (cf. Definition 1), then (1) all labelling functions returned by Algorithm 1 are legal; and (2) these are all the legal labelling functions for S.*

*Proof. Omitted, but soundness comes from the fact that lines 7 and 12 of Algo-
rithm 4 only add legal labelling functions and completeness from the facts that all
alternative solutions are tried in line 2 of Algorithm 1 and line 10 of Algorithm 4.*

4   Analysis and Comparisons with Other Work

In Sect. 2, we briefly described the Modgil-Caminada's algorithm for preferred
extensions and mentioned that it could behave very inefficiently. In fact, Charwat
et al. pointed out that for the class of argumentation frameworks $\langle \mathcal{A}, \mathcal{A}^2 \rangle$, the
algorithm produces $n!$ branches (where $n = |\mathcal{A}|$), all with the same extension
[8]. Since each node in each branch of execution corresponds to a transition
step, the total number of transition steps is at least twice as many. In fact, it
is $n! + \sum_{i=1}^{n-1} \frac{n!}{(n-i)!} \geq 2n!$, to be precise. Although arguably unrealistic, this

class of argumentation frameworks is particularly hard for Modgil-Caminada's algorithm, but it is dealt with trivially by our algorithm, requiring only n steps to identify that no nodes can be possibly labelled **in** and then producing the empty extension. This is because the higher the degree of attacks, the higher the degree of constraints and hence the lower the number of alternatives to check by our algorithm. Perhaps a more interesting class of frameworks to compare is what we call *bi-directed cycle graphs* involving a cycle with all nodes in both directions (see Fig. 6 (Start)). We now discuss the behaviour of both algorithms for this class of graphs. Modgil-Caminada's algorithm would start with the all-**in** labelling function and hence all nodes would be initially illegally labelled **in**. None is super-illegally labelled **in**, so the algorithm would iterate through *all* nodes, performing a transition step on each one and then recursively calling itself with the labelling functions resulting from the transitions. For the sake of argument, let us assume that the algorithm picks node A_1 first. A_1's label would be changed from **in** to **out**. As it would become legally labelled **out** and none of the nodes that it attacks is labelled **out**, the first transition step would result in the labelling TS1 of Fig. 6.[6] None of the nodes in TS1 are legally labelled **in** or super-illegally labelled **in**, so the algorithm would then again iterate through all nodes that remain illegally labelled **in** (4 in total). In the branches that pick a node adjacent to A_1, say A_2, the following would happen. The algorithm would change A_2's label to **out** (which is illegal), and then to **und**. A_1 is the only node that A_2 attacks and it is labelled **out**, but it is legally labelled so. The algorithm would then choose from one of the remaining illegally labelled nodes (of which there would be 3). If it agains picks an adjacent node, say A_3, it would change its label to **out** and then to **und**, and this process would continue until all nodes were re-labelled **und**. This sequence of transitions is depicted in graphs TS1, TS2,..., TS5 of Fig. 6. The algorithm would eventually pick A_3 or A_4 as an alternative choice to A_2 and in those branches it would eventually produce the preferred extensions. However, the number of recursive calls would still remain *close* to factorial (See Fig. 7). Our algorithm by contrast would start with all nodes labelled **und** and pick any initial seed node. In enumeration problems the choice is actually irrelevant as all eligible undecided nodes are attempted. In decision problems, we can start with an argument of interest and continue only if an appropriate extension can be constructed. If we start by propagating $A_1 = $ **in**, we are immediately forced to label A_2 and A_5 **out**, giving us only two further choices to generate the preferred extensions, i.e., either to label $A_3 = $ **in** or to label $A_4 = $ **in**. Figure 7 shows the number of transition steps performed by Modgil-Caminada's algorithm in bi-directed cycle graphs of up to 24 nodes and the number of recursive calls in our own algorithm (both implemented in EqArgSolver). For comparison, we included the factorial and 2^x functions.

In [6], Cerutti et al.'s proposed a meta-algorithm that decomposes the original argumentation framework into SCCs and uses a "base algorithm" at the base of the recursion to solve the original problem at the SCC level. As an illustration of the approach, the base algorithm employed a SAT solver. It should be possible

---

[6] There is an analogous branch for all other arguments $A_2,\ldots,A_5$.

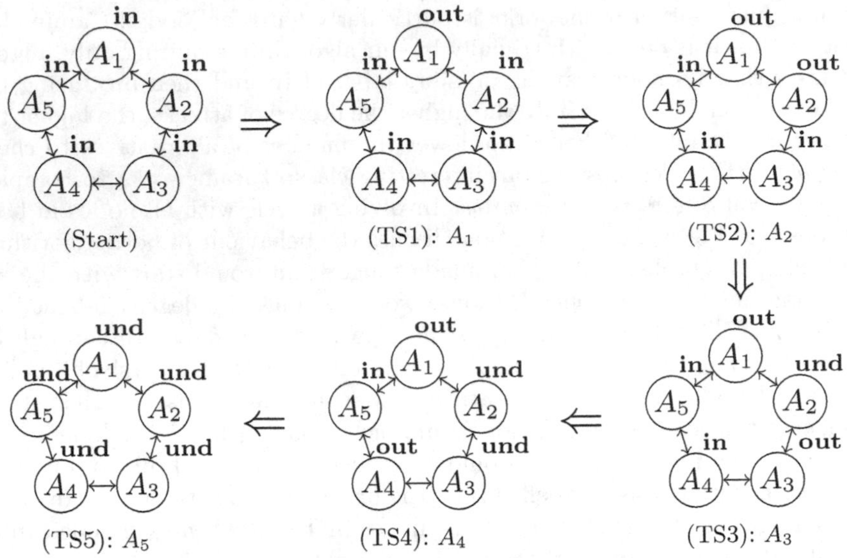

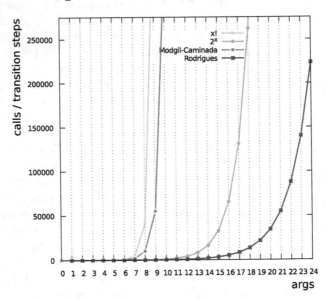

Fig. 6. Bi-directed cycle graph behaviour

Fig. 7. Transition steps × calls to `propagateIN` and `propagateOUT`

to swap the algorithm here proposed for the call to the SAT solver [6, Line 19, Algorithm 2] or vice-versa using an appropriate translation of the problem, since a conditioning solution simply constrains the set of possible models. This investigation will appear in a forthcoming paper. Finally, Nofal et al. proposed algorithms for decision problems in the preferred semantics [13]. The algorithm presented here is not restricted to this semantics only. We will however compare the approaches of these algorithms and ours in future work.

5   Empirical Evaluation

Apart from the special cases discussed above, we also conducted some experiments to compare Modgil-Caminada's algorithm with ours in randomly generated graphs. Our objective was not to conduct an extensive empirical evaluation between general solvers, as this will be done by the 2nd International Competition of Computational Models of Argumentation (ICCMA), but merely to provide a first-hand evaluation of the two labelling approaches. In order to eliminate any implementation factors that could directly affect the comparison between the two, they were both embedded within two versions of EqArgSolver which was invoked for the preferred semantics only. For further comparison we also recorded the results provided by Tweetysolver v1.2, which also uses decomposition into SCCs but uses a SAT solver for solutions. Tweetysolver was chosen because it is an off-the-peg easy-to-deploy solver and a "good enough" initial marker for the performance of SAT-based solvers *in this class of problems.*

We generated 3 datasets of 1,000 graphs each with maximum cardinality of 15, 25 and 35 nodes using `probo`'s SCC generator. The maximum number of SCCs in each graph was set to 2. Each dataset was divided into 10 sets of 100 graphs with probability $p = 0.1$, $p = 0.2$, $\ldots$, $p = 1$ of a node attacking another within an SCC. We submitted the 3,000 graphs thus generated to the solvers running on a PC with an Intel i7 4690 K processor and 32 Gb RAM. The left of Fig. 8 shows the comparative average time per graph successfully solved by each solver and the right shows the percentage of instances timed out within 180 s.

The graphs turned out to be rather too small to effectively stress test EqArgSolver using our algorithm. However, they clearly show the differences in performance between the two algorithms (and Tweetysolver). Both the version of EqArgSolver using our algorithm as well as Tweetysolver successfully solved all graphs submitted within the time limit. As expected, the version using Modgil-Caminada's algorithm timed out more frequently the more nodes the datasets contained. For graphs with up to 15 nodes, it timed out in roughly 10% of the problems, increasing to 40% of timeouts in graphs with up to 25 nodes; and then to 70% timeouts in graphs with up to 35 nodes. The actual average time per graph successfully solved varied rather erratically in the version using Modgil-Caminada's algorithm and this deserves further investigation. Our algorithm was clearly the fastest (just above 0 ms per graph on average). The execution times for Tweetysolver stayed relatively constant at around 1,000–1,250 ms per graph in all datasets. This shows some advancements in catching up with SAT reduction approaches.[7]

---

[7] A more robust SAT-based argumentation solver would employ special techniques to maximise the performance of the underlying SAT solver.

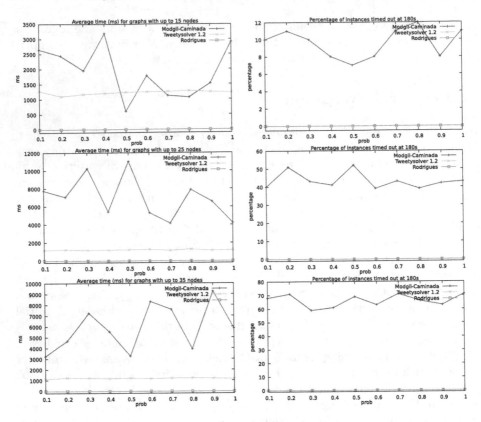

Fig. 8. Average execution time and % of time-outs for graphs with up to 15, 25 and 35 nodes.

6 Conclusions and Future Work

It is well known that the computation of grounded extensions is simply a matter of propagation of the **in** labels of unattacked arguments, which can be done very efficiently using the Discrete Gabbay-Rodrigues Iteration Schema [10]. In this paper we proposed a novel algorithm for the computation of all other complete extensions by looking for solutions to the SCCs of an argumentation framework. With minor modifications the algorithm can be used for the preferred and stable semantics as well.

The motivation for the development of this algorithm came from the following. In the solver GRIS [14], we used Modgil-Caminada's algorithm to compute the preferred extensions of an argumentation framework. However, Modgil-Caminada's algorithm proved very inefficient for all but the simplest graphs and can only compute the preferred extensions. We wanted a more efficient algorithm that could compute all complete extensions and that could also check argument acceptability without necessarily having to generate all extensions. The algorithm here proposed achieves all that and successfully replaced

Modgil-Caminada's algorithm in the solver EqArgSolver, which we submitted to the 2nd ICCMA (see http://argumentationcompetition.org/).

Given that solvers using reduction-based approaches to the computation of argumentation semantics took the top spots in the 1st ICCMA, the reader might ask if the development of direct algorithms and tools for argumentation semantics is worthwhile or whether we should simply concentrate on improving the reduction-based techniques. We would side with Cerutti et al.'s to argue that both approaches have a role to play [7] and combining them could be advantageous. In addition, we would claim that direct approaches are the only alternative in applications for which a translation to logic is either not possible at all or very cumbersome, e.g., in certain numerical argumentation networks.

We tested the new algorithm over tens of thousands of graphs of cardinality of up to 100,000 nodes. Rather than the number of nodes in the framework as a whole, it is the complexity and the number of SCCs involved that can stress a solver using the algorithm. Although some of these characteristics are unavoidable and intrinsic to the problem, the complexity could be reduced in our algorithm by avoiding multiple generation of the same solution arising in different search branches. As it stands, we attempt to label **in** every candidate argument in an SCC in order to guarantee the completeness of the set of solutions found, but this could be improved. Optimisations in this area are under investigation.

A further point to make is that within an SCC we can start the algorithm at an argument of interest to aid in decision problems of argument acceptability. It should also be possible to work backwards from a specific argument to see if an extension containing it can be constructed. This is work in progress.

Finally, each **in** labelling of a node forces the arguments that it attacks to be labelled **out**, which means that in each non-trivial SCC, each forward propagation reduces the complexity of the original problem by at least two arguments, but possibly many more in cases where the seed node attacks multiple arguments. We therefore expect the probability of attacks between nodes within an SCC to be inversely proportional to the execution time of our algorithm. This needs to be fully demonstrated and we also want to compare the performance of our algorithm with Nofal et al.'s [13] which, as for Modgil-Caminada's algorithm, can only generate the preferred extensions.

References

1. Baroni, P., Giacomin, M., Guida, G.: SCC-recursiveness: a general schema for argumentation semantics. Artif. Intell. **168**(1), 162–210 (2005)
2. Baumann, R.: Splitting an argumentation framework. In: Delgrande, J.P., Faber, W. (eds.) LPNMR 2011. LNCS (LNAI), vol. 6645, pp. 40–53. Springer, Heidelberg (2011). https://doi.org/10.1007/978-3-642-20895-9_6
3. Baumann, R., Brewka, G., Wong, R.: Splitting argumentation frameworks: an empirical evaluation. In: Modgil, S., Oren, N., Toni, F. (eds.) TAFA 2011. LNCS (LNAI), vol. 7132, pp. 17–31. Springer, Heidelberg (2012). https://doi.org/10.1007/978-3-642-29184-5_2

4. Caminada, M.: A labelling approach for ideal and stage semantics. Argum. Comput. **2**(1), 1–21 (2011)
5. Caminada, M., Gabbay, D.M.: A logical account of formal argumentation. Stud. Log. **93**(2–3), 109–145 (2009)
6. Cerutti, F., Giacomin, M., Vallati, M., Zanella, M.: A SCC recursive metaalgorithm for computing preferred labellings in abstract argumentation. In: 14th International Conference on Principles of Knowledge Representation and Reasoning (2014)
7. Cerutti, F., Vallati, M., Giacomin, M.: Where are we now? State of the art and future trends of solvers for hard argumentation problems. In: Baroni, P., Gordon, T., Scheffler, T. (eds.) Proceedings of COMMA, Frontiers in Artificial Intelligence and Applications, vol. 287, pp. 207–218. IOS Press (2016)
8. Charwat, G., Dvořák, W., Gaggl, S.A., Wallner, J.P., Woltran, S.: Methods for solving reasoning problems in abstract argumentation a survey. Artif. Intell. **220**, 28–63 (2015)
9. Dung, P.M.: On the acceptability of arguments and its fundamental role in nonmonotonic reasoning, logic programming and n-person games. Artif. Intell. **77**, 321–357 (1995)
10. Gabbay, D.M., Rodrigues, O.: Further applications of the Gabbay-Rodrigues iteration schema. In: Beierle, C., Brewka, G., Thimm, M. (eds.) Computational Models of Rationality, vol. 29, pp. 392–407. College Publications (2016)
11. Liao, B.: Toward incremental computation of argumentation semantics: a decomposition-based approach. Ann. Math. Artif. Intell. **67**(3), 319–358 (2013). http://dx.doi.org/10.1007/s10472-013-9364-8
12. Modgil, S., Caminada, M.: Proof theories and algorithms for abstract argumentation frameworks. In: Simari, G., Rahwan, I. (eds.) Argumentation in Artificial Intelligence, pp. 105–129. Springer, Boston (2009). https://doi.org/10.1007/978-0-387-98197-0_6
13. Nofal, S., Atkinson, K., Dunne, P.E.: Algorithms for decision problems in argument systems under preferred semantics. Artif. Intell. **207**, 23–51 (2014)
14. Rodrigues, O.: GRIS system description. In: Thimm, M., Villata, S. (eds.) System Descriptions of the 1st International Competition on Computational Models of Argumentation, pp. 37–40. Cornell University Library (2015)
15. Wu, Y., Caminada, M.: A labelling-based justification status of arguments. Stud. Log. **3**(4), 12–29 (2010)

The Formal Argumentation Libraries
of Tweety

Matthias Thimm[(✉)]

Institute for Web Science and Technologies (WeST),
University of Koblenz-Landau, Koblenz, Germany
thimm@uni-koblenz.de

Abstract. We provide an overview on the argumentation libraries of the
Tweety library collection to artificial intelligence and knowledge representation. These libraries comprise of implementations to abstract argumentation frameworks, as well as the most popular approaches to structured argumentation, and various further aspects. We briefly sketch the
functionalities of these libraries and give some pointers to how they can
be used.

1   Introduction

The *Tweety libraries for logical aspects of artificial intelligence and knowledge
representation*[1] [20,21] are a comprehensive collection of Java libraries for various
logical approaches to artificial intelligence. The Tweety libraries provide implementations of formalisms such as default logic [17], answer set programming
[9], belief revision [11], and, in particular, formal argumentation [1,3,6,8,12–
16,19,23].

The popularity of the *International Competition on Computational Models
of Argumentation*[2] (ICCMA) has shown that there is a growing interest in algorithmic approaches to formal argumentation. The formal argumentation libraries
of Tweety address this by providing a general and versatile collection of Java
classes to deal with various aspects of different approaches. The aim of this is not
to provide highly efficient implementations, but rather a simple and clear representation of argumentation concepts in an object-oriented manner that can be
easily understood and used by researchers and students not trained in algorithm
and software engineering.

The remainder of this paper gives a brief overview on the functionalities
provided within Tweety for the area of formal argumentation.

2   Overview

Tweety aims at providing a common framework for implementing different
approaches to artificial intelligence in general and knowledge representation in

---

[1] http://tweetyproject.org.

[2] http://argumentationcompetition.org.

© Springer International Publishing AG, part of Springer Nature 2018
E. Black et al. (Eds.): TAFA 2017, LNAI 10757, pp. 137–142, 2018.
https://doi.org/10.1007/978-3-319-75553-3_9

particular. It can be used by undergraduate students to better understand logical approaches to knowledge representation by actually working with them in a familiar object-oriented manner. Moreover, the main purpose of Tweety is to allow the easy implementation of new approaches by following a given strict framework and with the benefit of easily integrating concepts and methods of other formalisms. This allows for early testing of ideas and experimental evaluation in terms of feasibility studies.

Tweety is organized as a modular collection of Java libraries with a clear dependence structure. Each knowledge representation formalism has a dedicated Tweety library which provides implementations for both syntactic and semantic constructs of the given formalism as well as reasoning capabilities. Several libraries provide basic functionalities that can be used in other libraries. Among those is the *Tweety Commons* library which contains abstract classes and interfaces for all kinds of knowledge representation formalisms. Furthermore, the library *Math* contains classes for dealing with mathematical problems such as constraint satisfaction or optimization problems that often occur, in particular, in probabilistic approaches to reasoning. Most other Tweety projects deal with specific approaches to knowledge representation. Each Tweety library is organized as a Maven[3] project. Most libraries can be used right away as they only have dependencies to other Tweety libraries. Some libraries provide bridges to third-party libraries such as numerical optimization solvers which are not automatically found by Maven and have to be installed beforehand. However, all necessary third-party libraries can be installed by executing a single install file located within the Tweety distribution. We refer to [21] for a more detailed description of Tweety in general.

3   Argumentation Libraries

The package `net.sf.tweety.arg` is the general parent package for all approaches pertaining to formal argumentation. In the following, we briefly sketch the functionalities of the sub-package `net.sf.tweety.arg.dung` for abstract argumentation (Sect. 3.1), various sub-packages for structured argumentation (Sect. 3.2), and further approaches (Sect. 3.3).

3.1   Abstract Argumentation

Abstract argumentation frameworks (AAFs) due to Dung [6] are arguably the most investigated formalism for formal argumentation. An AAF is a tuple $AF = (A, R)$ where A is a set of arguments—atomic entities without inner structure—and R is a relation $R \subseteq A \times A$ modelling directed attack between arguments. Thus, an AAF can be represented as a directed graph. Semantics are given to these graphs using *extensions*, i.e. sets of arguments that are jointly acceptable according to some specific acceptance condition [1,6].

---

[3] http://maven.apache.org.

The Tweety package `net.sf.tweety.arg.dung` contains several classes for dealing with AAFs. The class `DungTheory`[4] models an AAF and provides several convenience methods for accessing the data structure and manipulate it. Abstract argumentation frameworks can be imported using the APX format [7] or programmatically using specific methods (see also Fig. 1). Tweety supports reasoning with AAFs using the extension-based approaches of grounded, stable, complete, preferred, ideal, semi-stable, CF2, and stage semantics as well as the ranking-based approaches of [10,22]. Finally, the package contains an implementation of the logic of dialectical outcomes of [13] that allows modelling and reasoning with extensions of subgraphs, and several factory classes for generating random AAFs.

```
DungTheory aaf = new DungTheory();
Argument a = new Argument("a"), b = new Argument("b"), c = new Argument("c");
aaf.add(a); aaf.add(b); aaf.add(c);
aaf.add(new Attack(a,b)); aaf.add(new Attack(b,a)); aaf.add(new Attack(b,c));

AbstractExtensionReasoner reasoner = new
                   StableReasoner(aaf, Semantics.CREDULOUS_INFERENCE);

System.out.println(reasoner.getExtensions());
```

Fig. 1. Code snippet for manually creating a simple AAF and determining its stable extensions.

Figure 1 shows a code snippet for creating a simple AAF and determining its stable extensions.

3.2   Structured Argumentation Approaches

Tweety contains implementations of the most popular approaches to structured argumentation, namely ASPIC+ [16], Assumption-based Argumentation (ABA) [23], Defeasible Logic Programming (DeLP) [8], and *deductive argumentation* [3]. In general, an approach to structured argumentation aims at providing an inner structure to arguments by allowing the representation of those through sets of formulas in some logic. For example, in the framework of *deductive argumentation* [3] classical logic—propositional and first-order logic—is used as the underlying knowledge representation formalism. Arguments are build from classical formulas by identifying a set of classical formulas as the *premise* and a single formula as the *conclusion* of an argument, such that the premise entails the conclusion. Therefore, arguments correspond to minimal proofs in the classic logical sense. If a knowledge base is inconsistent, arguments and counterarguments for different conclusions can be extracted from this knowledge base and put in relation to each other. While [3] bases its framework on classical logic, ASPIC+, ABA, and DeLP also incorporate aspects of non-classical formalisms that allow e.g. the use of default reasoning techniques for the construction of arguments.

---

[4] The class name `DungTheory` was chosen in favour of the class name `AbstractArgumentationFramework` in order to avoid confusion with the Java term `abstract` which is usually used as a prefix of an abstract class.

Tweety provides several functionalities for importing and working with knowledge bases in ASPIC$^+$, ABA, DeLP, and deductive argumentation. In particular, reasoning with these approaches can be reduced to reasoning with abstract argumentation frameworks by determining the corresponding AAF and using AAF reasoners as discussed above. Note that this is the standard semantical approach for ASPIC$^+$ and ABA. Note, however, that both DeLP and the deductive argumentation approach of [3] also provide proprietary reasoning mechanisms based on the construction of dialectical trees (or *argument graphs* in [3]) and their evaluation. Tweety provides implementation of these reasoning mechanisms as well, in particular the approach through knowledge base compilation for deductive argumentation from [2]. Finally, a web interface for the Defeasible Logic Programming approach is also available[5] and similar interfaces for other approaches are currently in development.

Figure 2 shows a small example using Tweety's ASPIC$^+$ implementation.

```
=> WearsRing                    | AspicArgumentationTheory<PropositionalFormula> t =
=> PartyAnimal                  |           parser.parseBeliefBaseFromFile(<File>);
                                | DungTheory aaf = t.asDungTheory();
d1: WearsRing => Married        |
d2: PartyAnimal => Bachelor     | AbstractExtensionReasoner reasoner =
                                |           new PreferredReasoner(aaf,
s1: Married -> ! Bachelor       |           Semantics.CREDULOUS_INFERENCE);
s2: Bachelor -> ! Married       | System.out.println(reasoner.getExtensions());
```

Fig. 2. The Tweety format of the classical ASPIC$^+$ example of the bachelor [16] (left) and a code snippet for reading this file into an `AspicArgumentationTheory`, inducing its abstract argumentation framework, and determining the latter's preferred extensions (right).

3.3 Further Approaches

Tweety also provides implementations to further approaches to formal argumentation, in particular to various approaches to probabilistic argumentation [12,13,15,19] and how those can be used for opponent modelling in strategies for persuasion [18]. Finally, Tweety provides an implementation of the approach of *social abstract argumentation* [14].

4 Conclusion

We gave a brief overview on the argumentation libraries of Tweety. In particular, we sketched the functionalities of libraries pertaining to abstract argumentation, structured argumentation, and further approaches.

Tweety is an active project and new approaches are added to the collection regularly. Current work is on an implementation for *Abstract Dialectical Frameworks* [5] as well as further approaches to ranking semantics [4].

---

[5] http://tweetyproject.org/w/delp.

References

1. Baroni, P., Caminada, M., Giacomin, M.: An introduction to argumentation semantics. Knowl. Eng. Rev. **26**(4), 365–410 (2011)
2. Besnard, P., Hunter, A.: Knowledgebase compilation for efficient logical argumentation. In: Proceedings of the 10th International Conference on Knowledge Representation (KR 2006), pp. 123–133. AAAI Press (2006)
3. Besnard, P., Hunter, A.: Elements of Argumentation. The MIT Press, Cambridge (2008)
4. Bonzon, E., Delobelle, J., Konieczny, S., Maudet, N.: A comparative study of ranking-based semantics for abstract argumentation. In: Proceedings of the 30th AAAI Conference on Artificial Intelligence (AAAI 2016) (2016)
5. Brewka, G., Ellmauthaler, S., Strass, H., Wallner, J.P., Woltran, S.: Abstract dialectical frameworks revisited. In: Proceedings of the 23rd International Joint Conference on Artificial Intelligence (IJCAI 2013) (2013)
6. Dung, P.M.: On the acceptability of arguments and its fundamental role in non-monotonic reasoning, logic programming and n-person games. Artif. Intell. **77**(2), 321–358 (1995)
7. Egly, U., Gaggl, S.A., Woltran, S.: ASPARTIX: implementing argumentation frameworks using answer-set programming. In: Garcia de la Banda, M., Pontelli, E. (eds.) ICLP 2008. LNCS, vol. 5366, pp. 734–738. Springer, Heidelberg (2008). https://doi.org/10.1007/978-3-540-89982-2_67
8. Garcia, A., Simari, G.R.: Defeasible logic programming: an argumentative approach. Theory Pract. Log. Program. **4**(1–2), 95–138 (2004)
9. Gelfond, M., Leone, N.: Logic programming and knowledge representation - the A-prolog perspective. Artif. Intell. **138**(1–2), 3–38 (2002)
10. Grossi, D., Modgil, S.: On the graded acceptability of arguments. In: Proceedings of the 24th International Joint Conference on Artificial Intelligence (IJCAI 2015), pp. 868–874 (2015)
11. Hansson, S.O.: A Textbook of Belief Dynamics. Kluwer Academic Publishers, Norwell (2001)
12. Hunter, A., Thimm, M.: On partial information and contradictions in probabilistic abstract argumentation. In: Proceedings of the 15th International Conference on Principles of Knowledge Representation and Reasoning (KR 2016), April 2016
13. Hunter, A., Thimm, M.: Optimization of dialectical outcomes in dialogical argumentation. Int. J. Approx. Reason. **78**, 73–102 (2016)
14. Leite, J., Martins, J.: Social abstract argumentation. In: Proceedings of the 22nd International Joint Conference on Artificial Intelligence (IJCAI 2011) (2011)
15. Li, H., Oren, N., Norman, T.J.: Probabilistic argumentation frameworks. In: Proceedings of the First International Workshop on the Theory and Applications of Formal Argumentation (TAFA 2011) (2011)
16. Modgil, S., Prakken, H.: The ASPIC+ framework for structured argumentation: a tutorial. Argum. Comput. **5**, 31–62 (2014)
17. Reiter, R.: A logic for default reasoning. Artif. Intell. **13**, 81–132 (1980)
18. Rienstra, T., Thimm, M., Oren, N.: Opponent models with uncertainty for strategic argumentation. In: Proceedings of the 23rd International Joint Conference on Artificial Intelligence (IJCAI 2013), August 2013
19. Thimm, M.: A probabilistic semantics for abstract argumentation. In: Proceedings of the 20th European Conference on Artificial Intelligence (ECAI 2012), August 2012

20. Thimm, M.: Tweety - a comprehensive collection of java libraries for logical aspects of artificial intelligence and knowledge representation. In: Proceedings of the 14th International Conference on Principles of Knowledge Representation and Reasoning (KR 2014), pp. 528–537, July 2014
21. Thimm, M.: The tweety library collection for logical aspects of artificial intelligence and knowledge representation. Künstliche Intelligenz **31**(1), 93–97 (2017)
22. Thimm, M., Kern-Isberner, G.: On controversiality of arguments and stratified labelings. In: Proceedings of the Fifth International Conference on Computational Models of Argumentation (COMMA 2014), September 2014
23. Toni, F.: A tutorial on assumption-based argumentation. Argum. Comput. **5**(1), 89–117 (2014)

Heureka: A General Heuristic Backtracking Solver for Abstract Argumentation

Nils Geilen and Matthias Thimm[✉]

Institute for Web Science and Technologies,
Universität Koblenz-Landau, Koblenz, Germany
thimm@uni-koblenz.de

Abstract. The HEUREKA solver is a general-purpose solver for various problems in abstract argumentation frameworks pertaining to complete, grounded, preferred and stable semantics. It is based on a backtracking approach and makes use of various heuristics to optimize the search.

> ευρηκα! ευρηκα! – *I have found it! I have found it!*
> – Archimedes of Syracuse (287–212 BC)

1 Introduction

An abstract argumentation framework (AAF) as defined by Dung [3] is a tuple $\Gamma = (\mathcal{A}, \mathcal{R})$ where $\mathcal{A}$ is a set of arguments and $\mathcal{R} \subseteq \mathcal{A}^2$ an attack relation between arguments. An attack $a \rightarrow b \in \mathcal{R}$ models that argument a defeats argument b. For any argument set $E \subseteq \mathcal{A}$, let E^+ be the set of arguments which are attacked by an element of E and let E^- be the set of arguments which attack an element of E. An AAF Γ is interpreted through the use of *extensions*, i.e., sets of arguments that provide a coherent view on the argumentation represented by Γ. An extension $E \subseteq \mathcal{A}$ is *conflict-free* iff there are no $a, b \in E$ with $a \rightarrow b$. An extension E is *stable* iff it is conflict-free and for every $b \in \mathcal{A} \setminus E$ there is $a \in E$ with $a \rightarrow b$. Other notions of extensions include complete, grounded, and preferred extensions, see [3] for the formal definitions.

HEUREKA is a software system that implements a direct backtracking approach for solving reasoning problems with respect to stable, complete, grounded, and preferred semantics. The backtracking approach makes use of a variety of heuristics to dynamically (re-)order the arguments to minimize the backtracking steps. HEUREKA is able to solve the problems of

- enumerating all extensions (EE),
- determining a single extension (SE),
- checking whether an argument is part of at least one extension, i.e., whether it is creduously justifiable (DC), and

E. Black et al. (Eds.): TAFA 2017, LNAI 10757, pp. 143–149, 2018.
https://doi.org/10.1007/978-3-319-75553-3_10

– checking whether an argument is part of every extension, i.e., whether it is sceptically justifiable (DS)

with respect to the four mentioned semantics. HEUREKA is written in C++ and available under the LGPL v3.0 licence on GitHub[1].

In the remainder of this paper, we describe the architecture of HEUREKA as it has been submitted to the *Second International Competition on Computational Models of Argumentation (ICCMA '17)*[2]. Note that a slightly shorter version of this paper has also been submitted as a system description to the competition.

2   Backtracking Algorithm

HEUREKA consists of a family of backtracking algorithms, one for each complete, preferred, and stable semantics which are similar to the algorithm defined in [5] but use dynamic heuristics to (re-)order how arguments are processed. The concrete algorithms differ only slightly so we focus our presentation here on the stable semantics and, in particular, on the task of computing all stable extensions.

At any time during the execution, a labelling function $\mathcal{L}ab$, which assigns to each argument either the value IN if it should be contained in the extension, OUT if it should be ruled out, or UNDEC if it is undecided, is maintained by the algorithm that keeps track of the current (partial) extension. A fourth label (BLANK) is used to indicated that an argument is not labelled yet. Let further IN($\mathcal{L}ab$) be the set of all arguments labelled IN by $\mathcal{L}ab$, and therefore the current solution. In a first step, the grounded extension E_{GR} is computed using a purely iterative algorithm which does not require backtracking [4] and an *intial labelling* is constructed. For an AAF $\Gamma = (\mathcal{A}, \mathcal{R})$ with the grounded extension E_{GR} let the *initial labelling* $\mathcal{L}ab_{init} : \mathcal{A} \rightarrow \{\text{IN}, \text{OUT}, \text{UNDEC}, \text{BLANK}\}$ be defined as

$$\mathcal{L}ab_{init}(a) = \begin{cases} \text{IN} & \text{if } a \in E_{GR} \\ \text{OUT} & \text{if } a \in E_{GR}^+ \\ \text{UNDEC} & \text{if } a \rightarrow a \\ \text{BLANK} & \text{otherwise} \end{cases}$$

Using a specific heuristic (see next section) a new argument a is selected and set to IN in $\mathcal{L}ab$. Setting this argument to IN may require that other arguments have to be rejected (because they are attacked by a) or need to be set to IN as well (because all attackers of them are now attacked by some IN-labelled argument), and so on, see [5] for the corresponding lookahead strategies. Those arguments are then marked correspondingly in $\mathcal{L}ab$. This step is repeated until either a stable extension has been determined or a contradiction occurs (an argument is labelled with two different labels). In the latter case, the algorithm backtracks and rejects an argument previously accepted. Algorithm 1 shows a

---

[1] https://github.com/nilsgeilen/heureka.
[2] http://www.dbai.tuwien.ac.at/iccma17.

Algorithm 1. Enumerate All Stable Extensions

Input: $\Gamma = (\mathcal{A}, \mathcal{R})$ AAF
                h heuristic
                $\mathcal{L}ab_{\text{init}}$ initial labelling
Output: $\mathcal{E}_{\text{ST}} \subseteq 2^{\mathcal{A}}$ stable extensions

1: ENUMERATE_EXTENSIONS($\mathcal{L}ab_{\text{init}}$)

2: **function** SET_IN($\mathcal{L}ab, a$)
3:     $\mathcal{L}ab(a) \leftarrow$ IN
4:     **for all** $b \in \{a\}^-$ **do**
5:         **if not** SET_UNDEC($\mathcal{L}ab, b$) **then**
6:             **return** false
7:     **for all** $b \in \{a\}^+$ **do**
8:         $\mathcal{L}ab(b) \leftarrow$ OUT
9:     **for all** $c \in (\{a\}^+)^!$ **do**
10:         **if** $\{c\}^- \subseteq \text{IN}(\mathcal{L}ab)^+$ **then**
11:             **if** $\mathcal{L}ab(c) = $ UNDEC **then**
12:                 **return** false
13:             **else if not** SET_IN($\mathcal{L}ab, c$) **then**
14:                 **return** false
15:     **if** IS_STABLE($\mathcal{L}ab$) **then**
16:         add IN($\mathcal{L}ab$) to $\mathcal{E}_{\text{ST}}$
17:         **return** false
18:     **else return** true

19: **function** SET_UNDEC($\mathcal{L}ab, a$)
20:     $\mathcal{L}ab(a) \leftarrow$ UNDEC
21:     **if** $|\{a\}^- \setminus \text{IN}(\mathcal{L}ab)^+| = 1$ **then**
22:         find $b \in \{a\}^- \setminus \text{IN}(\mathcal{L}ab)^+$
23:         **if not** SET_IN($\mathcal{L}ab, b$) **then**
24:             **return** false
25:     **return** true

26: **procedure** ENUMERATE_EXTENSIONS($\mathcal{L}ab$)
27:     let h choose next argument a, if there is none, **stop**
28:     **if** $\mathcal{L}ab(a) = $ BLANK **then**
29:         $\mathcal{L}ab' \leftarrow \mathcal{L}ab$
30:         **if** SET_IN($\mathcal{L}ab', a$) **then**
31:             ENUMERATE_EXTENSIONS($\mathcal{L}ab'$)
32:         **if** SET_UNDEC($\mathcal{L}ab, a$) **then**
33:             ENUMERATE_EXTENSIONS($\mathcal{L}ab$)
34:     **else** ENUMERATE_EXTENSIONS($\mathcal{L}ab$)

---

shortened version of this procedure. The functions SET_IN and SET_UNDEC set the labelling of the current argument to IN or undec, respectively, and propagate the changes following the mentioned lookahead strategies. For example all arguments attacked ba an argument labelled IN are set to OUT. At the end of SET_IN, the algorithm checks whether the current extension, i.e., the set of

IN-labelled arguments in $\mathcal{L}ab$, is stable, then it is reported as a stable extension and the algorithm backtracks as the current branch cannot contain any more extensions.

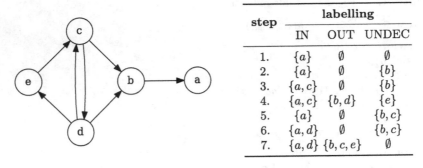

step	labelling		
	IN	OUT	UNDEC
1.	$\{a\}$	$\emptyset$	$\emptyset$
2.	$\{a\}$	$\emptyset$	$\{b\}$
3.	$\{a,c\}$	$\emptyset$	$\{b\}$
4.	$\{a,c\}$	$\{b,d\}$	$\{e\}$
5.	$\{a\}$	$\emptyset$	$\{b,c\}$
6.	$\{a,d\}$	$\emptyset$	$\{b,c\}$
7.	$\{a,d\}$	$\{b,c,e\}$	$\emptyset$

Fig. 1. AAF Γ (left) and algorithm steps (right) from Example 1; arguments not present in any set are BLANK

Example 1. Consider the AAF $\Gamma = (\mathcal{A}, \mathcal{R})$ depicted in Fig. 1 (left). Assume our heuristic function determines the following order of arguments: (a, b, c, d, e). In the first step, we determine that the grounded extension is empty and that there is no self-attacking argument, so we start with an empty labelling (all arguments are blank).

1. *decision:* a is picked by the heuristic and set to IN
2. as a consequence of step 1, all attackers/attackees of a are set to UNDEC/ OUT respectively, $\{a\}$ is not stable
3. *decision:* c is picked by the heuristic and set to IN
4. as a consequence of step 3, all attackers/attackees of c are set to UNDEC/ OUT respectively, $\{a,c\}$ is not stable
5. there are no more arguments which are still undecided, so the algorithm backtracks to the last decision in step 3 and sets c to OUT
6. *decision:* d is picked by the heuristic and set IN
7. as a consequence of step 6, all attackers/attackees of d are set to UNDEC/ OUT respectively, $\{a,d\}$ is stable $\Rightarrow$ stop

The backtracking algorithms for preferred and complete semantics are similar to the one for stable semantics. Reasoning problems pertaining to credulous/sceptical justification are solved by the same algorithms but with different termination criteria and slightly different initial steps.

3   Heuristics

While it is clear that the backtracking approach outlined before is a sound and complete procedure to enumerate extensions, its performance is highly dependent on the order in which arguments are processed. Observe that if this order is perfect, i.e., all arguments within the final extension are processed first, then no backtracking is needed and the algorithm has polynomial runtime. However, this runtime performance cannot, of course, be guaranteed but the choice of the heuristic used in ordering the arguments can deeply influence the runtime in general. HEUREKA comes with a series of different heuristics for this purpose.

In general, a heuristic h is a function $h : 2^{\mathcal{A}} \times \mathcal{A} \to \mathbb{R}$ that maps the current partial extension $E \subseteq \mathcal{A}$, i.e., the set of IN-labelled arguments in $\mathcal{L}ab$, and an argument $a \in \mathcal{A}$ to a real number $h(E, a)$. A large value $h(E, a)$ indicates that a should be likely included in the extension E and should be processed earlier than arguments with lower score. Some of our heuristics are defined independently of E and therefore need not to be recomputed after every modification of E. In general, however, HEUREKA allows for dynamic heuristics that are updated after every step.

A simple example of such a heuristic is the number of undefeated aggressors, i.e., the number of arguments which attack a but are not defeated by E. The number of undefeated aggressors $h_{\mathrm{UA}}(E, a)$ should be used as a negatively weighted component in a compound heuristic as every aggressor increases the vulnerability of an argument.

$$h_{\mathrm{UA}}(E, a) = \left| \{a\}^{-} \backslash E^{+} \right|$$

Another example which is independent of E is the ratio of an argument's in-degree and out-degree:

$$h_{\mathrm{deg}}^{\div}(E, a) = \frac{|\{a\}^{+}| + \epsilon}{|\{a\}^{-}| + \epsilon} \text{with } \epsilon \in \mathbb{R}$$

Path-based heuristics have proven useful in many cases. Let $d_i^{+}(a)$ be the number of paths of length i originating in a and let $d_i^{-}(a)$ be the number of paths of length i ending in a. The path-based components h_{path}^{+} and h_{path}^{-} map an argument to a combination of its outgoing paths or ingoing paths respectively.

$$h_{\mathrm{path}}^{+}(E, a) = \sum_{i=1}^{k} \alpha^{i} d_i^{+}(a)$$

$$h_{\mathrm{path}}^{-}(E, a) = \sum_{i=1}^{k} \beta^{i} d_i^{-}(a)$$

These heuristics can be combined into more complex path-based heuristics like $h_{\mathrm{path}}^{\Sigma} = h_{\mathrm{path}}^{-} + h_{\mathrm{path}}^{+}$ or $h_{\mathrm{path}}^{\Pi} = (-1) \cdot (h_{\mathrm{path}}^{-} + \epsilon) \cdot (h_{\mathrm{path}}^{+} + \epsilon)$.

Further heuristics have been implemented on top of well-known graph metrics such as betweenness centrality, eigenvector centrality, and matrix exponential. Another approach are SCC-based heuristics, which order arguments according to the ordering number of the strongly connected component, which they are part of, thus implementing ideas on SCC-recursiveness [1]. On top of the individual heuristics, HEUREKA also allows heuristics to be combined arithmetically.

For ICCMA'17, we fixed a heuristic for every problem based on a small experimental evaluation. For all tasks except SE-ST (enumerating some stable extension) we used the heuristic h_1, i.e., h_{path}^+ with fixed parameters $\alpha = 0.5$ and $k = 3$, defined as

$$h_1(E, a) = \sum_{i=1}^{3} \frac{d_i^+(a)}{2^i}$$

This heuristic shows the power of an argument to defend and defeat arguments. For the task SE-ST we used the heuristic h_2, which combines h_{path}^+ with h_{path}^- and h_{UA}.

$$h_2(E, a) = h_1(E, a) + \sum_{i=1}^{3} \frac{d_i^-(a)}{(-2)^i} - \frac{|\{a\}^- \setminus E^+|}{2}$$

This heuristic is influenced by the *matrix exponential* which has been suggested for this use in [2].

Later a systematic evaluation of the implemented heuristics has been conducted. During this evaluation $h_{\text{deg}}^{\pm}$ has proven most useful for solving problems under stable semantics while h_{path}^{Π} worked best when solving problems under complete or preferred semantics. For some graphs the performance could be substantially increased by adding an SCC-based component to the heuristic. In future work, heuristics could be explored, which also discriminate between OUT and UNDEC arguments intead of only analysing the partial extension.

4   Summary

We presented HEUREKA, a general-purpose argumentation solver based on the backtracking paradigm. The solver is backed by a number of heuristics that (dynamically) order the arguments of an abstract argumentation framework to minimize the number of necessary backtracking steps. During ICCMA'17, all results returned by HEUREKA have been correct. It landed in the center field for most tasks, while it was the fastest to find the grounded extension. Current and future work comprises analytical and empirical evaluation of the solver and its heuristics, as well as the development of new heuristics and combinations thereof.

References

1. Baroni, P., Giacomin, M., Guida, G.: SCC-recursiveness: a general schema for argumentation semantics. Artif. Intell. **168**(1–2), 162–210 (2005)
2. Corea, C., Thimm, M.: Using matrix exponentials for abstract argumentation. In: Proceedings of the First Workshop on Systems and Applications of Formal Argumentation (SAFA 2016), pp. 10–21, September 2016

3. Dung, P.M.: On the acceptability of arguments and its fundamental role in non-monotonic reasoning, logic programming and n-person games. Artif. Intell. **77**(2), 321–357 (1995)
4. Nofal, S., Atkinson, K., Dunne, P.E.: Algorithms for argumentation semantics: labeling attacks as a generalization of labeling arguments. J. Artif. Intell. Res. **49**, 635–668 (2014)
5. Nofal, S., Atkinson, K., Dunne, P.E.: Looking-ahead in backtracking algorithms for abstract argumentation. Int. J. Approx. Reason. **78**, 265–282 (2016)

EqArgSolver – System Description

Odinaldo Rodrigues[(✉)]

Department of Informatics, King's College London,
The Strand, London WC2R 2LS, UK
odinaldo.rodrigues@kcl.ac.uk

Abstract. This paper provides a general overview of EqArgSolver, a solver for enumeration and decision problems in argumentation theory. The solver is implemented from the ground up as a self-contained application in C++ without the use of any other external solver (e.g., SAT, ASP, CSP) or libraries.

1 Introduction

EqArgSolver is a computer application that can be used to solve *enumeration* and *decision* problems in argumentation theory. EqArgSolver builds and expands on the prototype GRIS [9] submitted to the 1st International Competition on Computational Models of Argumentation (ICCMA, [1]). It includes two technical advances that result in significant improvements in performance [7] and functionality. Firstly, EqArgSolver uses the discrete version of the Gabbay-Rodrigues Iteration Schema (dGR-iteration schema) [4], which can be implemented in a much more efficient way than its full-fledged counterpart [3]. Secondly, the component in GRIS responsible for computing preferred extensions (and based on Modgil and Caminada's algorithm for the computation of preferred labellings [6]) has been replaced by a novel and more efficient algorithm [7] that can compute *all* complete extensions. This allows EqArgSolver to handle the following two types of problems: *(i)* Given an argumentation network $\langle S, R \rangle$, to produce one or all of the extensions of the network under the grounded, complete, preferred or stable semantics; and *(ii)* Given an argument $X \in S$, to decide whether X is accepted credulously or sceptically according to one of those semantics.

The solver follows the general process of computation described in [5], which requires the decomposition of the framework into SCCs and the arrangement of these into layers following the direction of attacks between the arguments.

The dGR-iteration schema is employed in what we call a *grounding* module that propagates a (conditioning) solution $\boldsymbol{f}$ (under a particular semantics) to an attacked SCC in a subsequent layer. Provided $\boldsymbol{f}$ is an assignment that gives the correct label for every argument in the graph according to Dung's semantics (see [6])[1] the result of the propagation will also be legal. The numerical computations of the dGR-iteration schema are optimised by using the integer

---

[1] Henceforth, these assignments will be referred to as *legal assignments*.

© Springer International Publishing AG, part of Springer Nature 2018
E. Black et al. (Eds.): TAFA 2017, LNAI 10757, pp. 150–158, 2018.
https://doi.org/10.1007/978-3-319-75553-3_11

values 0 = **out** (rejected), 1 = **und** (undecided), and 2 = **in** (accepted) which are more efficient than the real values $\{0, \frac{1}{2}, 1\}$. It is worth emphasising, that although the dGR-iteration schema is employed in these computational tasks, EqArgSolver actually uses a *direct approach* to the problem (in the sense of [2]), i.e., argumentation problems are solved via operations performed directly on the graph. We will refer to the labels 0/**out**, 1/**und** and 2/**in** interchangeably.

The newly proposed algorithm [7] ensures that all arguments left undecided by the propagation of solutions to SCCs are systematically tried for inclusion in some extension (this is explained in more detail in Sect. 2).

EqArgSolver has been submitted to the 2nd iteration of ICCMA, whose results will be announced at the 2017 International Workshop on Theory and Applications of Formal Argument (TAFA-17).

2 System Overview

EqArgSolver accepts problems submitted according to **probo**'s syntax (see [1]). The problem specification is fully validated before the computation proceeds. Algorithm 1 gives a high-level overview of the computation process, which we now briefly describe. Some shortcuts allowing early termination are described in Sect. 4.

The framework is first divided into SCCs and arranged into layers (line 3). The starting point is an initial partial solution labelling all arguments as undecided (all-**und**, line 4). The solutions to each layer expand on the previous layers' solutions to include the new labelling assignments for the layer's arguments.

The composition of a typical layer is shown in Fig. 2(L). It consists of a block of trivial SCCs that are mutually dependent and operated on in one step, and a set of non-trivial SCCs that are independent from each other (and can be computed in parallel). Before working on a layer, each partial solution generated for the preceding layer is propagated to the layer's SCCs in order to *condition* its argument values—a process that we call *grounding*. Grounding will fully determine all of the values of the arguments in the trivial block (line 9) but not all of the values of the arguments in the non-trivial SCCs. Some of these arguments will be left undecided although they could potentially be labelled **in** in a larger extension (line 12). A newly proposed algorithm [7] ensures that all such arguments are systematically tried for inclusion generating all solutions for the SCC (line 13). These solutions are only partial to the argumentation framework as a whole and are combined with each other in a process called by Liao [5] the *horizontal and vertical combinations of partial solutions* (lines 14 and 16, respectively).[2]

This systematic process is repeated until all relevant layers are processed. The resulting solutions are then output as extensions (line 20). This may involve removing some of the extensions that are not relevant to the problem at hand (i.e., non-preferred or non-stable extensions in problems in the preferred and stable semantics, respectively).

---

[2] Some filtering to eliminate solutions not leading to maximal extensions in preferred/stable semantics problems is also done, although this is not shown in Algorithm 1. For full details, refer to [7].

The computation of the solutions to the problems in the grounded semantics does not require the decomposition of the framework into layers. The dGR-iteration schema mentioned previously can be applied to the entire argumentation framework (without decomposition) to compute the grounded extension. However, since the decomposition of the network into SCCs and their arrangement into layers can be performed very efficiently, the extra decomposition cost is offset by the performance gain obtained through the computation by layers in all but a few special cases, and is therefore our preferred choice for all semantics. Further optimisation here is possible but left as future work.

2.1 How Grounding Works

Propagation of the conditioning values of a solution is done using the dGR-iteration schema, whose behaviour we can only outline due to space limitations. Each node $X \in S$ gets an equation describing its value at iteration $i + 1$ ($V_{i+1}(X)$) based on the nodes' values at iteration i. Let $Att(X)$ denote the attackers of X, then the general format of the equations for a node X is

$$V_{i+1}(X) = 1 - \max_{Y \in Att(X)} \{V_i(Y)\}$$

with the values $0 = $ **out**, $\frac{1}{2} = $ **und**, $1 = $ **in**. However, by multiplying the equations by 2 we can take advantage of integer operations, which are faster. Therefore, in EqArgSolver we use the equivalence

0	out
1	und
2	in

All nodes get initial value **und** (i.e., we set $V_0(X) = 1$, for all $X \in S$). The sequence of values $V_0, v_1, \ldots$ of all nodes will converge in time linear to the cardinality of the set of nodes involved in the system of equations. The final values in the sequence will correspond to an extension (see [4] for details). Even though each iteration involves the computation of the values of n nodes, in practice a topological sorting of the nodes can be used to achieve convergence in linear time.[3]

Figure 1, depicts a sample argumentation framework, its associated system of equations, and the behaviour of the schema in several grounding scenarios until convergence is achieved. In layer 0 the schema produces the solution $X = 2$, $A = 0$, and $W, Y = 1$. It is easy to see that besides X, arguments W and Y could also be included in (distinct) complete extensions. The new algorithm is invoked at this point considering all candidate arguments that could potentially

---

[3] The version of EqArgSolver submitted to the 2nd ICCMA did not take advantage of this simple optimisation, but since submission it has been incorporated offering a huge performance improvement in certain classes of large graphs.

be labelled **in**. This imposes some constraints on any additional candidate partial solutions. In this example, it will produce the two remaining partial solutions to layer 0: $W = 2, Y = 0$ and $W = 0, Y = 2$. Propagating layer 0's three conditioning solutions to layer 1 is done again by grounding SCC_4 with the solutions. The result of these groundings can be seen in Fig. 1(R). It produces extensions $\{X\}$, $\{X, W\}$ and $\{X, Y, D\}$.

Full details of the dGR iteration schema can be found in [4].

The behaviour loosely described above is implemented in a component that we refer to as the *grounding module*, whose main function is to propagate the results of a conditioning solution through the nodes of an SCC of interest.

2.2   Generating All Complete Extensions

As mentioned, the grounding module may leave some nodes with label **und** which could potentially be labelled **in** yielding a larger extension (e.g., nodes W, Y in layer 0 or the nodes in layer 1 of Fig. 1). Our algorithm attempts to label **in** all such undecided nodes, propagating the results as required. When this is employed judiciously, it not only generates all remaining complete extensions, but also offers significant performance gains because the legal labelling of an argument imposes constraints that help prune the search space of feasible solutions. Full details are given in [7].

Input: Graph G
Output: Extensions of G

1 **EqArgSolver**
2 | Read and validate graph G
3 | Decompose G into SCCs and arrange them into layers $L_0, L_1, \ldots, L_{k-1}$
4 | $Sols \leftarrow \{$all-**und**$\}$
5 | **for** $i \leftarrow 0$ **to** $k - 1$ **do**   /* Iterate through layers */
6 | | $newSols \leftarrow \varnothing$
7 | | **foreach** $f \in Sols$ **do**
8 | | | $\lambda \leftarrow$ **GR-ground**(L_i, f); $TSB \leftarrow$ trivial SCC block of L_i
9 | | | $LayerSols \leftarrow \{\lambda \downarrow TSB\}$
10 | | | $S \leftarrow$ non-trivial SCCs in L_i
11 | | | **foreach** $S \in \mathcal{S}$ **do**
12 | | | | $possIns \leftarrow$ candidate **in**-nodes of S according to λ
13 | | | | SCC-sols$\leftarrow$**findExtsFromArgs**$(possIns, S, f, \lambda \downarrow S)$
14 | | | | Horizontally combine SCC-sols with solutions in $LayerSols$
15 | | | **end foreach**
16 | | | Add vertical combination of f with each $\gamma \in LayerSols$ to $newSols$
17 | | **end foreach**
18 | | $Sols \leftarrow newSols$
19 | **end for**
20 | Output $Sols$
21 **end**

Algorithm 1. EqArgSolver's overall processing sequence.

Argumentation Framework:

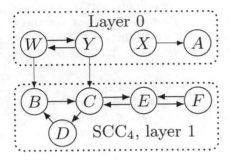

Equations:

$$V_{i+1}(X) = 2$$
$$V_{i+1}(A) = 2 - V_i(X)$$
$$V_{i+1}(W) = 2 - V_i(Y)$$
$$V_{i+1}(Y) = 2 - V_i(W)$$
$$V_{i+1}(B) = 2 - \max\{V_i(W), V_i(D)\}$$
$$V_{i+1}(C) = 2 - \max\{V_i(B), V_i(Y), V_i(E)\}$$
$$V_{i+1}(E) = 2 - \max\{V_i(C), V_i(F)\}$$
$$V_{i+1}(D) = 2 - \max\{V_i(C)\}$$
$$V_{i+1}(F) = 2 - \max\{V_i(E)\}$$

Results of grounding:

Layer 0

	X	A	W	Y
V_0	1	1	1	1
V_1	2	1	1	1
V_2	2	0	1	1
	in	**out**	**und**	**und**

Layer 1, sol $W=2$, $Y=0$

	B	C	D	E	F
V_0	1	1	1	1	1
V_1	0	1	1	1	1
	out	**und**	**und**	**und**	**und**

Layer 1, sol $W=1$, $Y=1$

	B	C	D	E	F
V_0	1	1	1	1	1
	und	**und**	**und**	**und**	**und**

Layer 1, sol $W=0$, $Y=2$

	B	C	D	E	F
V_0	1	1	1	1	1
V_1	1	0	1	1	1
V_2	1	0	2	1	1
V_3	0	0	2	1	1
	out	**out**	**in**	**und**	**und**

Fig. 1. Examples of grounding invocations.

3   Functionality and Design Choices

EqArgSolver can tackle enumeration and decision problems (sceptical and cred-ulous) of the grounded, complete, preferred and stable semantics. EqArgSolver can also provide solutions for the *Dung's Triathlon*, i.e., to compute in sequence the grounded extension, all stable extensions, and all preferred extensions of an argumentation framework. Graphs must be supplied as a *trivial graph format* text file, consisting of a sequence of argument designators one per line, followed by the separator "#" in its own line, followed by a list of pairs of argument designators, a pair per line, where the first element of the pair is the attacking argument and the second element is the attacked argument.

Each argument in EqArgSolver is assigned an internal identifier (an unsigned integer) and the argumentation graph is represented internally as an enhanced adjacency list. The argument data structure used is shown in Fig. 2(C) and (R). `layer` is the graph layer assigned by the decomposition; `extArgId` is the exter-nal argument identifier (the string given in the graph's input file); and `attsIn` and `attsOut` give, respectively, the list of incoming and outgoing attacks of the argument. This data structure is associated with the internal argument identifier using C++'s associative container `unordered_map`.

Storing both directions of the attack relation in the vectors `attsIn` and `attsOut` makes it more efficient to traverse the graph as needed. Similarly, a second associative container is created using the external node identifier as key and the internal node identifier as value (this is useful in decision problems). A (partial) solution is just a mapping from node identifiers to unsigned integers.

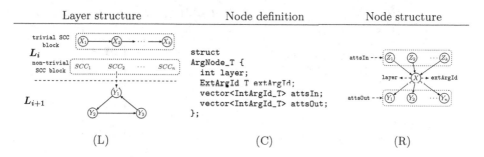

Layer structure	Node definition	Node structure
(L)	(C)	(R)

Fig. 2. Data representation in EqArgSolver

In order to avoid resizing of the associative container at creation time (which in large graphs can be very inefficient), EqArgSolver looks ahead in the input graph file to count the total number of arguments. It then creates a hash map with a sufficiently large number of buckets to represent the entire graph. This ensures that even graphs with many hundreds of thousands of nodes can be created in a just a few seconds.

4   Optimisations in Decision Problems

We now give a brief overview of some "shortcuts" that can be taken during the computation of some decision problems. The presentation here is not comprehensive but it serves to illustrate some of the details omitted in Algorithm 1. The shortcuts can in many cases avoid the full computation of all (or some of the) extensions of the argumentation framework. The left hand side of Fig. 3 contains the general schema of computation using layers and its right hand side contains an example that will be used to explain the main concepts involved.

As mentioned in Sect. 2, the decomposition of the argumentation framework will divide it into layers, where each layer is composed of a block containing the trivial SCCs of the layer and a block containing the layer's non-trivial SCCs. For example, in the sample argumentation framework shown on the right of Fig. 3, layer 0 contains only the non-trivial SCC block (NTSB) containing the SCC $A \leftrightarrow B$; layer 1 contains the trivial SCC block (TSB) containing the single trivial SCC with argument U and the NTSB containing the non-trivial SCCs

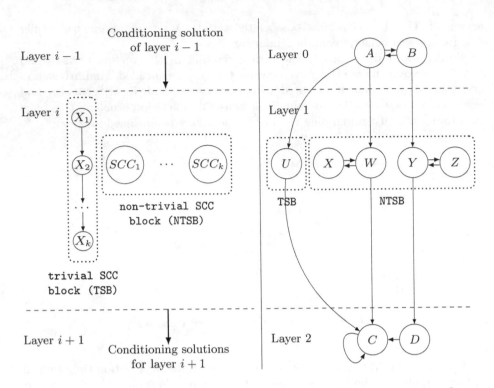

Fig. 3. Relationship between partial solutions and layers of the argumentation framework.

$X \leftrightarrow W$ and $Y \leftrightarrow Z$; and layer 2 contains only the NTSB containing the non-trivial SCC $C \leftrightarrow D$.

Let us start our considerations with the *grounded semantics*, which produces one single solution and is therefore simpler. From our discussion in Sect. 2.1, it should be clear that an answer to an argument acceptability decision problem under this semantics depends only on the computation of the solution up to the layer containing the argument in question. So for instance, if the problem is to decide whether U belongs to the grounded extension, there is no need to compute the solution to layer 2, as U belongs to layer 1 and the nodes in layer 2 cannot affect the acceptability of U in any way. Similarly, as U is in the TSB of layer 1, we can safely restrict the propagation of the solution of layer 0 to this block alone ignoring all SCCs in the NTSB. If the propagation labels U **in**, then the answer is positive, otherwise it is negative. In our example, the grounded solution to layer 0 is $A = B = $ **und**. When this is propagated to the TSB of layer 1 we get $U = $ **und** and we can stop there with a negative answer.

Consider now the *credulous approach* to the *complete semantics*. A positive answer must be given if any complete extension contains the argument in question. For example, if the problem is to determine whether X is credulously accepted under the complete semantics, then it should also be clear that the computation can be stopped at layer 1 of the graph. Moreover, the solutions to the SCCs in the NTSB are independent of each other, so we can restrict the computation in layer 1 to the SCC $X \leftrightarrow W$. There are three solutions to layer 0, all of which condition layer 1: $f_1 = \{A = \textbf{in}, B = \textbf{out}\}$; $f_2 = \{A = \textbf{out}, B = \textbf{in}\}$ and $f_3 = \{A = \textbf{und}, B = \textbf{und}\}$. In this case a (positive) answer can be obtained when solution f_1 is propagated to the SCC $X \leftrightarrow W$. This solution will force W to be labelled $\textbf{out}$, and hence X to be labelled $\textbf{in}$. In general, a negative answer to the credulous approach can only be given when all options have been exhausted and no solution includes the argument. This means generating all partial solutions for the SCC under all conditioning solutions. The situation with the sceptical approach is symmetrical to this. If the decision problem is sceptical, then we can provide a negative answer as soon as we obtain a solution that assigns a label other than $\textbf{in}$ to X. An example of this is when we propagate f_3 to $X \leftrightarrow W$, leaving both X and W undecided (which is a solution corresponding to an extension that does **not** contain X). A positive answer can only be given once all solutions have been considered and they all include the argument in question.

The considerations for the *preferred semantics* are slightly more complex as only the solutions that maximise the nodes labelled $\textbf{in}$ will yield preferred extensions. The *stable semantics* on the other hand will require the full computation of the framework because in general one cannot guarantee that a partial solution will form the basis of a stable extension. For example, a partial solution may include the argument X; label all arguments up to the current layer $\textbf{in}$ or $\textbf{out}$; and then have an expansion through propagation to a subsequent layer that leaves some arguments undecided (therefore not yielding a stable extension). Consider the partial solution $f_{11} = \{A = \textbf{in}, B = \textbf{out}, U = \textbf{out}, X = \textbf{in}, W = \textbf{out}, Y = \textbf{in}, Z = \textbf{out}\}$. As far as layers 0 and 1 are concerned, no arguments are undecided. However, no stable extension of the argumentation framework in Fig. 3 contains $\{A, X, Y\}$, even though it is itself a preferred extension.

5   Conclusions and Discussion

Many further improvements continue to be made to EqArgSolver. After the submission to the 2nd ICCMA, we improved the internal data representation of solutions and employed topological sorting of the TSB improving the performance of the grounding module considerably in some frameworks containing very large acyclic chains of attacks. Topological sorting of TSBs allows for the exclusion of nodes from the dGR iteration schema computation as soon as their values converge.

At the time of writing, further data representation alternatives are being considered to speed up some critical operations and the forward propagation algorithm is being optimised in two ways: (1) we want to avoid the multiple generation of the same solution (see [7]); and (2) we want to "learn" from failed forward propagations to help prune the search space of solutions further.

Finally, we have identified a number of randomly generated graphs with problem instances that were particularly difficult to solve. Work is under way to understand why these graphs are so challenging.

It is worth emphasising that the implementation aspects of a particular technique or algorithm play a fundamental role in the efficiency of any solver employing them. In the large graphs used in the 2nd ICCMA (some of which have millions of nodes), even the efficiency of reading, representing the graph internally, and outputting the solutions have a direct impact on the solver's performance.

The remarkable achievement of argumentation solvers employing SAT-reduction techniques can be largely (but not solely) attributed to recent advances in SAT-solving techniques, notably *conflict-driven clause learning* [8]. On top of that, any efficient reduction-based solver will exploit whatever optimisation parameters the underlying solver has to offer.

As SAT solvers operate on the logical translations of constraints arising from the argumentation context and learns from irreconcilable clauses to prune unfeasible solutions early, it should be possible to do the same in algorithms operating directly on the argumentation graph itself.

References

1. Cerutti, F., Oren, N., Strasse, H., Thimm, M., Vallati, M.: The First International Competition on Computational Models of Argumentation (ICCMA 2015) (2015). http://argumentationcompetition.org/2015/index.html
2. Charwat, G., Dvořák, W., Gaggl, S.A., Wallner, J.P., Woltran, S.: Methods for solving reasoning problems in abstract argumentation - a survey. Artif. Intell. **220**, 28–63 (2015)
3. Gabbay, D.M., Rodrigues, O.: Equilibrium states in numerical argumentation networks. Log. Univers. **9**(4), 1–63 (2015)
4. Gabbay, D.M., Rodrigues, O.: Further applications of the Gabbay-Rodrigues iteration schema in argumentation and revision theories. In: Beierle, C., Brewka, G., Thimm, M. (eds.) Computational Models of Rationality, vol. 29, pp. 392–407. College Publications (2016)
5. Liao, B.: Efficient Computation of Argumentation Semantics. Academic Press, Cambridge (2014)
6. Modgil, S., Caminada, M.: Proof theories and algorithms for abstract argumentation frameworks. In: Simari, G., Rahwan, I. (eds.) Argumentation in Artificial Intelligence, pp. 105–129. Springer, Boston (2009). https://doi.org/10.1007/978-0-387-98197-0_6
7. Rodrigues, O.: A forward propagation algorithm for semantic computation of argumentation frameworks. In: Black, E., et al. (eds.) TAFA 2017. LNAI, vol. 10757, pp. 120–136. Springer, Cham (2017)
8. Silva, J.P.M., Sakallah, K.A.: GRASP: a new search algorithm for satisfiability. In: Proceedings of the 1996 IEEE/ACM International Conference on Computer-aided Design, ICCAD 1996, pp. 220–227. IEEE Computer Society, Washington, D.C. (1996)
9. Thimm, M., Villata, S.: System descriptions of the 1st International Competition on Computational Models of Argumentation (ICCMA 2015). CoRR, abs/1510.05373 (2015)

Team Persuasion

David Kohan Marzagão[1](✉) ⓘ, Josh Murphy[1], Anthony P. Young[1] ⓘ,
Marcelo Matheus Gauy[2], Michael Luck[1], Peter McBurney[1],
and Elizabeth Black[1]

[1] Department of Informatics, King's College London, London, England
{david.kohan,josh.murphy,peter.young,michael.luck,peter.mcburney,
elizabeth.black}@kcl.ac.uk
[2] Institut für Theoretische Informatik, ETH Zürich, Zürich, Switzerland
marcelo.matheus@inf.ethz.ch

Abstract. We consider two teams of agents engaging in a debate to persuade an audience of the acceptability of a central argument. This is modelled by a bipartite abstract argumentation framework with a distinguished topic argument, where each argument is asserted by a distinct agent. One partition defends the topic argument and the other partition attacks the topic argument. The dynamics are based on flag coordination games: in each round, each agent decides whether to assert its argument based on local knowledge. The audience can see the induced sub-framework of all asserted arguments in a given round, and thus the audience can determine whether the topic argument is acceptable, and therefore which team is winning. We derive an analytical expression for the probability of either team winning given the initially asserted arguments, where in each round, each agent probabilistically decides whether to assert or withdraw its argument given the number of attackers.

1 Introduction

Argument-based persuasion dialogues provide an effective mechanism for agents to communicate their beliefs and reasoning in order to convince other agents of some central topic argument [11]. In complex environments, persuasion is a distributed process. To determine the acceptability of claims, a sophisticated agent or audience should consider multiple, possibly conflicting, sources of information that can have some level of agent-hood. In this paper, we consider teams of agents that work together in order to convince some audience of a topic argument. While strategic considerations have been investigated for one-to-one persuasion (e.g. [15]), and for one-to-many persuasion (e.g. [9]), the act of persuading as a team is a largely unexplored problem.

Consider a political referendum, where two campaigns seek to persuade the general public of whether or not they should vote for or against an important proposition. Each campaign consists of separate agents, where each agent is an

D. Kohan Marzagão—Supported by CNPq (206390/2014-9)
M. M. Gauy—Supported by CNPq (248952/2013-7).

expert in a single argument. For example, an environmentalist might argue how a favourable outcome in the referendum would reduce air pollution. Each agent can assert its argument to the public, and each agent is aware of counterarguments that other agents can make. However, no agent can completely grasp all aspects of the campaign, for example the environmentalist may be ignorant of relevant economic issues. If the agent thinks there are no counterarguments to its argument, then it should keep asserting its argument, as it is beneficial for its team. While each agent wishes to further their team's persuasion goal, they do not want to risk having their argument publicly defeated by counterarguments.

From this example, we consider a team of agents to have three key properties that differentiate them from an individual agent when persuading. Firstly, each agent may have localised knowledge which is inaccessible and non-communicable to other agents in the same team. Secondly, agents may not be wholly benevolent, potentially acting in their own interest before that of their team; reconciling this conflict between individual and team goals makes strategising more complex. Thirdly, there is no omniscient or authoritative agent able to determine the actions of the other agents in the team, meaning each agent must act independently, making the problem a distributed one. This problem is distinct from that of an individual persuader, and therefore requires a different approach to model the outcomes of persuasion.

We approach the problem of modelling team persuasion by exploring a particular team persuasion game, in which two opposing teams attempt to convince an audience of whether some central issue, termed the *topic*, is acceptable or not. For simplicity, we assume that each agent in a team is individually responsible for one argument in the domain, being an expert on that particular argument. As such, each agent must independently decide whether to actively assert its argument to the audience, or to hold back from asserting its argument. The persuasion game proceeds in rounds, where in each round an agent decides whether to assert its argument. An agent can decide to stop asserting its argument even if in previous rounds they had asserted it. Teams aim to reach a state in which the topic is acceptable or unacceptable according to the audience (depending on whether the agent is defending or attacking the topic), and in which no individual agent will change its decision of whether to assert its argument; in such a state the topic is guaranteed to retain its (un)acceptability indefinitely. When deciding whether to assert its argument, an agent takes into account whether the other agents are currently asserting their arguments. It aims to have a positive effect on its team's persuasion goal, but may also wish to avoid having its own argument publicly defeated (since this may, for example, negatively affect their public standing or reputation). When deciding whether to assert its argument, the agent must therefore balance the potential positive effect of this on its team's persuasion goal with the risk of its own argument being publicly defeated.

The audience determines whether they find the topic argument acceptable in a particular round by considering the set of arguments that are currently asserted. Note that the audience has no knowledge of which arguments were

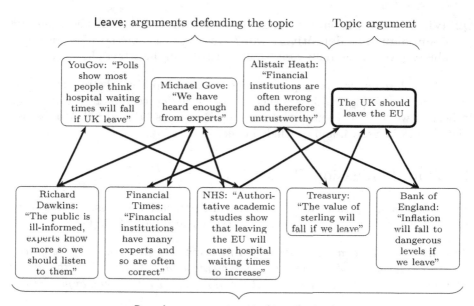

Leave; arguments defending the topic   Topic argument

Remain; arguments attacking the topic

Fig. 1. An instantiated example of a bipartite argumentation framework.

asserted in previous rounds; we consider the audience to be memoryless, only considering the arguments that are asserted in the current round.

For example, consider the arguments in Fig. 1, in which the directed edges represent conflict between arguments. The topic argument in this example is that the United Kingdom should leave the European Union, with three arguments defending the topic and five arguments attacking the topic (some indirectly). Each argument is controlled by a particular individual or institution. The agents are organised into two teams, those defending the topic (the Leave campaign), and those attacking the topic (the Remain campaign). Consider the argument that might be asserted by the Treasury: the Treasury is motivated to assert their argument as it directly attacks the topic argument (which they are seeking to dissuade the audience of). If they are aware of the argument possibly asserted by Alistair Heath, they may decide not to assert their own argument to avoid the risk of being publicly defeated. The public decides whether leaving the European Union is acceptable depending on which arguments are currently being asserted. The contribution of this paper is the application of *team persuasion games* [10] to model public debates of this form. We answer the following:

Q1 How do we formalise the situation where one team has *definitively won*? We define such a situation to be a state where agents that are asserting their arguments will continue to do so, and agents not asserting their arguments will never do so.

Q2 What is the probability that a particular team (e.g. the Remain Campaign) has definitively won? We prove an expression for this probability, given the initially asserted arguments and the attacks between them.

In Sect. 2 we define a team persuasion game on a bipartite abstract argumentation framework [6], which is a special case of a *flag coordination game* [10]. In Sect. 3, we use our framework to answer Questions Q1 and Q2. We discuss related work in Sect. 4, and conclude in Sect. 5.

2   Team Persuasion Games

In this section we present our model of team persuasion games. We begin by briefly reviewing the relevant aspects of abstract argumentation [6].

Definition 1. *An **argumentation framework** is a directed graph (digraph) $AF := \langle A, R \rangle$ where A is the set of arguments and $R \subseteq A \times A$ is the attack relation, where $(a, b) \in R$ denotes that the argument a attacks the argument b.*

Figure 1 is an example argumentation framework. We will only consider *finite*, non-empty argumentation frameworks, i.e. where $A \neq \varnothing$ is finite. Given an argumentation framework, we can determine which sets of arguments (*extensions*) are justified given the attacks [6]. There are many ways (*semantics*) to do this, each based on different intuitions of justification. We do not assume a specific semantics in this paper, only that all agents and the audience use the same semantics.

Definition 2. *Let AF be an argumentation framework. The set $\mathrm{Acc}(AF) \subseteq A$ is **the set of acceptable arguments of** AF, with respect to some argumentation semantics under credulous or sceptical inference. An argument a is said to be **acceptable** with respect to AF iff $a \in \mathrm{Acc}(AF)$.*

We model team persuasion as an instance of a *flag coordination game* over an argumentation framework [10]. A flag coordination game consists of a network of agents and an index representing discrete time. Each agent has a set of flags of different colors (representing e.g. choices or states) and a set of other agents it can see. In each round, each agent raises a colored flag synchronously and independently, as the output of some (possibly random) decision procedure given what the agent sees other agents doing in the preceding round. Such models have been studied in the context of the adoption of new technology standards, voting and achieving consensus [10, Section 1]. We now adopt a specific instance of a flag coordination game for our purposes.

Definition 3. *A **team persuasion framework** is a tuple $\langle AF, X, \beta, \Gamma, \phi, \Lambda \rangle =: \mathcal{F}$. Let $AF = \langle A, R \rangle$ be an argumentation framework, where the nodes represent arguments, each owned by distinct agents.[1] Let $\phi : A \to \mathcal{P}(A)$ be **the***

---

[1] As each *argument* is owned by a distinct *agent*, we use the terms interchangeably.

visibility function,[2] *i.e.* $\phi(a) \subseteq A$ *is the set of arguments that a can see. Let* $X := \{on, off, topic\}$ *denote* **the set of states.** *Let* $t \in A$ *be a distinguished argument called* **the topic (argument).** *Define* **the state function** $\beta : A \to X$ *such that* $\beta(t) := topic$ *and* $(\forall a \in A - \{t\}) \beta(a) \in \{on, off\}.$[3]

Let $\mathcal{S} := X^A$ *be the space of functions that assigns a state to each argument, which defines a* **configuration.** *Let* $\Gamma \subseteq \mathcal{S}$ *be* **the set of goal states.** *For* $a \in A$ *let* λ_a *be* **the decision algorithm of agent** *a, that takes input* β *and* ϕ *and outputs* $s(a) \in X$, *for* $s \in \mathcal{S}$. *We define* Λ *as the set of algorithms for all* $a \in A$.

The team persuasion framework is such that each agent asserts a single argument, which can attack and be attacked by other asserted arguments, so it is isomorphic to an argument framework. Each of the agents can assert their argument (turning it *on*) or not assert their argument (turning it *off*). The topic is a special argument that is labelled *topic* throughout the duration of the game.

Definition 4. *Let* $\mathcal{F}$ *denote a team persuasion framework. Let* $i \in \mathbb{N}$ *denote discrete time. Consider the sequence of configurations* $[s_0, s_1, ...]$, *indexed by* i. *We call* s_0 **the initial configuration,** *and* s_i *is the* i^{th} **configuration.** *The update rule is such that for all* $a \in A - \{t\}$, $s_{i+1}(a) \in X$ *is the output of* λ_a *given* $s_i(b) \in X$ *for all* $b \in \phi(a)$ *and possibly* $\beta(a)$. *Further,* $(\forall i \in \mathbb{N}) s_i(t) := topic$. **A team persuasion game with initial configuration** s_0 *is the tuple* $\langle \mathcal{F}, s_0 \rangle$.

Initially, the agents start in some initial configuration defined by whether each agent asserts its argument. In each subsequent round, the agents decide using their own decision procedure whether to assert or stop asserting their argument in the next round, given the actions of other agents they see.

Both teams are presenting their arguments to an audience who are assumed to be memoryless across rounds and can only see the topic and the arguments that are being currently presented. This prompts the following definition.

Definition 5. *Given a team persuasion game,* **the set of arguments that are on in round** i *is* $A_i^{on} := \{a \in A \,|\, s_i(a) = on\} \cup \{t\}$. *The* **induced argument framework** *is* $AF_i^{on} := \langle A_i^{on}, R_i^{on} \rangle$, *where* $R_i^{on} := R \cap [A_i^{on} \times A_i^{on}]$.

The audience will therefore see a sequence of argument frameworks $(AF_i^{on})_{i \in \mathbb{N}}$ as the teams debate each other about the topic. The audience can determine which team is winning based on whether the topic is justified in a given round.

Definition 6. *In a given round* $i \in \mathbb{N}$ *of a team persuasion game, we say that* **the team of defenders are winning** *iff* $t \in \text{Acc}(AF_i^{on})$ *iff* **the team of attackers are not winning.**

In each round the acceptability of the topic may change, and hence the winner can change. We are interested in definitively winning states, as defined in Q1 in Sect. 1. We explore the existence of such states in Sect. 3.

---

[2] If X is a set, then $\mathcal{P}(X)$ is its power set.

[3] We further assume that ϕ is such that if $b \in \phi(a)$ then a can also see $\beta(b) \in X$.

Since we are modelling the arguments between two teams, each trying to persuade or dissuade an audience of the topic, we specialise to bipartite argumentation frameworks because no agent should attack an argument of another agent in its own team. Further, the framework is weakly connected because all arguments asserted are relevant to the debate. Further, we assume that every argument has a counterargument, and that the topic is not capable of defending itself, so it does not directly attack any argument.

Definition 7. *Our team persuasion frameworks* $\mathcal{F} = \langle AF, X, \beta, \Gamma, \phi, \Lambda \rangle$ *have an underlying argument framework* $AF = \langle A, R \rangle$ *that is bipartite and weakly connected, with the requirements that* $(\forall a \in A)\, a^- \neq \varnothing$ *and* $t^+ = \varnothing$.[4] *As AF is bipartite, let U and W be the two partitions of A such that* $t \in U$. *We call* **U the set of defenders of the topic,** *and* **W the set of attackers of the topic.** *The* **set of goal states** *is* $\Gamma := \{\gamma_u, \gamma_w\}$, *where* $\gamma_u(U - \{t\}) = \{on\}$ *and* $\gamma_u(W) = \{off\}$, *and* $\gamma_w(U - \{t\}) = \{off\}$ *and* $\gamma_w(W) = \{on\}$.[5]

The goal states indicate that each team has the goal of unilaterally asserting their arguments and making the opposing team unilaterally withdraw their arguments. See Fig. 2 for an example of γ_u, and Fig. 3 for an example of γ_w. In our figures, white (resp. black) nodes are arguments that are *on* (resp. *off*).

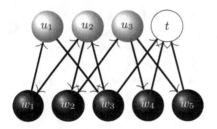

Fig. 2. The defenders' goal state γ_u; all defenders are asserting their argument.

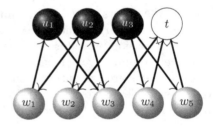

Fig. 3. The attackers' goal state γ_w; all attackers are asserting their argument.

2.1 Agent Visibility

There are several possible forms of the agents' visibilities, ϕ, for example:

V1 $(\forall a \in A)\, \phi(a) = a^- := \{b \in A \,|\, (b, a) \in R\}$,
V2 $(\forall a \in A)\, \phi(a) = a^+ := \{b \in A \,|\, (a, b) \in R\}$, or
V3 $(\forall a \in A)\, \phi(a) = a^- \cup a^+$ (both).

---

[4] Recall that for an AF $\langle A, R \rangle$ where $a, t \in A$, $a^- := \{b \in A \,|\, (b, a) \in R\}$, and $t^+ := \{b \in A \,|\, (t, b) \in R\}$.
[5] Recall that for a function $f : X \to Y$ and $A \subseteq X$ the *image set of A under f* is $f(A) := \{y \in Y \,|\, (\exists x \in A)\, f(x) = y\}$.

Recall from (see footnote 3) that if $b \in \phi(a)$, then a can also see the state $s(b)$ of b. It is possible to define $\phi(a) \subseteq A$ to be completely arbitrary, beyond the immediate neighbours of a. However, it is not currently clear how the behaviour of an agent might be influenced by knowledge of the states of arguments beyond the immediate neighbours especially if there is to be localised knowledge (see Sect. 3). In this paper, we focus on V1, leaving the other cases for future work.

2.2 The Agents' Decision Algorithm

We claim that agents with visibility of a^- can be motivated by two factors: their desire to make the topic acceptable/unacceptable to the audience (the goal of the team), and their desire not to have their argument publicly defeated (the goal of the individual). An individual does not want to have its argument publicly defeated (that is, its argument is asserted, but is not considered acceptable by the audience in the current round), as it is somehow a challenge to the agent's authority, and therefore reflects negatively on its ego. An agent can estimate how likely it is that their argument will be publicly defeated by considering how many attacking arguments the agent could see are being asserted: the more attackers that are asserted, the more likely one of the attacks will be successful, and therefore the higher the chance its argument is defeated.

- **Altruistic:** An agent which is only motivated by the team goal of making the topic (un)acceptable would always assert its argument a, regardless of the state of the arguments in a^-. We call such selfless agents *altruistic*.
- **Timid:** An agent which is only motivated by its individual goal of not having its argument being publicly defeated would never assert its argument, regardless of which arguments in a^- are being asserted. If the agent never asserts its argument, it can never be defeated, and therefore will always achieve its individual goal.
- **Balanced:** An agent motivated by both factors must find a way to balance these two goals. Such an agent is certain to assert its argument when none of its attackers are asserted, because the chance of a successful defeat is minimal. Similarly, the agent is least likely to assert when all of its attackers are asserted because the chance of successful defeat is maximised.

As a starting point for our analysis we will consider balanced agents. We define the probability of the agent not asserting its argument when all of its attackers are *on* as 1, and conversely the probability of the agent not asserting its argument when all of its attackers are *off* as 0. To begin with, we assume this probability increases linearly, proportional to the number of arguments a^- that are *on*.

Definition 8. *Let $\mathcal{F}$ be a team persuasion framework on an argument framework AF as defined in Definition 7. An agent $a \in A - \{t\}$ is **balanced** iff λ_a (Definition 3) is defined as follows. For $i \in \mathbb{N}$, λ_a outputs $s_{i+1}(a) = off$ with conditional probability*

$$\mathbb{P}\left(s_{i+1}(a) = \mathit{off}\,|\,s_i\right) := \frac{|a^- \cap A_i^{on}|}{|a^-|} \in [0,1]. \tag{1}$$

Further, λ_a *outputs* $s_{i+1}(a) = on$ *given* s_i *with probability* $1 - \mathbb{P}\left(s_{i+1}(a) = \mathit{off}\,|\,s_i\right)$. *We will assume that for all* $a \in A - \{t\}$, a *is balanced.*

Example 1. Consider Fig. 4, which represents the situation in Fig. 1 as a team persuasion framework with the initial configuration where the *on* arguments are u_2, u_3, w_2, and w_3, with the rest of the arguments being *off.* Consider the argument w_3. It is attacked by u_1 and u_2, which are respectively *off* and *on.* Therefore, the probability of w_3 remaining *on* in the next round is $\frac{1}{2}$.

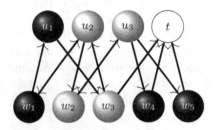

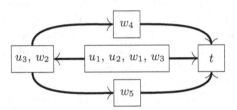

Fig. 4. An initial configuration $(\mathcal{F}, s_0)$ for the example in Fig. 1.

Fig. 5. Condensation graph of Fig. 4, showing strongly connected components.

3   Results

From the setup described in Sect. 2, we can now answer more precise versions of the two questions posed at the end of Sect. 1.

F1 Are there any states of the arguments (*on* or *off*) in which no agent is going to change their state in any future round according to λ_a as defined in Eq. 1? We call such a state a *state-stable configuration.*[6]

F2 What is the probability of a particular team winning, i.e. achieving a state-stable configuration, where the topic is either acceptable or unacceptable?

3.1   State-Stable Configurations

We now answer Question F1, which concerns state-stable configurations.

Definition 9. *A **state-stable configuration** is a function $s \in S$ such that, if attained at round $i \in \mathbb{N}$ of the team persuasion game following Eq. 1, will also be the state of the game in all subsequent rounds.*

---

[6] This is to avoid confusion with the notion of *stable semantics* [6].

This formalises the intuition that no agent is going to change their state in any future round once a state-stable configuration is reached.

Proposition 1. *Given the setup of Sect. 2, the two goal states, γ_u and γ_w (Definition 7) are the only state-stable configurations.*

Proof. Please refer to Appendix A for all proofs in this paper.

3.2   Probabilities for State-Stable Configurations

We now answer Question F2. We first translate our team persuasion game into a *consensus game*. In a consensus game, the update is such that in round $i+1$, every digraph node a *copies* the color of a randomly (uniformly) sampled neighbour in a^-, rather than adopting the opposite color as in Eq. 1 [10].

The translation is as follows. We consider the finite, weakly connected, bipartite digraph $G := \langle V, E \rangle$ which is the induced subgraph of $\langle A, R \rangle$ with nodes $V := A - \{t\}$. For each configuration $s : A - \{t\} \to X := \{on, off\}$ we define a *coloring function* $s' : V \to X' := \{0, 1\}$ such that

$$s'(a) := 1 \text{ if } [(a \in U \text{ and } s(a) = \text{on}) \text{ or } (a \in W \text{ and } s(a) = \text{off})]. \qquad (2)$$
$$s'(a) := 0 \text{ if } [(a \in U \text{ and } s(a) = \text{off}) \text{ or } (a \in W \text{ and } s(a) = \text{on})]. \qquad (3)$$

We intuitively associate the color 1 with the state on and similarly, 0 with off, but notice how this association is swapped for $a \in W$. Thus, the correspondence $s \mapsto s'$ is well-defined and bijective.

Example 2. Consider the digraph in Fig. 4.[7] Given this initial configuration s_0 such that $s_0 (u_1) = $ off, $s_0 (u_2) = $ on... etc. (see Example 1), we get a corresponding s' where $s' (\{u_2, u_3, w_1, w_4, w_5\}) = \{1\}$ and $s' (\{u_1, w_2, w_3\}) = \{0\}$, by (see footnote 5). If we arrange $V = \{u_1, \ldots, u_3, w_1, \ldots, w_5\}$, we can represent s'_0 as the boolean vector $(0, 1, 1, 1, 0, 0, 1, 1)$.

We now give some definitions and results for consensus games on digraphs.

Definition 10. *Let $G = \langle V, E \rangle$ be a finite digraph. Given some fixed order of the nodes $V = \{a_1, \ldots, a_{|V|}\}$,[8] the **(row-normalised) in-matrix** of G is the $|V| \times |V|$ matrix $H := (h_{ij})$, where*

$$\text{if } (v_j, v_i) \in E \text{ then } h_{ij} = \frac{1}{|v_i^-|}, \text{ else } h_{ij} = 0. \qquad (4)$$

The intuition of Eq. 4 is that the ith node $v_i \in V$ has a probability $h_{ij} > 0$ to copy the color of v_j when $(v_j, v_i) \in E$.

---

[7] Note that Fig. 5 will be relevant for a following proof.

[8] In the context of team persuasion games, we write all nodes in U first and then the nodes in W, as in Example 2.

Definition 11. *Let $G = \langle V, E \rangle$ be a digraph. Its **condensation** is the digraph $\langle \mathcal{K}, \mathcal{E} \rangle$ such that $\mathcal{K} \subseteq \mathcal{P}(V)$ is the set of strongly connected components (SCCs) of G and $(K, K') \in \mathcal{E} \subseteq \mathcal{K}^2$ iff $[(\exists a \in K)(\exists b \in K')(a, b) \in E$ and $K \neq K']$. A **source component** is a component with no in degree.*

Example 3. The condensation of Fig. 4 is Fig. 5. The only source component is $\{u_1, u_2, w_1, w_3\}$.

The following theorem answers Question F2 with an analytic expression of the probability of a particular team winning. We then apply this to solve our motivating example in Example 4. Intuitively, we first look at the condensation of a given bipartite AF. Since source components are not going to be influenced by any external argument, the probability of them reaching either one of the state-stable configurations is independent of the eventual state of the rest of the AF. Also, non-source components have no influence over the final outcome, since once the source components stabilise, they will be a constant influence in either defending or attacking the topic. Thus, we need all source SCCs to converge to the same state-stable configuration, otherwise a global state-stable configuration will not be reached. Finally, in order to calculate the probability of either the defender or the attacker to win in each source SCC, we find each individual agents' influence in the network.

Theorem 1. *Consider a team persuasion game on a bipartite $AF = \langle A, R \rangle$ with initial coloring s_0'. Let $\mathcal{K} = \{\{t\}, K_1, \ldots, K_m\}$ be the set of SCCs of AF (for some $m \in \mathbb{N}^+$), where $\{t\}$ is the component that contains only the topic argument. We also define **source**$_\mathcal{K} \subseteq \mathcal{K}$ as the set of source SCCs in the condensation of AF. Let $\mathcal{K}_{\{t\}} \subseteq$ **source**$_\mathcal{K}$ denote the set of SCCs for which there is a $\mathcal{E}$-path in the condensation of AF to $\{t\}$.*

Let μ_K be the stationary distribution of H_K, where H_K[9] is the in-matrix of the subgraph of AF induced by $K \in \mathcal{K}$ (Definition 10). Let $\mu(K) = \sum_{a \in K} \mu(a)$ for $K \subset A$. Finally, each set $K \in \mathcal{K}_{\{t\}}$ has a value g that stands for the greatest common divisor (gcd) of the lengths of all cycles in K. This generates a g-partite AF with partitions $\{K^1, \ldots, K^g\}$ as in Lemma 2 (Appendix A). We have that[10]

$$\mathbb{P}(\gamma_u \text{ is reached} \mid s_0) = \prod_{K \in \mathcal{K}_{\{t\}}} \prod_{i=1}^{g} \left(\frac{1}{\mu(K^i)} \sum_{a \in K^i} \mu_{K^i}(a) s_0'(a) \right). \tag{5}$$

Example 4. Consider the bipartite $AF = \langle V, E \rangle$ in Fig. 1 and s_0 as in Fig. 4. The condensation graph can be seen in Fig. 5, so $\mathcal{K} = \{\{t\}, K_1, K_2, K_3, K_4\}$, where $K_1 = \{u_1, u_2, w_1, w_3\}$, $K_2 = \{u_3, w_2\}$, $K_3 = \{w_4\}$ and $K_4 = \{w_5\}$. K_1 is the only source component. Since K_1 (indirectly) influences the acceptability of the

---

[9] Recall the row vector μ is the stationary distribution of H iff $\mu H = \mu$.

[10] We have abused notation here: we have considered γ_u to be a state configuration not on the entire AF, but just on the subgraph induced by the arguments that have a path to the topic. In other words, we exclude arguments that do not even indirectly influence the acceptability of the topic.

topic, we have $\mathcal{K}_{\{t\}} = \{K_1\}$. We now need to evaluate $\mu = \mu_{K_1}$, a stationary distribution of the in-matrix H_{K_1}, the induced subgraph of AF generated by K_1. Then, we have

$$\mu H_{K_1} = \mu \Leftrightarrow \mu \begin{pmatrix} 0 & 0 & 1 & 0 \\ 0 & 0 & \frac{1}{2} & \frac{1}{2} \\ 0 & 1 & 0 & 0 \\ \frac{1}{2} & \frac{1}{2} & 0 & 0 \end{pmatrix} = \mu \Rightarrow \mu = \frac{1}{10}(1, 4, 3, 2). \tag{6}$$

Note that $g = 2$. We now use the initial configuration s_0 and the translation to s_0' according to Eqs. 2 and 3. We have $s_0'(u_1) = 0$, $s_0'(u_2) = 1$, $s_0'(w_1) = 1$, $s_0'(w_3) = 0$, therefore, by Theorem 1, we have

$$\mathbb{P}(\gamma_u \text{ is reached} \mid s_0) = \prod_{K \in \mathcal{K}_{\{t\}}} \prod_{i=1}^{g} \left(\frac{1}{\mu(K^i)} \sum_{a \in K^i} \mu_{K^i}(a) s_0'(a) \right)$$

$$= \left(\frac{1}{\mu(K_1^1)}(4) \right) \left(\frac{1}{\mu(K_1^2)}(3) \right) = \frac{12}{25} = 48\%. \tag{7}$$

Therefore, the probability of the topic being accepted is 48%. Analogously, the probability of the topic being rejected is given by

$$\mathbb{P}(\gamma_w \text{ is reached} \mid s_0) = \left(\frac{1}{\mu(K_1^1)}(1) \right) \left(\frac{1}{\mu(K_1^2)}(2) \right) = \frac{2}{25} = 8\%. \tag{8}$$

The probability for this game not reaching a state-stable configuration is 44%.

4   Related Work

In this paper we have presented and analysed an argumentation model for a very common form of public debate. Our work has made two novel contributions. The first contribution is the formalisation using argumentation frameworks of public policy debates where multiple parties with only local information propose arguments to support (or attack) claims of interest to a wider audience, seeking to persuade that audience of a claim (or not, as the case may be). The second contribution is the use of flag coordination games, specifically its analysis of the dynamics of graph coloring, to understand the properties of this formal framework. Analogues of graph coloring have been used in argumentation, for example, in labelling semantics to determine acceptability of arguments [3]. However, to the best of our knowledge, interpreting such colorings as the argument having been asserted or not, and the dynamics of how such a coloring changes, have not previously been used in argumentation theory.

The general problem of two parties with contradictory viewpoints, each seeking to persuade an impartial third party of their viewpoint, has been investigated in economics, e.g. using game theory [13,14] or mechanism design [7,8]. Applying argumentation theory to study multi-agent persuasion with two teams, in which one is arguing for the acceptability of a topic and the other against, has been investigated in the work by Bonzon and Maudet [2]. They focus specifically on the problem with respect to the kinds of dialogue that occur on social websites, specifying that agents "vote" on the attack relations between arguments. One of the main differences between their work and ours is that they assume that each agent has a total view of the argumentation framework, where as we assume agents have a specific area of expertise and thus, in general, do not have complete knowledge of the structure of the argumentation framework. Furthermore, agents in their formulation do not have any motivation to act in a way that might be detrimental to their team's goal, whereas agents in our work may also be motivated by their own individual goals.

Dignum and Vreeswijk developed a testbed that allows an unrestricted number of agents to take part in an inquiry dialogue [5]. The focus of their work is on the practicalities of conducting a multi-party dialogue, concerned with issues like turn-taking, rather than in the strategising of agents participating in such a dialogue. Bodanza et al. [1] survey work on how multiple argumentation frameworks may be aggregated into a single framework. While this direction of work considers how frameworks from multiple agents might be merged, it removes the strategic aspect of persuasion which we are interested in here.

5 Conclusion and Future Work

We have shown how to determine the probability of each team winning in a team persuasion game (Theorem 1). However, we have shown that not all games become state-stable (Appendix A, Theorem 1), having no definite winner. Considering games which do not become state-stable, we are interested in determining (1) in what proportion of rounds is the topic acceptable, and (2) what is the probability the topic being acceptable at a specific round in the future. With respect to the first question, we might determine the winning team to be the one who makes the topic acceptable/unacceptable in the majority of rounds. The second question is particularly interesting in the context of referendum-like domains, in which there is a set date (round n) in which the audience determines whether the topic is acceptable (and thus which team wins): in this case it does not matter whether there is state-stability, only that the topic is acceptable in round n.

Future work will apply the techniques of [10] to the situation investigated in this paper. Specifically, if the team persuasion game will reach a goal state, we can calculate the expected number of rounds until that happens [10, Proposition 4]. Further, we can study the game-theoretic implications of some knowledgeable external agent "bribing" a specific agent to either assert or stop asserting its argument [10, Section 3]. We will also investigate different generalisations of the team persuasion game. There are various assumptions on the digraph that we can modify. For example, generalising from bipartite to multipartite argumentation frameworks where many teams seek to persuade the audience. Additionally, we can lift the assumption that no agent attacks its fellow agents of the same team. Such a team seems quite unlikely (and thus is not considered here), but occasionally this may occur, e.g. a campaigner who wishes to leave the EU because their environmental laws are too restrictive on UK businesses, and a campaigner who wishes to leave the EU because they do not have strong enough environmental laws; both campaigners would be on the same team, but their arguments are seemingly in conflict. Further generalisations include: consideration of the different visibility functions ϕ for each agent, or the more realistic case where each agent can assert more than one argument, or when each agent has a non-linear version of Eq. 1, or the consideration of heterogeneous agents in the same framework that can also be altruistic or timid. We will show that the results also apply to the case when the attacking arguments are weighted differently by agents in Eq. 1, which we will articulate in future work. Ultimately, we hope such generalisations can give insight into situations where individual goals and societal goals conflict to a greater extent, and how this conflict can be resolved.

A   Lemmas and Proofs

Proof. (of Proposition 1). To show that γ_u is a state stable configuration, notice that in round $i \in \mathbb{N}$, if γ_u is attained, then for $a \in U - \{t\}$, the probability (Eq. 1) a will be off in round $i + 1$ is zero, because $a^- \subseteq W$ and all attackers of a are off. Therefore, a will still be on in round $i + 1$. Similarly, we can show that the probability of being off for all $b \in W$ in round $i + 1$ is one. Therefore, in round $i + 1$, the state is still γ_u. A similar argument to this proves that if γ_w is attained in round i, then it will also be the state for round $i + 1$. By induction over i, γ_u and γ_w satisfy Definition 9.

We now show that both γ_u and γ_w are the only state stable configurations. Assuming the contrary. Then, we have a configuration different from γ_u and γ_w in which no argument has a positive probability of changing their state. In this case, we would have two nodes, say u_1 and u_2, in the same partition, say U, that have different colors (otherwise we have γ_u and γ_w). Since G is weakly connected, there is a path that ignores edges' directions from u_1 to u_2. This path has even length and, therefore, since u_1 to u_2 are different, there must be at least two consecutive nodes in this path with the same color. One it attacking the other, therefore, the attacked one has a positive probability of changing their color. We have a contradiction. Thus γ_u and γ_w must be the only state-stable configurations in a bipartite AF.   ∎

We now answer a more general version of Question F2 using the framework of consensus games and colors. We derive a formula for a color to win the consensus game on a strongly connected digraph, given that consensus will be achieved. We then investigate the necessary conditions for consensus to be achieved, and derive an expression for the probability of failing to achieve consensus that depends on s_0'. We then generalise to the case of weakly connected graphs, and answer Question F2 via our translation back into team persuasion games.

The in-matrix H of the digraph G can be seen as a transition matrix of a time homogeneous Markov chain, where the each node v represents a state and the reversed edges represent the transitions. If the Markov chain is irreducible and finite, there is a unique *stationary distribution*, which is a row vector $\mu \in \mathbb{R}_+^V$ that satisfies $\mu H = \mu$.

Proof. (of Theorem 1). The theorem follows from the following lemmas. Note that these lemmas are considering a general digraph $G = (V, E)$ and colors 0 and 1. We also denote **0** and **1** as the consensus on color 0 and 1 respectively.

Lemma 1. *A consensus game on a strongly-connected digraph* $G = \langle V, E \rangle$ *reaches consensus with probability for all initial configurations* 1 *iff* $\gcd C = 1$, *where* $C \subseteq \mathbb{N}$ *is the set of the lengths of all cycles in* G. *In the case* $\gcd C = g > 1$, *then* G *is* g-*partite with parts* $V_1, \dots, V_g$ *where all edges go from* V_i *to* V_{i+1}.

Proof. (of Lemma 1). ($\Leftarrow$) Assuming $\gcd C = 1$. Then, given an initial configuration, a game has already reached consensus or it has not. If not, we can assume, WLOG, that there is at least one $v \in V$ colored 0. We note that the $\gcd C_v = 1$, where C_v is the set of the lengths of the cycles passing through v. This follows from the fact that G is strongly connected. We can then show that there is a large enough $n_0 > 0$ such that for any $n \geq n_0$, we have $\mathbb{P}(s_n(u) = 0 \mid s_0) > 0$ for all $u \in V$. For that it is enough to show that there is finite n_0, such that for every $n \geq n_0$ there is a directed path from v to u of length n. The existence of such n_0 follows from Lemma 2.1 of [12]. Thus, if the game runs long enough, it will reach consensus (either **0** or **1**) with probability 1.

($\Rightarrow$) We now want to prove that if the game reaches consensus with probability 1, then $\gcd C = 1$. We are going to prove this by showing that if $\gcd C > 1$, then there is a positive chance that the game never reaches consensus. Let $\gcd C = g > 1$. We start by showing that the graph must be not only a g-partite graph, but also of the form that every edge from a node in partition i points to a node in partition $i + 1 \pmod{g}$. Let $v \in V$. For all $w \in V$, we define the partition that w belongs to by taking the $x \pmod{g}$, where x is the length of any path from v to w.

We show that this is well defined. First, the existence of such a path is guaranteed by the strongly connectivity of G. Also, the lengths of all paths from v to w must coincide modulo g. If not, by concatenating both paths to the same returning path from w to v, we would have created two cycles from v to v that differ in length modulo g (by assumption, all cycles must be $0 \pmod{g}$).

We now observe that, if the game reaches a configuration in which a partition is all 0 and another all 1, consensus will never be reached. Thus it can not be reaching consensus for sure from all possible initial configurations. We will show that no non-consensual initial configuration reaches consensus with probability 1. ∎

Lemma 2. *Consider a consensus game in a strongly connected and direct graph* $G = \langle V, E \rangle$ *in which* $\gcd C = g$. *Then, we know by Lemma 1 that* G *is* g-*partite*[11]

---

[11] By 1-partite, we mean $\gcd C = 1$ and $V_1 = V$.

and we denote the partitions $V_1, \ldots, V_g$. *We further denote* $\mu(U) = \sum_{v \in U} \mu(v)$
for $U \subset V$. *In these conditions,*

$$\mathbb{P}(\text{Colour } 1 \text{ wins in } G \mid s_0) = \prod_{i=1}^{g} \left(\frac{1}{\mu(V_i)} \sum_{v \in V_i} \mu(v)s_0(v) \right) \qquad (9)$$

Proof. (of Lemma 2). We use a similar approach to the one in [10] and apply
Theorem 1 of [4]. Note that the state of vertices of V_{i+1} in the round $n + 1$,
depends only in the state of vertices of V_i in the round n. We can then consider
g parallel consensus games on g copies of G, where in the i-th consensus game
we set the initial state of the vertices in V_i to their original initial state in the
consensus game, but set the state of all other vertices to 1. Denote by p_i the
probability of the i-th consensus game reaching a 1 winning state. It is then easy
to see that $\mathbb{P}(1 \text{ wins in } G \mid s_0) = \prod_{i=1}^{g} p_i$.

We are left to show that $p_i = \dfrac{1}{\mu(V_i)} \sum_{v \in V_i} \mu(v)s_0(v)$. For that end, over
the i-th consensus game define the random variable $X_n = \sum_{v \in V_j} \mu(v)s_n(v)$,
where $j = n + i - 1 (\mathrm{mod}\, g)$. We show that the process $(X_n)_{n \in \mathbb{N}}$ is a martingale
with respect to the sequence s_n. We need to show that $\mathbb{E}(X_{n+1} \mid s_n) = X_n$.
By linearity of expectation $\mathbb{E}(X_{n+1} \mid s_n) = \sum_{v \in V_{j+1}} \mu(v)\mathbb{E}(s_{n+1}(v) \mid s_n)$. Note
that $\mathbb{E}(s_{n+1}(v) \mid s_n) = \sum_{u \in V_j} h_{vu}s_n(u)$ and by changing the order of summa-
tion we get that: $\mathbb{E}(X_{n+1} \mid s_n) = \sum_{u \in V_j} s_n(u) \sum_{v \in V_{j+1}} \mu(v)h_{vu}$. Due to station-
arity of μ and the fact that h_{vu} is non-zero only for $v \in V_{j+1}$, we have that
$\sum_{v \in V_{j+1}} \mu(v)h_{vu} = \mu(u)$, which implies that $\mathbb{E}(X_{n+1} \mid s_n) = X_n$.

Now, it is easy to see that $\mu(V_i)p_i = \mathbb{E}(X_\infty \mid X_0) = \mathbb{E}(X_0)$ and this proves
that $p_i = \dfrac{1}{\mu(V_i)} \sum_{v \in V_i} \mu(v)s_0(v)$, which concludes the result. ∎

Lemma 3. *Consider a consensus game played in a weekly connected digraph*
$G = \langle V, E \rangle$ *and let* $\mathcal{K} = \{K_1, \ldots, K_n\}$ *be the set of strongly connected compo-*
nents (SCC) of G. *We define* **source**$_\mathcal{K}$ *as the set of SCCs that have no attack*
coming from the outside, i.e., if $K \in \mathcal{K}$, *then* $K \in$ **source**$_\mathcal{K}$ *if for every* $(a, b) \in E$
such that $b \in K$, *we have* $a \in K$. *Then,*

$$\mathbb{P}(\text{Colour } 1 \text{ wins in } G \mid s_0) = \prod_{K \in \text{source}_\mathcal{K}} \mathbb{P}(\text{Colour } 1 \text{ wins in } K \mid s_0) \qquad (10)$$

Proof. (of Lemma 3). First note that each $K \in$ **source**$_\mathcal{K}$ is independent of each
other, since they are independent from anything outside each of these SCCs.
Then, we cannot have consensus if they reach different consensus. It remains
now to observe that, in the case they reach same consensus colors, then all the
other SCCs will eventually stabilise in the same color. That happens because of
the influence they receive from components in **source**$_\mathcal{K}$, so consensus cannot be
achieved by any other color. Finally, for every node not in a source component,
there is a path from a source node to it, therefore there is a non-zero probability
that the game achieves the sources' color. ∎

References

1. Bodanza, G., Tohme, F., Auday, M.: Collective argumentation: a survey of aggregation issues around argumentation frameworks. J. Argum. Comput. **8**(1), 1–34 (2016)
2. Bonzon, E., Maudet, N.: On the outcomes of multiparty persuasion. In: McBurney, P., Parsons, S., Rahwan, I. (eds.) ArgMAS 2011. LNCS (LNAI), vol. 7543, pp. 86–101. Springer, Heidelberg (2012). https://doi.org/10.1007/978-3-642-33152-7_6
3. Caminada, M.: On the issue of reinstatement in argumentation. Log. Artif. Intell. **4160**, 111–123 (2006)
4. Cooper, C., Rivera, N.: The linear voting model: consensus and duality (2016)
5. Dignum, F.P.M., Vreeswijk, G.A.W.: Towards a testbed for multi-party dialogues. In: Dignum, F. (ed.) ACL 2003. LNCS (LNAI), vol. 2922, pp. 212–230. Springer, Heidelberg (2004). https://doi.org/10.1007/978-3-540-24608-4_13
6. Dung, P.M.: On the acceptability of arguments and its fundamental role in non-monotonic reasoning, logic programming and n-person games. Artif. Intell. **77**, 321–357 (1995)
7. Glazer, J., Rubinstein, A.: Debates and decisions: on a rationale of argumentation rules. Games Econ. Behav. **36**(2), 158–173 (2001)
8. Glazer, J., Rubinstein, A.: On optimal rules of persuasion. Econometrica **72**(6), 1715–1736 (2004)
9. Hunter, A.: Toward higher impact argumentation. In: Proceedings of the The 19th American National Conference on Artificial Intelligence, pp. 275–280. MIT Press (2004)
10. Kohan Marzagão, D., Rivera, N., Cooper, C., McBurney, P., Steinhöfel, K.: Multi-agent flag coordination games. In: Proceedings of the 16th International Conference on Autonomous Agents & Multiagent Systems, pp. 1442–1450. International Foundation for Autonomous Agents and Multiagent Systems (2017)
11. Prakken, H.: Formal systems for persuasion dialogue. Knowl. Eng. Rev. **21**(02), 163–188 (2006)
12. Rosales, J.C., García-Sánchez, P.A.: Numerical Semigroups, vol. 20. Springer Science & Business Media, New York (2009). https://doi.org/10.1007/978-1-4419-0160-6
13. Shin, H.: The burden of proof in a game of persuasion. J. Econ. Theory **64**, 253–264 (1994)
14. Shin, H.: Adversarial and inquisitorial procedures in arbitration. RAND J. Econ. **29**, 378–405 (1998)
15. Thimm, M.: Strategic argumentation in multi-agent systems. Künstl. Intell. **28**, 159–168 (2014)

Towards a General Framework for Dialogues That Accommodate Reasoning About Preferences

Sanjay Modgil[(⊠)]

Department of Informatics, King's College London, London, UK
sanjay.modgil@kcl.ac.uk

Abstract. Argumentation theory provides foundations for distributed non-monotonic reasoning in the form of inter-agent dialogues. However current dialogue models do not accommodate reasoning about possibly conflicting preferences used in arbitrating amongst attacking arguments. We provide a framework for persuasion dialogues that accommodates such reasoning. Agents exchange locutions that implicitly define an $ASPIC^+$ theory consisting of rules and premises. The theory's defined arguments instantiate an extended argumentation framework (EAF) that accommodates arguments claiming preferences over other arguments, so that evaluation of the EAF's justified arguments determines the outcome of the dialogue. We also evaluate the outcome of a dialogue based on the dialectical status of moves in the dialogue, propose restrictions on dialogue moves and conjecture correspondences between the two outcome definitions.

Keywords: Argumentation · Dialogue · Preferences · $ASPIC^+$

1 Introduction

In Dung's theory of argumentation [8], arguments and attacks are defined by a belief base ($\mathcal{B}$) of logical formulae. An argument X may then be said to successfully attack (defeat) Y if Y is not strictly preferred to X (assuming a given strict ordering $\prec$ over arguments [1,3,18]). Preferences can thus be used to arbitrate amongst attacking arguments. The claims of justified arguments in the Dung framework (AF) of arguments related by defeats, identify the non-monotonic inferences from $\mathcal{B}$, where these claims may correspond to non-monotonic inference relations defined directly over $\mathcal{B}$.

The dialectical characterisation of non-monotonic inference paves the way for distributed non-monotonic reasoning in the form of argumentation-based dialogues in which agents persuade interlocutors as to the truth of a claimed belief or deliberate over a choice of action (see [13] for a review). Dialogue protocols sanction when locutions are legal replies to other locutions. At any stage in a dialogue, an outcome in favour of a topic (e.g., the claimed belief or proposed action) can be affirmed if the topic is non-monotonically inferred from the

© Springer International Publishing AG, part of Springer Nature 2018
E. Black et al. (Eds.): TAFA 2017, LNAI 10757, pp. 175–191, 2018.
https://doi.org/10.1007/978-3-319-75553-3_13

contents of exchanged locutions (e.g., [7,10]) or the claim of a justified argument in the AF incrementally constructed from the contents of locutions (e.g., [9,19]). However current formalisms assume a fixed exogenously given preference relation over arguments (which in turn may be based on preferences over the arguments' constituents.) that is assumed to be agreed upon by the agents. Agents cannot therefore justify, reason about, and resolve conflicts amongst preferences, so limiting the range of applicability of these dialogue models.

The main contribution of this paper is a framework for formalising persuasion dialogues accommodating argumentation based reasoning about possibly conflicting preference information; information that is now part of the domain of discourse. We focus on persuasion dialogues as these are often embedded in dialogues of other types. Locutions define an $ASPIC^+$ argumentation theory [18,21], whose defined arguments are subsequently evaluated in *Extended Argumentation Frameworks (EAFs)* [15] which extend AFs to include arguments claiming preferences over other arguments, rather than assume a single exogenously given preference ordering. We choose $ASPIC^+$, as this framework for structured argumentation is shown in [18,21,24] to capture a range of argumentation formalisms (e.g., [4,11,22]) and non-monotonic logics (e.g., [5,6]), so bestowing a considerable degree of generality to our dialogical framework.

In Sect. 2 we review the $ASPIC^+$ framework, EAFs and the instantiation of EAFs by $ASPIC^+$ arguments. We modify $ASPIC^+$ so as to accommodate dialogue protocols that have a 'public semantics' in that no reference to the contents of participating agents' beliefs bases (argumentation theories) is made. Rather, it is the contents of locutions that incrementally define an argumentation theory. Section 3 then presents our main contributions. Firstly, we define a protocol that regulates use of some typical dialogue locutions, *as well as* locutions that include arguments claiming preferences over other arguments. A key challenge is to accommodate the ubiquitous use of 'why' locutions in dialogues. For example, an agent submits 'α since β' (A) and then when questioned "why β", submits 'β since γ' (B), where B 'backward extends' A to define the argument $A' = $ 'α since β and β since γ'. Such backward extensions usually limit the types of preferences assumed in dialogues in which counter-arguments are required to defeat their targets. For example (assuming a given fixed $\prec$), if A were moved as a defeat on C given that $A \not\prec C$, then one must assume that A is not weakened when backward extended to define A' (e.g., see [19]; note that this assumption precludes use of the *weakest link* principle for evaluating the strength of arguments) as it may then be that $A' \prec C$, and so the legality of moving A as a defeat on C is negated. However, we will see that this problem does not arise when agents are able to reason and argue about preferences as part of the dialogue. Secondly, we define how the outcome of a dialogue is determined. The $ASPIC^+$ arguments defined by the contents of exchanged locutions are evaluated in an EAF. If the dialogue topic is the claim of a justified argument, then the proponent of the topic is said to be winning the dialogue. We additionally formalise an approach taken in [9,19], whereby an outcome in favour of a topic is affirmed by reference to the 'dialectical status' of moves in the tree of locutions generated by the dialogue. We then propose restrictions on moves, adapting those used in argument game proof theories for establishing membership of an argument in a preferred extension of an AF [16], and conjecture a correspondence between the dialectical status of moves made under these restrictions and the justified status of arguments in the EAF defined by the dialogue. We conclude in Sect. 4, pointing to directions for future research.

2   Background

$ASPIC^+$ arguments are inference trees constructed from an agent's 'axiom' and 'ordinary' premises, and defeasible and strict inference rules. Only the fallible ordinary premises and fallible consequents of defeasible rules can be attacked (axiom premises are infallible). For example, (informally) an argument concluding γ constructed from a premise α by chaining the inference rules β if α and then γ if β. However, in this paper we are interested in distributed agents exchanging $ASPIC^+$ arguments that are defined without explicit reference to the premises and rules of these agents; rather the contents of locutions incrementally define the premises and rules from which arguments are constructed and evaluated to determine who is currently 'winning' the dialogue. An agent might thus move an 'incomplete' argument α since β, where β is not a premise in the agent's knowledge base. Only on being challenged as to why β is the case, might the agent then backward extend his initial argument by moving β since γ. Hence, in what follows we define arguments without reference to a specific agent's belief base (premises and rules), and such that we refer to the leaves of an $ASPIC^+$ inference tree simply as 'leaves' and not as 'premises'.

All agents are assumed to share: (1) a language $\mathcal{L}$ (lower case greek letters will refer to arbitrary formulae in $\mathcal{L}$); (2) a naming function for defeasible rules that allows agents to undercut attack an argument on a defeasible rule, and; (3) a function $^-$ that generalises negation, and specifies the set of wffs in conflict with any $\psi \in \mathcal{L}$. Formally:

Definition 1. $^-$ *is a function from $\mathcal{L}$ to $2^{\mathcal{L}}$, such that: φ is a* **contrary** *of ψ if $\varphi \in \overline{\psi}$, $\psi \notin \overline{\varphi}$; φ is a* **contradictory** *of ψ (denoted '$\varphi = -\psi$'), if $\varphi \in \overline{\psi}$, $\psi \in \overline{\varphi}$.*

Definition 2. *We assume the* **universal** *argumentation system* $(\mathcal{L}, \mathcal{R}, n, ^-)$ *where $\mathcal{R} = \mathcal{R}_s \cup \mathcal{R}_d$ is a set of strict ($\mathcal{R}_s$) and defeasible ($\mathcal{R}_d$) inference rules which are respectively of the form:*

$$\varphi_1, \ldots, \varphi_n \rightarrow \varphi \text{ and } \varphi_1, \ldots, \varphi_n \Rightarrow \varphi$$

(where φ_i, φ are meta-variables ranging over wff in $\mathcal{L}$), and $\mathcal{R}_s \cap \mathcal{R}_d = \emptyset$, and n is a partial function such that $n : \mathcal{R}_d \longrightarrow \mathcal{L}$.

We assume a set of agents $\{Ag_1, \ldots, Ag_n\}$ where each agent is equipped with an *argumentation theory* (AS_i, KB_i) consisting of an argumentation system $AS_i = (\mathcal{L}, \mathcal{R}_i, n, ^-)$, $\mathcal{R}_i \subseteq \mathcal{R}$ and a knowledge base $KB_i \subseteq \{\alpha | \alpha \in \mathcal{L}\}$ consisting of disjoints sets of ordinary (KB_i^p) and axiom premises (KB_i^n).

We now define $ASPIC^+$ arguments as in [18], but without reference to a given set of inference rules and premises in an argumentation theory. Hence, unlike [18] we do not refer to the leaves of an argument as premises, and nodes are labelled f for fallible and if for infallible. Intuitively, a leaf node formula labelled f indicates that either the formula is an ordinary premise or inferred using a defeasible rule in the agent's theory, and if indicates either an axiom premise or inferred using a strict rule in the agent's theory. A non-leaf formula labelled f (if) indicates that the formula is inferred from a defeasible (strict) inference rule whose antecedents are the children of the non-leaf node.

Definition 3. *An* **argument** *A is either:*
(1) a single node $\phi \in \mathcal{L}$, labelled f *or* if, *in which case A is said to be* elementary*:*
$\texttt{Leaves}(A) = \{\phi\}$; $\texttt{Conc}(A) = \phi$; $\texttt{DefRules}(A) = \texttt{StRules}(A) = \emptyset$; $\texttt{Sub}(A) = \{\phi\}$, *or:*
(2) a tree of nodes (in which case A is said to be complex*) with root node* $\texttt{Conc}(A) = \phi$, *child nodes $\phi_1, \ldots, \phi_n$ of ϕ, where each ϕ, $\phi_{i=1\ldots n}$ is labelled* f *or* if, *and for $i = 1 \ldots n$, ϕ_i is the root node $\texttt{Conc}(A_i)$ of an argument A_i. We say that:*

$\texttt{Leaves}(A) = \texttt{Leaves}(A_1) \cup \ldots \cup \texttt{Leaves}(A_n)$; $\texttt{Sub}(A) = \texttt{Sub}(A_1) \cup \ldots \cup \texttt{Sub}(A_n) \cup \{A\}$; $\texttt{DefRules}(A) = \texttt{DefRules}(A_1) \cup \ldots \cup \texttt{DefRules}(A_n) \cup \{\phi_1, \ldots, \phi_n \Rightarrow \phi\}$ *if ϕ is labelled* f. $\texttt{StRules}(A) = \texttt{StRules}(A_1) \cup \ldots \cup \texttt{StRules}(A_n) \cup \{\phi_1, \ldots, \phi_n \rightarrow \phi\}$ *if ϕ is labelled* if.
Finally, for any argument A, $\texttt{Concs}(A) = \{\texttt{Conc}(A') | A' \in \texttt{Sub}(A)\}$ denotes the set of all nodes in the argument A.

Note that in [18], arguments are defined as above, but with reference to an argumentation theory (AS, KB), so that in (1) an argument is a ϕ that is an ordinary or axiom premise in KB, and in (2) any rule in an argument must be in $\mathcal{R}$ in AS, and the notation $\texttt{Prem}(A)$ replaces $\texttt{Leaves}(A)$.

We define here arguments extending an argument A on leaf nodes, to define A'.

Definition 4. *Let $\texttt{Leaves}(A) = \{\phi_1, \ldots, \phi_n\}$, and let A' be the argument A where for each $\phi_j \in \{\phi_i, \ldots, \phi_k\} \subseteq \{\phi_1, \ldots, \phi_n\}$, ϕ_j is replaced by a complex argument A''_j such that $\texttt{Conc}(A''_j) = \phi_j$. Then A'* **extends** *A on $\phi_i, \ldots, \phi_k$ with $A''_i, \ldots, A''_k$.*

$ASPIC^+$ attacks include undercuts on applications of defeasible rules and rebut attacks on the conclusions of defeasible rules or undermine attacks on ordinary premises. Since in this paper the leaves of an argument exchanged in a dialogue are not necessarily premises, we group rebut and undermining attacks under the term 'formula attacks'.

Definition 5. *A* **attacks** *B* **on** *B' iff A undercuts or (contrary) formula attacks B on B', where:*

- *A* **undercuts** *B (on B') iff $\texttt{Conc}(A) \in \overline{n(r)}$ for some $B' \in \texttt{Sub}(B)$ such that B''s top rule r is defeasible.*

– A **formula attacks** B on B' iff $\text{Conc}(A)$ *is a contradictory of* φ *for some* $B' \in \text{Sub}(B)$ *such that* $\text{Conc}(B') = \varphi$, *and* φ *is labelled* f. A **contrary formula attacks** B on B' iff $\text{Conc}(A)$ *is a contrary of* φ *for some* $B' \in \text{Sub}(B)$ *such that* $\text{Conc}(B') = \varphi$, *and* φ *is labelled* f.

Given a strict preference relation $\prec$ over arguments, one can determine the success of the 'preference-dependent' formula attacks (as defeats). Undercuts and contrary formula attacks succeed as defeats independently of preferences (see [18]).

Definition 6. A *defeats* B if A *undercuts, or contrary formula attacks* B, *or* A *formula attacks* B *on* B' *and* $A \nprec B'$.

Example 1. Figure 1(i) shows $ASPIC^+$ arguments A and A'' exchanged in a dialogue, where A' extends A on β with A''. B formula attacks A' on A''. Note the argument B in which ϵ is labelled *if*, indicating that ϵ is infallible.

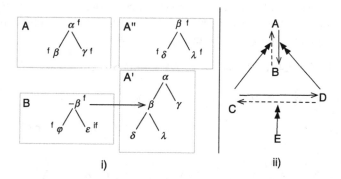

Fig. 1. (i) A' extends A on β with A''. B formula attacks A'; (ii) EAF for weather example; dashed arrows are attacks invalidated by preference arguments.

Extended Argumentation Frameworks (EAFs). [15] extend AFs to include arguments that express preferences over other arguments, thus providing for instantiation by formalisms that accommodate reasoning about possibly conflicting preference information. For example, consider the following dialogue between agents P and O:

P_1 "Today will be dry in London since CNN forecast sunshine" $= A$
O_2 "Today will be wet in London since BBC forecast rain" $= B$
P_3 "But CNN are statistically more accurate than the BBC" $= C$
O_4 "However the BBC are more trustworthy than CNN" $= D$
P_5 "But statistics basis for comparison than your instincts about their relative trustworthiness" $= E$.

A and B attack each other since they express contradictory conclusions. C is an argument justifying the preference $B \prec A$, and so attacks (invalidates) the attack from B to A. Similarly, D attacks the attack from A to B. Since C and D express contradictory preferences, they attack each other. However E justifies a preference for C over D and so attacks the attack from D to C. Hence C and so A (at the expense of B) is justified.

Definition 7. *An* Extended Argumentation Framework *(EAF) is a tuple $(\mathcal{A}, \mathcal{C}, \mathcal{D})$ such that $\mathcal{A}$ is a set of arguments, $\mathcal{C} \subseteq (\mathcal{A} \times \mathcal{A})$ is the attack relation $(A \to B$ denotes $(A,B) \in \mathcal{C})$, and:*

- *$\mathcal{D} \subseteq (\mathcal{A} \times \mathcal{C})$ $(C \twoheadrightarrow (A \to B)$ denotes $(C,(A,B)) \in \mathcal{D})$*
- *If $(X,(Y,Z))$, $(X',(Z,Y)) \in \mathcal{D}$ then (X,X'), $(X',X) \in \mathcal{C}$*

$S \subseteq \mathcal{A}$ is conflict free iff $\forall A, B \in S$, if $(A,B) \in \mathcal{R}$ then $(B,A) \notin \mathcal{R}$ and $\exists C \in S$ s.t. $(C,(A,B)) \in \mathcal{D}$. Defeats are parameterised by a set of arguments S: if A attacks B then A **defeats** B **w.r.t.** S if there is no argument C in S that claims a preference for B over A. An argument A is then acceptable w.r.t. a set S if every argument B defeating A (w.r.t. S) is defeated (w.r.t. S) by some $C \in S$ and there is a 'reinstatement set' for this latter defeat. The extensions of an *EAF* are then defined as for Dung frameworks:

Definition 8. *Let $(\mathcal{A}, \mathcal{C}, \mathcal{D})$ be an (EAF), and $S \subseteq \mathcal{A}$.*

- *A defeats B w.r.t. S (denoted $A \to^S B$) iff $(A,B) \in \mathcal{C}$ and $\neg\exists C \in S$ s.t. $(C,(A,B)) \in \mathcal{D}$.*
- *$R_S = \{X_1 \to^S Y_1, \ldots, X_n \to^S Y_n\}$ is a reinstatement set for $C \to^S B$, iff:*
  1. *$C \to^S B \in R_S$*
  2. *for $i = 1 \ldots n$, $X_i \in S$*
  3. *$\forall X \to^S Y \in R_S, \forall Y'$ s.t. $(Y',(X,Y)) \in \mathcal{D}$, there is a $X' \to^S Y'$ in R_S*
- *A is acceptable w.r.t. S iff $\forall B$ s.t. $B \to^S A$, there is a C in S s.t. $C \to^S B$ and there is a reinstatement set for $C \to^S B$.*

Let S be conflict free. Then S is: an admissible extension iff every argument in S is acceptable w.r.t. S; complete iff admissible and each argument which is acceptable w.r.t. S is in S; preferred iff a set inclusion maximal complete extension; stable iff $\forall B \notin S$, $\exists A \in S$ such that $A \to^S B$

In Example 1, $\{E, C, A\}$ is the single complete extension. For arbitrary finitary[1] *EAF*s, the grounded extension is defined by the fixed point reached by iteration of an *EAF*s characteristic function $\mathcal{F}$, beginning with $\emptyset$, where $\mathcal{F}(S) = \{X | X \text{acceptable w.r.t. } S\}$. This is because in general $\mathcal{F}$ is not monotonic and so one cannot guarantee existence of a least fixed point. However, $\mathcal{F}$ is monotonic for *hierarchical EAF*s [15] in which case one can identify the grounded extension as the least fixed point of $\mathcal{F}$.

---

[1] Each argument (attack) is attacked by a finite number of arguments.

We now instantiate EAFs by $ASPIC^+$ arguments and attacks [17], whereby we assume a function $\mathcal{P}$ that maps the conclusion of an individual argument to strict preferences over other arguments; e.g., given A and B with respective sets of defeasible rules $\{r_1\}$ and $\{r_2, r_3\}$ then if C concludes $(r_1 < r_2) \wedge (r_1 < r_3)$, then $\mathcal{P}(\mathrm{Conc}(C)) = A \prec B$ (under the *Elitist* set ordering [18]):

Definition 9. *Let $\mathcal{A}$ be a set of* ASPIC^+ *arguments, $\mathcal{C}$ the attack relation defined over $\mathcal{A}$, and $\mathcal{P} : \mathcal{L} \mapsto 2^{\mathcal{A} \times \mathcal{A}}$. Then $(\mathcal{A}, \mathcal{C}, \mathcal{D})$ is defined as in Definition 7, where $(C, (A, B)) \in \mathcal{D}$ iff A formula attacks B on B' and $A \prec B' \in \mathcal{P}(\mathrm{Conc}(C))$.*

3   A Framework for Dialogues

3.1   Defining the Dialogue Moves and Protocol

We formalise a framework for two party persuasion dialogues in which each agent can construct arguments from their own argumentation theories and the contents of the locutions submitted during the course of the dialogue. The proponent P starts a dialogue by submitting either an elementary or complex argument, whose claim is the 'topic' of the dialogue about which she wishes to persuade her opponent. A dialogue is then a sequence of moves consisting of locutions, where each agent replies to a move of her interlocutor. Since agents can move multiple replies to an interlocutor's move (either all at once or on backtracking), a dialogue can be represented as a tree in which each path from root to leaf consists of alternating moves by P and O.

Definition 10. *Let $(\mathcal{L}, \mathcal{R}, n, {}^-)$ be the universal argumentation system (recall Definition 2) and $\mathcal{A}$ the set of all arguments whose nodes are formulae in $\mathcal{L}$ and whose strict and defeasible rules are in $\mathcal{R}$. A **locution** is of the form $pf(c)$ where pf is the performative* argue *or* prefer*, in which case c is an argument $X \in \mathcal{A}$, else pf is the performative* why *or* concede*, in which case c is a formula $\phi \in \mathcal{L}$.*

Definition 11. *A **move** m is a tuple $< i, ag, l, j >$ where $id(m) = i \in \mathbb{N}$ is the identifier of the move, $pl(m) = ag \in \{P, O\}$ is the player of the move (henceforth $\overline{ag} = O$ if $ag = P$, $\overline{ag} = P$ if $ag = O$), $s(m) = l$ is the locution, and $t(m) = j \in \mathbb{N}$ is the identifier of the target of m (i.e., the id of the move that m replies to). $\mathcal{M}$ denotes the set of all possible moves, and for any m, if $s(m) = pf(c)$ we say m is a 'pf move'.*
We may refer to a move m by its locution $s(m)$, and instead of writing 'm is the move s.t. $t(m') = i$, $id(m) = i$', we may simply write 'm $= t(m')$' or 'm' replies to m'.
*A **dialogue** D is a sequence $m_1, \ldots, m_i, \ldots$ s.t. each ith move has identifier i, $t(m_1) = 0$, and for $i > 1$, $t(m_i) = j$ for some $j < i$.*
*A finite dialogue $D = m_1, \ldots, m_i, \ldots$ can be represented as a **dialogue tree** T_D consisting of a set of **disputes** $\{d_1, \ldots, d_l\}$ where each dispute is a sequence of moves $m_1, \ldots, m_n$[2], m is a move in D iff m is a move in some dispute, and for $j = 1 \ldots n-1$, $t(m_{j+1}) = id(m_j)$ (i.e., each move in a dispute is a reply to the preceding move).*

---

[2] In this representation each $i = 1 \ldots n$ does not denote the identifier of the move.

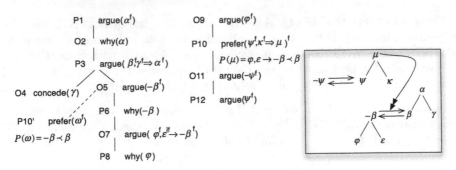

Fig. 2. Dialogue tree beginning with P1, continuing from O9 shown on right. Inset the *EAF* defined by the dialogue.

Consider the example dialogue tree in Fig. 2, showing the locutions and the order in which they are moved by P or O. Notice the two disputes generated by the successive moves O4 and O5. Notice also the potential additional dispute generated by P backtracking by moving P10' to reply to O5 after O9 (we will see later that this move is prohibited by the dialogue protocol).

The locutions in Definition 10 are common to many argumentation based models of dialogues, apart from the prefer locution which we introduce to enable moving arguments claiming preferences over other arguments. Why moves account for the fact that agents may: construct arguments for conclusions that are then added to their premises (cf. lemmas); often submit 'incomplete' arguments that are not fully backward extended, or; assume premises that are in need of further justification. For example, suppose an argument X instantiating the scheme for practical reasoning [2]: 'In circumstances S doing action A will have effect E so achieving goal G and promoting value V' (e.g., S might be a patient diagnosis warranting a medical treatment A). One of the scheme's critical questions 'why is S true ?' can be addressed as a *why* move, eliciting a reply providing an argument for S, so effectively backward extending X on S (note that the agent might be able to both construct a complex argument for S – possibly having had to first acquire information in order to do so – and have S included as a premise, c.f. a lemma as described above). Hence, we define sequences of moves that successively backward extend an argument:

Definition 12. *Let T_D be a dialogue tree, m_i a move in some dispute $d = m_1, \ldots, m_i, \ldots$ in T_D s.t. $s(m_i) = argue(X)$, m_{i+1} is not a why move. Let $j < i$ be the smallest j s.t. m_j is an argue or prefer move, m_{j-1} is not a why move, and for $k = j+1 \ldots i$, m_k is either a why move replying to an argue move or an argue move replying to a why move. Then $m_j, \ldots, m_i$ is an **argument extension sequence** (**aes**) in d and in D, that begins with m_j and terminates with m_i.*

An *aes* therefore begins with $argue(X)$ or $prefer(X)$, and thereafter consists of alternate why and argue moves that terminates in an argue move.

Definition 13. *Let $pf(X_1)$, $why(\phi_1)$, ..., $why(\phi_n)$, $argue(X_{n+1})$ be an* aes *where $pf \in \{argue, prefer\}$. Suppose for $i = 1 \ldots n$, $\phi_i \in$ Leaves(X_i), and for $1 < i \leq n+1$, Conc$(X_i) = \phi_{i-1}$. Let $X_1' = X_1$, and (recalling Definition 4) define for $i = 1 \ldots n$:*

$$X_{i+1}' \ extends \ X_i' \ on \ \phi_i \ with \ X_{i+1}$$

We say that the aes *defines the argument X_{n+1}'.*

In Fig. 2, P1–P3 is an *aes* that defines the argument A in Fig. 1, and O5–O7 and O5–O9 are *aess* that define B in Fig. 1.

The preference $Z \prec Y$, claimed by an argument in a prefer locution, may refer to a Z moved as an attack on some Y, where Z is defined by a *aes*.

Definition 14. *Let $d = m_1, \ldots, m_n$ be a dispute in a dialogue tree T_D. Then a sub-dispute $m_i, \ldots, m_k$ of d is an* **attack pair** (Z, Y) **on** Y' **in** d *iff Z attacks Y on Y', and: $m_i = prefer(Y)$ or $argue(Y)$, and either:*

1. *$k = i+1$, $m_k = argue(Z)$, and m_k does not begin an* aes *in d, or;*
2. *$m_{i+1} = argue(Z_1), \ldots, m_k = argue(Z_n)$ is an* aes *in d that defines the argument Z.*

If $\neg \exists d' \neq d \in T_D$ such that $m_i, \ldots, m_k, \ldots, m_j$ is an aes *then $m_i, \ldots, m_k$ is a* **maximal attack pair** (Z, Y) **on** Y' **in** d. *We say that m_k terminates the (maximal) attack pair, and 'the attack pair is moved in d by $pl(m_k)$'.*

Letting A and B be the arguments in Fig. 1. P3, O5 is an attack pair $(-\beta, A)$ on β in the dispute $P1, \ldots, P10'$ in Fig. 2, whereas it is not a maximal attack pair. P3, ..., O9 is a maximal attack pair (B, A) on β in the dispute $P1, \ldots, P12$.

We define a dialogue protocol by defining the set of all legal dialogues, which in turn are defined by the conditions under which a move can be a legal reply to another move. Since these conditions make no reference to the beliefs of the participating agents, we give a 'public semantics' for the protocol [20]. The protocol allows a considerable degree of freedom as to the moves that agents can make, and can be considered a 'core protocol' to which further rules and restrictions can be added depending on specific requirements (as we illustrate later).

Definition 15. *$\mathcal{D}$ is set of all possible* **legal dialogues**, *s.t.:*

1. *$\forall m \in \mathcal{M}$ s.t. $pl(m) = P$, $s(m) = argue(X)$: $m_1 = m$ is a dialogue in $\mathcal{D}$.*
2. *If $D = m_1, \ldots, m_{n-1} \in \mathcal{D}$ then $D' = m_1, \ldots, m_{n-1}, m_n \in \mathcal{D}$ iff*

   2.1 For $i = 2 \ldots n$, $pl(m_i) = P$ or $pl(m_i) = O$;
   2.2 If m_n replies to m_i, then $pl(m_n) = \overline{pl(m_i)}$ and there is no reply $m_{j \neq n}$ to m_i such that $s(m_n) = s(m_j)$;
   2.3 If $s(m_n) = argue(X)$, $t(m_n) = m_i$, then either:
      2.3.1 $s(m_i) = argue(Y)$ or $prefer(Y)$, and X attacks Y on Y', or;

2.3.2 $s(m_i) = prefer(Y)$, $\mathcal{P}(\text{Conc}(Y)) = A < B$, $\mathcal{P}(\text{Conc}(X)) = B <$
    A, or;

2.3.3 $s(m_i) = why(\phi)$ and $\text{Conc}(X) = \phi$.

2.4 *If $s(m_n) = prefer(X)$, then letting T_D be the dialogue tree for D and d the dispute $m_1, \ldots, m_n \in T_D$:*

  $m_i, \ldots, m_{n-1}$ *($i \geq 1$) is a* maximal attack pair *(Z, Y) on Y' in d such that the attack (Z, Y) is a formula attack, and $\mathcal{P}(X) = Z < Y'$. We say m_n is a reply to an attack pair.*

2.5 *If $s(m_n) = why(\phi)$, $t(m_n) = m_i$, then $s(m_i) = argue(X)$ or $prefer(X)$, and $\phi \in \text{Leaves}(X)$, and there is no m in D that replies to m_i such that $s(m) = concede(\phi)$.*

2.6 *If $s(m_n) = concede(\phi)$, $t(m_n) = m_i$, then $s(m_i) = argue(X)$, $\phi \in \text{Concs}(X)$.*

The first condition states that every dialogue begins with an argue move by P. 2.1 allows P (O) to make multiple moves in one turn (e.g., O4 and O5 in Fig. 2), and 2.2 prohibits players replying to their own moves or repeating a locution in reply to a move.

An argue move can be used to attack another argument Y on Y' (2.3.1). An argument X can also be moved against a Y claiming a preference that has been used to invalidate the success of an attack, if X claims a contradictory preference (2.3.2). For example O4 replying to P3 in the dialogue in Fig. 3(i). An agent can also move an argument concluding ϕ as a reply to a why move questioning ϕ (2.3.3).

A prefer move submits an argument that declares a strict preference for some Y' over Z, where (Z, Y) is a *maximal* attack pair, so that Z is either an argument moved in a single move, or defined by a *aes* consisting of a series of backward extensions, and Z formula attacks Y (on Y') (recall that undercuts and contrary formula attacks succeed as defeats independently of preferences). Thus the attack is rendered un-successful by the preference argument. Note that we avoid the problem (described in Sect. 1) with approaches (e.g., [19]) that need to assume arguments are not weakened on backward extending. In Fig. 2, O5 moves $-\beta$ to

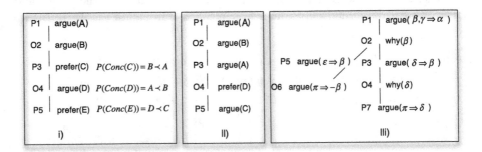

Fig. 3. Weather dialogues (i) and (ii). Two argue moves in reply to same why move shown in (iii) .

attack A in P3 on β. O5,...,O9 backward extends $-\beta$ to define the argument B, and P10 then moves an argument claiming $B \prec \beta$, so invalidating B's attack on A. Also note that P10′ would not be a valid reply given that O5 begins the *aes* O5,...,O9 that defines B in another dispute (i.e., P3,O5 is *not* a maximal attack pair). However, what if P10′ was moved prior to P6 and the subsequent backward extension of $-\beta$? We show later that P10′ will not then affect the outcome of the dialogue.

An agent can *at any point* concede (2.6) some ϕ that is the conclusion of any sub-argument (i.e,. $\texttt{Concs}(X)$) of a moved argument X (for example she earlier questions ϕ and then when presented with an argument for ϕ concedes ϕ to explicitly indicate that she is persuaded). Although an agent may concede the conclusion ϕ of an argument, she may still question or attack a premise or attack an intermediate conclusion of the argument, indicating that although persuaded as to the truth of ϕ, she is not persuaded as to reasons (i.e., the argument) given for ϕ. Of course if every $\phi \in \texttt{Concs}(X)$ is conceded, she must be persuaded as to the line of reasoning concluding ϕ. Indeed, 2.5 precludes questioning a ϕ that has been conceded. However, 2.3.1 does not require that no move $concede(\texttt{Conc}(Y'))$ replies to m_i in order that one can move an argument attacking Y on Y'. Thus, although γ is conceded in Fig. 2 (O4), O may subsequently acquire information to construct an argument for $-\gamma$. Such information may be acquired from the contents of arguments submitted by P. The use of premises/rules supplied by an interlocutor is illustrated by P's use (in P7) of O's premise π (in O6) in Fig. 3(iii).

3.2   Commitments and Dialogue Outcomes

During a dialogue, the contents of locutions are added to the participants' commitment stores. These commitments may be used to enforce an agent's dialogical consistency (e.g., requiring his commitments to be consistent at all times), enable agents to use the beliefs of their interlocutors, and attach dialogical obligations to the contents of commitment stores [23]. Commitments can also be used to determine the termination and outcome of a dialogue. For example, the proponent wins as soon as the opponent concedes the topic. However, since our focus is on providing dialogical characterisations of non-monotonic reasoning, we want that a dialogue is won just in case there is a justified $ASPIC^+$ argument for the topic in the EAF instantiated by the agents' commitment stores. Moreover, this allows for an *any-time outcome* definition [19]; at any stage in the dialogue the current winner can be identified based on the commitments.

We now define updates to agents' commitment stores (CSs). Unlike standard accounts, the update is not defined based only on the moved locution, but also accounts for the other locutions thus far moved. To illustrate, observe that in Fig. 3(iii), P1 commits the defeasible rule $\beta, \gamma \Rightarrow \alpha$. If the dialogue were not to proceed further, then by default one assumes that P constructs this argument given premises β and γ. However, O2's $why(\beta)$ obliges P3 to backward extend on β, so committing to $\delta \Rightarrow \beta$ in place of β as a premise. Now δ is committed to as a premise, but when O4's $why(\delta)$ challenges δ, the burden of proof on P to justify why δ is the case has thus far not been met, and so one no longer includes δ as a premise. P5 backtracks to provide an alternative argument for β, which is attacked by O6, and then P7 uses O's premise π to backtrack and provide an argument for δ. In general then, at any stage of the dialogue the rules in any argue move are committed, and a leaf node ϕ is added as a premise just in case at least one $why(\phi)$ move is replied to by an $argue(\phi)$.

Definition 16. *Let* $D = m_1, \ldots, m_n$ *be a dialogue, and for* $ag \in \{P, O\}$ *let* $arg(ag, D) = \{X | \exists m_i \ s.t. \ pl(m_i) = ag, s(m_i) = argue(X) \ or \ prefer(X)\}$. *Then:*

$CS(ag, D) = \mathtt{Rl}(ag, D) \cup \mathtt{Pr}(ag, D)$, *where:*

$\mathtt{Rl}(ag, D) = \bigcup_{X \in arg(ag, D)} \mathtt{DefRules}(X) \cup \mathtt{StRules}(X)$

$\mathtt{Pr}(ag, D) = \bigcup_{X \in arg(ag, D)} \{\phi | \phi \in \mathtt{Leaves}(X), \ if \ \exists m \ in \ D \ s.t. \ s(m) = why(\phi)$

   then for some m_j *in* D *s.t.* $s(m_j) = why(\phi)$, $\exists m_k$ *that replies to* m_j,

   $pl(m_k) = ag, \ s(m_k) = argue(\phi)^3\}$.

In Fig. 3(iii), $\mathtt{Pr}(P, D) = \{\delta, \gamma\}$ after move P3, $\{\gamma\}$ after O4, and $\{\epsilon, \pi, \gamma\}$ after P7. We now define the argumentation theory defined by a dialogue:

Definition 17. *Let* D *be a dialogue and* $CS = CS(ag, D) \cup CS(\overline{ag}, D)$. *Then:*

- $\mathcal{R}_D = \{r | r \ is \ either \ a \ strict \ or \ defeasible \ rule \ in \ CS\}$;
- $KB^p = \{\phi | \phi \in CS, \phi \ is \ a \ leaf \ labelled \ 'f' \ in \ an \ argument \ moved \ in \ D\}$;
- $KB^n = \{\phi | \phi \in CS, \phi \ is \ a \ leaf \ labelled \ 'if' \ in \ an \ argument \ moved \ in \ D\}$.

Then $AT_D = (AS_D = (\mathcal{L}, \mathcal{R}_D, n, ^-), (KB^p, KB^n))$ *is the argumentation theory defined by* D *(equivalently* T_D*)*

Arguments are then defined as in [18]; that is, as in Definition 3, but now with reference to AT_D as described after Definition 3. The EAF defined by the dialogue is then defined as in Definition 9. An any-time outcome for the dialogue can now be defined:

Definition 18. *Let* D *be a dialogue and* $(\mathcal{A}, \mathcal{C}, \mathcal{D})$ *the EAF defined by* AT_D. *Let* X *be an* initial argument *defined by a maximal aes that begins with* m_1, *else if* m_1 *does not begin an aes,* $m_1 = argue(X)$ *is the* initial argument. *Then* ***P wins under s** semantics* ($s \in \{preferred, stable, grounded\}$) *iff* $\exists X'$ *that extends* X *on* $\Omega \subseteq \mathtt{Leaves}(X)$ *s.t.* X' *is in some* s *extension of* $(\mathcal{A}, \mathcal{C}, \mathcal{D})$, *else O wins.*

Note there may be many initial arguments; e.g., $m_1 = argue(\alpha)$, $m_2 = why(\alpha)$ and P replies with $argue(\beta \Rightarrow \alpha)$ *and* $argue(\gamma \Rightarrow \alpha)$, which are both initial arguments. Moreover, suppose these two *aes*s are maximal, and a $why(\beta)$ moved *elsewhere* in the dialogue is replied to with $argue(\delta \Rightarrow \beta)$. Hence β would

---

[3] Note that if agents play 'logically perfectly' whereby agents make all move that are legal, $argue(\phi)$ would be moved as a reply against *all* $why(\phi)$ moves, including m.

not be a premise in $KB^{p(n)}$, and P is the winner only if the argument extending $\beta \Rightarrow \alpha$ with $\delta \Rightarrow \beta$ is justified[4].

Example 2. For Fig. 2's dialogue: $CS(P, D) = \{(\beta, \gamma \Rightarrow \alpha), (\psi, \kappa \Rightarrow \mu), \beta, \gamma, \psi, \kappa\}$ and $CS(O, D) = \{(\phi, \epsilon \rightarrow -\beta), \phi, \epsilon, -\psi\}$. The AT_D contains the rules and ordinary premises $\{\beta, \gamma, \psi, \kappa, \phi, -\psi\}$ and axiom premises $\{\epsilon\}$ in these commitment stores, and the EAF defined by AS_T is shown (inset) in Fig. 1. P wins under the preferred semantics (the arguments concluding α and μ are in an admissible and hence preferred extension of the EAF), whereas P loses under the grounded semantics.

Following Prakken [19], we now define the dialectical status of moves – *in* or *out* – in a dialogue tree D_T, so as to determine the winner of the dialogue. Concede moves cannot be replied to, and are effectively 'surrendering' replies [19] that do not affect the dialectical status of their targets. Hence these moves are not assigned any status. However $why(\phi)$ attacks its target argument since the burden of proof is on the agent moving the argument to justify ϕ. An argue reply also attacks its target if the target is an argument, or a why move (in the latter case by fulfilling the burden of proof), and a prefer reply attacks its target argument by invalidating an attack from the target argument. However, in the latter case the preference $Z \prec Y'$ should only invalidate an attack from Z to Y on Y', if (Z, Y) is a *maximal* attack pair (recall Definition 14).

Definition 19. *m is an **attacking reply** iff m is an argue, why or prefer move, where if $s(m) = prefer(X)$ then m is a reply to a maximal attack pair. An argue, prefer or why move m is then said to be **in** iff if m' is an attacking reply to m then m' is **out**. Otherwise m is out.*

In Fig. 2, Pi is *in* for $i = 1, 3, 6, 8, 10, 12$ and Oi is *out* for $i = 2, 5, 7, 9, 11$. *Suppose P10' had been moved before O7* (which is allowed by the protocol). P10' would then attack reply O5 so that P10' would be *in* and O5 *out*. If the dialogue then continued and terminated at O7, O5 would then be *in* since P10' would no longer attack reply O5 (as it would be a prefer move that does not reply to a maximal attack pair) and so would no longer be *in*. In Fig. 3(iii), P7, P3, P1 and O6 are *in*, and O2, O4 and P5 are *out*.

In [19], Prakken shows that if P and O play logically perfectly (see Footnote 3), the topic of persuasion ϕ is *in* (i.e., P wins) iff there is an argument concluding ϕ in the grounded extension of the AF instantiated by the theory defined by the dialogue. We now augment Definition 15's protocol so as to define a dialogue for the *preferred* semantics. We then conjecture that an initial argument X is *in* iff X is in an admissible (and hence preferred) extension of the EAF defined by the dialogue. Proof of this conjecture will then be established in future work, as a result that applies to this paper's protocol extended to allow moves that *retract* arguments.

---

[4] Under 'logically perfect' play, $why(\beta)$-$argue(\delta \Rightarrow \beta)$ would extend the *aes* that terminates with $argue(\beta \Rightarrow \alpha)$ and so it would suffice to check that the now initial argument – $(\delta \Rightarrow \beta, \beta \Rightarrow \alpha)$ – is justified in the EAF.

We adapt the rules in [14] in which an argument game proof theory is defined for a *given EAF*. In [14], a game tree consists of all possible proponent (*pro*) and opponent (*opp*) arguments that attack their adversary's arguments or attacks (as indicated by the given *EAF*). *Pro*'s initial argument is in an admissible extension of the *EAF* iff there is a winning strategy (a sub-tree of the game tree) in which every *opp* argument or attack is attacked by a *pro* argument, and the *pro* arguments in the winning strategy (i.e., the candidate admissible extension) are conflict free. *Opp* is restricted so that if in a dispute d, an *opp* argument or attack has already been replied to (attacked) by *pro*, then *opp* cannot repeat the argument/attack. To see why these restrictions are needed, consider an *EAF* consisting of two symmetrically attacking arguments A and B. A is in an admissible extension, but if *opp* can repeat, one might have an infinite dispute $A - B - A - B \ldots$. The non-repetition restriction on *opp* means that $A - B - A$ cannot be continued, and defines a winning strategy. We now adapt this non-repetition rule to disputes in a dialogue tree (we will later consider the repetition of why moves as a condition of logically perfect play which applies to P and O).

Definition 20. *Let d be a dispute $m_1, \ldots, m_n$.* **An attack pair (X, Y) on Y' by $ag \in \{P, O\}$ in d is said to fail** *if the attack pair terminates in m_k, $pl(m_k) = ag$, $s(m_{k+1}) = prefer(Z)$, $\mathcal{P}(\texttt{Conc}(Z)) = X \prec Y'$.*

The dispute $d' = m_1, \ldots, m_n, m_{n+1}$ where $pl(m_{n+1}) = ag$, is legal under non-repetition for ag, iff:

$s(m_{n+1}) = argue(X)$ or $prefer(X)$ implies there is no attack pair (Z, X') on X'' moved in d by $\overline{ag}$ such that $X'' \in \texttt{Sub}(X)$, and if $s(m_{n+1})$ terminates an attack pair (X', Y) on Y', then there is no attack pair (X', Y) on Y' by ag in d that fails.

Note we do not require above that the attack pair (X', Y) on Y' by ag in d is maximal. Hence if P10$'$ were moved prior to O7, and the dispute $P1, \ldots, P10'$ was extended by further moves to $P1, \ldots, P10', m, \ldots, m'$, then if amongst $m, \ldots, m'$, P moves an argument with fallible premise or defeasible conclusion β, the non-repetition rule applied to O would prohibit O from moving $argue(-\beta)$ as an attacking reply. This is despite the fact that the attack by O5 ($-\beta$) on P3 (β) does not define a maximal attack pair. However, recall (Definition 19) that P10$'$ is not an attacking reply and does not have an effect on the dialogical status of O5, and so any restrictions on O in a dispute that extends P10$'$ will make no difference to the outcome of the dialogue.

A protocol for the preferred semantics is defined as follows.

Definition 21. *$\mathcal{D}$ is the set of all possible **legal dialogues under the preferred semantics** if all dialogues in $\mathcal{D}$ satisfy 1, 2.1, 2.2, and 2.6 in Definition 15, and if $D = m_1, \ldots, m_{n-1} \in \mathcal{D}$, then $D' = m_1, \ldots, m_{n-1}, m_n \in \mathcal{D}$, where:*

- *if $pl(m_n) = P$ then D' satisfies 2.3, 2.4 and 2.5 in Definition 15.*
- *if $pl(m_n) = O$ then D' satisfies 2.3, 2.4 and 2.5 in Definition 15, and the dispute d in T_D that terminates in m_n is legal under non-repetition for O.*

To illustrate the non-repetition rule, suppose a continuation of the dispute ending in P12. O cannot make argue moves that define the argument $X = \phi, \epsilon \Rightarrow -\beta$ and such that X attacks β in this continuation. This is because P10 already invalidates this attack with a preference. However, O could repeat X if it is not used to attack β (e.g., in reply to a $why(-\beta)$ move), as X is not directly attacked by P in the dispute. However, in the dialogue in Fig. 3(ii), O cannot move $argue(D)$ as a reply to P5.

We now define a dialogue outcome that declares P the winner just in case P's initial move is in, and the contents of P's moves define a conflict free set of arguments.

Definition 22. *Let $D = m_1, \ldots, m_n$, $in(D) = \{m | m$ is $in, pl(m) = P\}$ and $CS(P, in(D))$ be defined as in Definition 16 with '$in(D)$' replacing 'D'. Let S be the set of all $ASPIC^+$ arguments that can be constructed from the premises $\{\phi | \phi \in CS(P, in(D))\}$ and the defeasible and strict rules in $CS(P, in(D))$. Then if m_1 is in and S is a conflict free set in the EAF defined by D, then P is the winner of D, else O is the winner.*

Definition 23. *Let $(\mathcal{A}, \mathcal{C}, \mathcal{D})$ be the EAF defined by the dialogue $D = m_1, \ldots, m_n$, and T_D the dialogue tree for D. D is **logically perfect** iff:*

- *For any $X \in \mathcal{A}$ if m is a legal reply to some m_i, where $s(m) = argue(X)$ or $s(m) = prefer(X)$, then m is a reply to m_i in D, and*
- *If $m = why(\phi)$ is a legal reply to some $m_i = argue(X)$ or m_i terminating an aes defining X, then m is a reply to m_i in D, unless $why(\phi)$ appears in the dispute $d = m_1, \ldots, m_i$ in T_D as a reply to some $m_{j<i} = argue(X)$ or an aes in d that terminates in $m_{j<i}$ and defines X.*

Note the second condition above excludes either player from moving a $why(\phi)$ to an argument whose leaf node he has already challenged. This prevents filibustering by both players; e.g., $argue(\phi) - why(\phi) - argue(\phi) - why(\phi) \ldots$.

Conjecture 1. P is the winner (according to Definition 22) of a logically perfect dialogue D played under the preferred semantics protocol iff P wins under preferred semantics (according to Definition 18).

Example 3. In Fig. 3(iii), logically perfect play would entail P repeating P5 as a reply to O6. In Fig. 2, logically perfect play would entail P moving $why(\epsilon)$ in reply to O7, and O replying $why(\beta)$ to P3, and $why(\psi)$, $why(\kappa)$ to P10. Note that after these why moves, none of ϵ, β, ψ or κ remain in the commitment stores, so that the corresponding elementary arguments would not be in the dialogue's defined EAF and would not be moved under logically perfect play.

4   Conclusions

To the best of our knowledge, this paper is the first to formalise dialogues that accommodate argumentation-based reasoning about preferences over arguments.

In [12], prefer locutions express an ordering over proposals in deliberation dialogues, but reasoning about preferences is not accommodated. Other dialogue models that formalise distributed reasoning through relating the dialogue outcome to the justified arguments defined by the contents of the locutions, include [9,19]. The former define assumption based argumentation (ABA) frameworks [4] and do not accommodate preferences. The general framework for persuasion in [19] does not assume $ASPIC^+$ arguments, and assumes a fixed exogenously given preference relation. Moreover, [19] requires that if A is used to defeat an argument B, then A is not weakened on being backward extended (recall the discussion in Sect. 1).

In future work we will further develop this paper's proposed framework. We intend extending this paper's protocols to accommodate *retract* moves, and will then define a grounded semantics protocol that essentially 'flips' the non-repetition restriction so that it applies to P rather than O. We will then formally prove correspondence theorems of the type described in Conjecture 1, so fully establishing formal frameworks for distributed non-monotonic reasoning that accommodate reasoning about preferences.

Acknowledgements. This work was funded by the EPSRC project CONSULT: Collaborative Mobile Decision Support for Managing Multiple Morbidities (EP/P010105/1).

References

1. Amgoud, L., Cayrol, C.: A reasoning model based on the production of acceptable arguments. Ann. Math. Artif. Intell. **34**(1–3), 197–215 (2002)
2. Atkinson, K., Bench-capon, T., Mcburney, P.: Computational representation of practical argument. Synthese **152**, 157–206 (2006)
3. Bench-Capon, T.J.M.: Persuasion in practical argument using value-based argumentation frameworks. J. Logic Comput. **13**(3), 429–448 (2003)
4. Bondarenko, A., Dung, P., Kowalski, R., Toni, F.: An abstract, argumentation-theoretic approach to default reasoning. Artif. Intell. **93**, 63–101 (1997)
5. Brewka, G.: Preferred subtheories: an extended logical framework for default reasoning. In: International Joint Conference on Artificial Intelligence, pp. 1043–1048 (1989)
6. Brewka, G.: Adding priorities and specificity to default logic. In: MacNish, C., Pearce, D., Pereira, L.M. (eds.) JELIA 1994. LNCS, vol. 838, pp. 247–260. Springer, Heidelberg (1994). https://doi.org/10.1007/BFb0021977
7. Brewka, G.: Dynamic argument systems: a formal model of argumentation processes based on situation calculus. J. Logic Comput. **11**, 257–282 (2001)
8. Dung, P.M.: On the acceptability of arguments and its fundamental role in non-monotonic reasoning, logic programming and n-person games. Artif. Intell. **77**, 321–357 (1995)
9. Fan, X., Toni, F.: A general framework for sound assumption-based argumentation dialogues. Artif. Intell. **216**, 20–54 (2014)
10. Gordon, T.F.: The pleadings game. Artif. Intell. Law **2**(4), 239–292 (1993)
11. Gorogiannis, N., Hunter, A.: Instantiating abstract argumentation with classical logic arguments: postulates and properties. Artif. Intell. **175**(10), 1479–1497 (2011)

12. Kok, E.M., Meyer, J.-J.C., Prakken, H., Vreeswijk, G.A.W.: A formal argumentation framework for deliberation dialogues. In: McBurney, P., Rahwan, I., Parsons, S. (eds.) ArgMAS 2010. LNCS (LNAI), vol. 6614, pp. 31–48. Springer, Heidelberg (2011). https://doi.org/10.1007/978-3-642-21940-5_3
13. McBurney, P., Parsons, S.: Chapter 13: dialogue games for agent argumentation. In: Simari, G., Rahwan, I. (eds.) Argumentation in Artificial Intelligence, pp. 261–280. Springer, Boston (2009). https://doi.org/10.1007/978-0-387-98197-0_13
14. Modgil, S.: Labellings and games for extended argumentation frameworks. In: Proceedings of 21st International Joint Conference on Artifical Intelligence, pp. 873–878 (2009)
15. Modgil, S.: Reasoning about preferences in argumentation frameworks. Artif. Intell. **173**(9–10), 901–934 (2009)
16. Modgil, S., Caminada, M.: Chapter 6: Proof theories and algorithms for abstract argumentation frameworks. In: Simari, G., Rahwan, I. (eds.) Argumentation in Artificial Intelligence, pp. 105–129. Springer, Boston (2009). https://doi.org/10.1007/978-0-387-98197-0_6
17. Modgil, S., Prakken, H.: Reasoning about preferences in structured extended argumentation frameworks. In: Proceedings of COMMA 2010, pp. 347–358 (2010)
18. Modgil, S., Prakken, H.: A general account of argumentation with preferences. Artif. Intell. **195**, 361–397 (2013)
19. Prakken, H.: Coherence and flexibility in dialogue games for argumentation. J. Logic Comput. **15**, 1009–1040 (2005)
20. Prakken, H.: Formal systems for persuasion dialogue. Knowl. Eng. Rev. **21**(2), 163–188 (2006)
21. Prakken, H.: An abstract framework for argumentation with structured arguments. Argum. Comput. **1**(2), 93–124 (2010)
22. Walton, D.N.: Argument Schemes for Presumptive Reasoning. Lawrence Erlbaum, London (1996)
23. Walton, D.N., Krabbe, E.C.W.: Commitment in Dialogue: Basic Concepts of Interpersonal Reasoning. State University of New York Press, New York (1995)
24. Young, A.P., Modgil, S., Rodrigues, O.: Prioritised default logic as rational argumentation. In: Proceedings of AAMAS 2016, pp. 626–634 (2016)

Enumerating Preferred Extensions:
A Case Study of Human Reasoning

Alice Toniolo[1](✉), Timothy J. Norman[2], and Nir Oren[3]

[1] School of Computer Science, University of St Andrews,
St Andrews, Scotland, UK
a.toniolo@st-andrews.ac.uk
[2] Department of Electronics and Computer Science,
University of Southampton, Southampton, UK
[3] Department of Computing Science,
University of Aberdeen, Aberdeen, Scotland, UK

Abstract. This paper seeks to better understand the links between human reasoning and preferred extensions as found within formal argumentation, especially in the context of uncertainty. The degree of believability of a conclusion may be associated with the number of preferred extensions in which the conclusion is credulously accepted. We are interested in whether people agree with this evaluation. A set of experiments with human participants is presented to investigate the validity of such an association. Our results show that people tend to agree with the outcome of a version of Thimm's probabilistic semantics in purely qualitative domains as well as in domains in which conclusions express event likelihood. Furthermore, we are able to characterise this behaviour: the heuristics employed by people in understanding preferred extensions are similar to those employed in understanding probabilities.

Keywords: Argumentation · Probabilistic semantics · User evaluation

1 Introduction

One of the strengths of argumentation theory is its qualitative nature. For example, in Dung's theory, arguments are either within, or outside an extension, and no notion of argument strength is required in order to obtain desirable features—such as reinstatement—from the system. More recently, researchers have begun considering more quantitative frameworks, particularly in the context of probabilistic argumentation (e.g., [8,10,11,18]), through weighted argumentation systems [2,7] and graduality within argumentation [4]. The immediate question then arises as to whether such quantitative representations appropriately capture human reasoning and intuitions, as well as questions regarding the relationship between formal qualitative representations and human quantitative (or semi-quantitative) reasoning. As a concrete example—which we focus on in this paper—one could view multiple extension semantics, such as the preferred

© Springer International Publishing AG, part of Springer Nature 2018
E. Black et al. (Eds.): TAFA 2017, LNAI 10757, pp. 192–210, 2018.
https://doi.org/10.1007/978-3-319-75553-3_14

semantics, as capturing different possible worlds. This would then suggest that even qualitative argumentation can capture some notion of uncertainty.

This view can be further extended by considering situations where the arguments within an extension are themselves about uncertain facts, effectively changing the likelihood of each extension. If this is the case, then even in purely qualitative domains (represented through logical argumentation), where no quantified information exists, the degree of acceptability of a conclusion is associated with the number of preferred extensions in which the conclusion is credulously accepted. This paper investigates the validity of this claim, by means of an experiment with human participants.

The remainder of the paper is structured as follows. In Sect. 2, we expand the motivations of this work. In Sect. 3, we introduce an ASPIC-like argumentation framework followed by an overview of its use and key assumptions underpinning our experiments (Sect. 4). Section 5 details our experimental settings. In Sect. 6, methodology, hypotheses and results are discussed. We present our conclusions in Sect. 8.

2   Background and Motivation

Haenni [8] considers uncertainty as being an evaluation of probability on the premises which propagates throughout the argumentation system. Similarly, other studies such as [15,18] model uncertainty on the premises as being associated with the uncertainty of the sources, in the latter case due to the different degrees of trustworthiness of the sources themselves. Li et al. [11] consider a different take on probability, namely that the probability of an argument represents a prediction on how likely it is that the argument is justified.

In this work, we are interested in studying the links between the preferred extensions as used in argumentation, and how these are interpreted as probabilities by people with regards to the acceptability of a conclusion. Let us consider a conclusion of an argument within a structured argumentation framework. Generally, argumentation frameworks presented in the literature use extensions to decide whether a conclusion is accepted. In purely qualitative argumentation frameworks, this acceptance is either credulous (when there is at least one extension in which the argument under consideration is accepted), or sceptical (when the argument is accepted in all extensions) [13]. As dictated by the nature of qualitative frameworks, the enumeration of extensions in which a conclusion is accepted does not influence the decision as to whether a conclusion is accepted. However, here we claim that the number of extensions in which a conclusion is accepted has an effect on deciding whether the conclusion is to be considered justified, even if the argumentation framework is fully qualitative[1].

---

[1] Note that we use the terms argument and conclusion somewhat interchangeably as in the work we describe, a specific conclusion was the result of a unique argument. In future work, we will consider situations where multiple arguments may lead to the same conclusion, c.f., the so called universal semantics [6].

The problem of understanding the role of enumeration of extensions has been studied by Thimm [16] in abstract argumentation. Thimm presents a novel argumentation framework in which a probabilistic semantics is used to associate an argument with a degree of belief. This belief is computed as function of the number of extensions in which the argument appears to be justified. In our work, we use a similar approach where we consider the enumeration of preferred extensions in evaluating the believability of a conclusion. Thimm claims that this assessment provides a degree of confidence when selecting an option. Here we want to understand whether this is the case, i.e., whether people actually do use a similar heuristic to make a decision on what conclusions are the most believable. In Thimm's work, a probability is associated with each extension, and this influences the degree of belief placed in an argument. In our study we want to understand whether doing so is comparable to human reasoning with probability.

Unlike Thimm's work, we use structured argumentation frameworks, as we are interested in the believability of conclusions rather than arguments. Our core research question is then as follows: *do people agree with the evaluation given by probabilistic interpretation of argumentation semantics?* To address this question, we define an ASPIC-like structured argumentation framework from which we can formalise the problem.

3 An ASPIC-Like Framework with Probabilistic Semantics

In order to identify plausible conclusions, we use a simplified ASPIC-like argumentation framework with ordinary premises and defeasible rules without preferences or undercuts [13,14]. We derive the degree of belief in a conclusion obtained by applying argumentation semantics to arguments obtained from the framework, and then considering a probabilistic interpretation of the results.

3.1 Argumentation Framework

Definition 1. *An argumentation system AS is a tuple $\langle \mathcal{L}, ^-, \mathcal{R} \rangle$ where $\mathcal{L}$ is a logical language, $^-$ is a contrariness function, and $\mathcal{R}$ is a set of defeasible rules. The contrariness function $^-$ is defined from $\mathcal{L}$ to $2^{\mathcal{L}}$, such that given $\varphi \in \bar{\phi}$ with $\varphi, \phi \in \mathcal{L}$, if $\phi \notin \bar{\varphi}$, φ is called the contrary of ϕ, otherwise if $\phi \in \bar{\varphi}$ they are contradictory (including classical negation $\neg$). A defeasible rule is $\varphi_0, \ldots, \varphi_j \Rightarrow \varphi_n$ where $\varphi_i \in \mathcal{L}$.*

Definition 2. *A knowledge-base K in AS is a subset of the language $\mathcal{L}$. An argumentation theory is a pair $AT = \langle K, AS \rangle$.*

An *argument* A is derived from K of theory AT. Let $Prem(A)$ indicate the premises of A, $Conc(A)$ the conclusion, and $Sub(A)$ the subarguments:

Definition 3. *Given a set of arguments Arg, argument $A \in Arg$ is defined as:*

- $A = \{\varphi\}$ *with* $\varphi \in K$ *where* $Prem(A) = \{\varphi\}$, $Conc(A) = \varphi$, $Sub(A) = \{\varphi\}$.
- $A = \{A_1, \ldots, A_n \Rightarrow \phi\}$ *if there exists a defeasible rule in AS s.t. $Conc(A_1)$, $\ldots, Conc(A_n) \Rightarrow \phi \in \mathcal{R}$ with $Prem(A) = Prem(A_1) \cup \cdots \cup Prem(A_n)$, $Conc(A) = \phi$ and $Sub(A) = Sub(A_1) \cup \cdots \cup Sub(A_n) \cup A$.*

Attacks are defined as those arguments that challenge others, while defeats are those attacks that succeed:

Definition 4. *Given two arguments A_A and A_B:*

- A_A *rebuts* A_B *on* $Arg_{B'}$ *iff* $Conc(A_A) \in \bar{\varphi}$ *for* $A_{B'} \in Sub(A_B)$ *such that $A_{B'} = \{A_{B1"}, \ldots, A_{Bn"} \Rightarrow \varphi\}$.*
- A_A *undermines* A_B *on* φ *iff* $Conc(A_A) \in \bar{\varphi}$ *such that* $\varphi \in Prem(A_B)$.

Definition 5. *Defeat is a binary relationship $Def : Arg \times Arg$ where a defeat is represented as $(A_A, A_B) \in Def$. An argument A_A defeats an argument A_B iff: (i) A_A rebuts A_B on $A_{B'}$; or (ii) A_A undermines A_B on φ.*

Definition 6. *An abstract argumentation framework $AF = (Arg, Def)$ corresponding to an AT contains the set of arguments Arg as defined in Definition 3 and a set of defeats Def as in Definition 5.*

Sets of acceptable arguments (i.e., extensions ξ) in an AF can be computed according to a semantics. Here we use the preferred semantics. The set of credulous preferred extensions is $\hat{\xi}_P = \{\xi_1, \ldots, \xi_n\}$, where every ξ_i is a maximal set of arguments (with respect to set inclusion) that is conflict free and admissible.

Definition 7. *Given an abstract argumentation framework $AF = (Arg, Def)$, a set of arguments $S \subseteq Arg$ is conflict-free iff there is no $A_A, A_B \in S$ such that $(A_A, A_B) \in Def$. An argument $A_A \in S$ is admissible iff for every A_B such that $(A_B, A_A) \in Def$, there is a $A_C \in S$ such that $(A_C, A_B) \in Def$.*

3.2 Probabilistic Semantics for an Argument Theory

Having described a simple ASPIC-like framework, we now describe how Thimm's probabilistic semantics [16] is used to associate probabilities with conclusions.

The set of all possible sets of arguments is referred to as $\mathcal{K} = 2^{Arg}$, and the set of preferred extensions $\hat{\xi}_P$ is a subset of $\mathcal{K}$. A probability function of the form $P : 2^{\mathcal{K}} \to [0, 1]$ assigns a probability to each set of possible extensions of AF. For $\xi \in \mathcal{K}$, $P(\xi)$ is the probability that ξ is an extension. For now, we make the assumption that extensions are equiprobable. Then the probability of ξ is:

$$P(\xi) = \begin{cases} 1/|\hat{\xi}_P| & \xi \in \hat{\xi}_P \\ 0 & \xi \notin \hat{\xi}_P \end{cases} \tag{1}$$

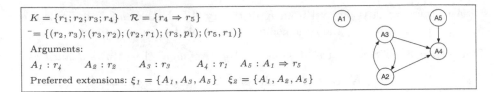

$K = \{r_1; r_2; r_3; r_4\}$ $\mathcal{R} = \{r_4 \Rightarrow r_5\}$

$\bar{} = \{(r_2, r_3); (r_3, r_2); (r_2, r_1); (r_3, p_1); (r_5, r_1)\}$

Arguments:

$A_1 : r_4$ $A_2 : r_2$ $A_3 : r_3$ $A_4 : r_1$ $A_5 : A_1 \Rightarrow r_5$

Preferred extensions: $\xi_1 = \{A_1, A_3, A_5\}$ $\xi_2 = \{A_1, A_2, A_5\}$

Fig. 1. Example of argumentation theory

For $P(\xi)$ and argument $A \in Arg$:

$$\hat{P}(A) = \sum_{A \in \xi \subseteq Arg} P(\xi) \tag{2}$$

Given the probability function P, $\hat{P}(A)$ represents the degree of belief that an argument A is in an extension according to P.

As Thimm suggests we now have an indication of the degree of belief of each argument that gives a characterisation of the uncertainty which is inherent in the AF. We must define several additional concepts in order to describe the acceptability of conclusions within the argumentation framework.

From [13] we know that a wff $\varphi \in \mathcal{L}$ is sceptically justified if φ is the conclusion of a sceptically justified argument, and credulously justified if φ is not sceptically justified and is the conclusion of a credulously justified argument. Hence we define a *justification ratio* μ of a conclusion φ as follows.

Definition 8. *Given a set of arguments $\mathcal{A} = \{A_1, \ldots, A_n\}$ such that for any A_i, $Conc(A_i) = \varphi$, we define the* justification ratio *as $\mu(\varphi) = \sum_{A_i \in \mathcal{A}} \hat{P}(A_i)$.*

The justification ratio $\mu(\phi)$ captures the probability of a conclusion being justified based on the likelihood of the arguments which justify it. If equiprobable extensions are assumed—as well as unique conclusions for each argument—then we obtain:

$$\mu(\varphi) = \hat{P}(A) = \sum_{A \in \xi \subseteq Arg} 1/|\hat{\xi}_P| \quad \text{where } \varphi \in Conc(A)$$

Example 1. We now illustrate the framework with the following example. Consider the AT presented in Fig. 1. We obtain two preferred extensions ξ_1, ξ_2 with $P(\xi_1) = P(\xi_2) = 0.5$. The justification ratios are then as follows:

$$\mu(r_1) = 0 \quad\quad \mu(r_2) = \mu(r_3) = 0.5 \quad\quad \mu(r_4) = \mu(r_5) = 1$$

4   Characterising Reasoning with Extensions

In the previous section, we explored a method to assign a degree of belief to a conclusion (which we denoted as the justification ratio) in relation to the enumeration of extensions by adapting Thimm's probabilistic semantics. Our main objective is to determine whether people agree with these probabilistic semantics; i.e., whether the justification ratio has a correlation with people's opinion of the believability of a conclusion. We believe that this is the case on the basis of the assumption that *people's reasoning with extensions may be understood in relation to reasoning with the rules of classical probability*. This assumption leads us to a second objective, namely characterising how people rate the believability of a conclusion.

Our analysis is based on the following observations:

- Classical probability assigns a likelihood to a piece of information φ on the basis of the ratio between the number of favourable and unfavourable cases which support or attack the information. Hence, consider a set of possible worlds W and a subset of the worlds $V \subseteq W$ in which a proposition $r_i \subset \mathcal{L}$ holds, the probability of r_i is as follows.

$$ p(r_i) = \frac{\# \text{ of worlds where } r_i \text{ holds}}{\text{total } \# \text{ possible worlds}} = \frac{|V|}{|W|} $$

- Similarly, we can consider the set of preferred extensions $\hat{\xi}_P$ as the set of possible explanations of a world, and the degree of belief of a conclusion r_i as given by the justification ratio $\mu(r_i)$. Let us refer to the subset of extensions in which r_i is acceptable as $\hat{\xi}_P^{r_i}$. From Definition 8 we obtain the following.

$$ \mu(r_i) = \sum_{A \in \xi \subseteq Arg} 1/|\hat{\xi}_P| = \frac{\# \text{ extensions in which } r_i \text{ is acceptable}}{\text{total } \# \text{ extensions}} = \frac{|\hat{\xi}_P^{r_i}|}{|\hat{\xi}_P|} $$

In the above situation, we assume that the information is purely qualitative. However, the information may refer to the likelihood of an event or a fact [1]. For example, an event E described in r_i can be subject of a proposition $r_j = $ "there is a ω chance that event E may occur". Continuing with the similarity between reasoning with extensions and reasoning with probability, we also seek to understand the behaviour in the case in which the user is presented with information that is about the likelihood of events, as well as the uncertainty introduced via the possibility of some information being, or not being, inferred. In this case, the believability of a conclusion may be explained by two heuristics depending on whether people consider these as dependent or independent events. The similarity with an argumentation framework outcome can then be established in the former case through the use of conditional probability, or in the latter by using the multiplication law of probability. For this research, we assume that the second heuristic is adopted, resulting in the following observations:

- ω indicates the probability of the event $p(E)$. Given $p(r_i)$, the probability of r_j using the multiplication law for independent events is: $p(r_j) = \omega * p(r_i)$
- Similarly, in an argumentation framework with probabilistic semantics, given the justification ratio $\mu(r_i)$, the justification ratio of r_j is: $\mu(r_j) = \omega * \mu(r_i)$.

We are now in a position to describe our experiments, designed to determine (1) whether the probabilistic interpretation of argumentation semantics described above represents human reasoning, and (2) whether the similarities observed between probabilistic and argument based reasoning are valid.

5   Experiment Design

Our overall objective is to understand whether people agree with the outcome of Thimm's probabilistic semantics. In our experiments, we asked a participant to rate the believability of a proposition under different experimental conditions α, as defined below. While considering different experimental conditions, we posed the following question to our subjects: "Given the condition α, how likely is that you believe r_i"? The subjects were asked to respond on a 5-points Likert scale, a commonly used scale for user studies, recorded as user evaluation $u(r_i)$ of a conclusion r_i (with 1: Extremely Unlikely – 5: Extremely Likely). Our hypothesis is that there is a positive correlation between the user rating $u_\mu(r_i)$ and the justification ratio $\mu(r_i)$. We also hypothesise that there is a positive correlation between the user evaluation of the likelihood of a piece of information r_i—$u_p(r_i)$—and its associated probability $p(r_i)$. Finally, we show that there is a similarity between the two ratings $u_p(r_i)$ and $u_\mu(r_i)$.

Definition 9. *An experimental condition α is a tuple $\alpha = \langle Domain, Scenario, Proposition, Interpretation, Percentage, Fraction, Ratio \rangle$.*

We now define the components of an experimental condition α.

5.1   Two Types of Information

As discussed in Sect. 4, information—represented via propositions—can be classified into two categories, or domain types in the context of the experiment.

Domain 1: Purely qualitative propositions $r_i \in \mathcal{L}$ in which the text is about a piece of information.

Domain 2: Propositions $r_j \in \mathcal{L}$ in which the text is about a piece of information and its probability of occurring.

In the former, we want to demonstrate that even in purely qualitative scenarios, people agree with the outcome of Thimm's probabilistic semantics: that the believability of a conclusion is related to the number of extensions in which that conclusion is accepted. With the latter domain, we want to demonstrate that in scenarios in which conclusions are about the probability of some information, the outcome of the probabilistic semantics is still an important factor in assessing the believability of a conclusion. The two types of propositions lead to two sets of experiments.

5.2   Scenarios and Propositions

In the experiments we use seven base scenarios within a social inference domain—inferences drawn from social media information and corroborated with background knowledge to draw potentially unwanted conclusions [12]. While our work is generalisable to other domains, this seemed to lend itself well to the design of the experiments. The base scenarios are derived from reported incidents in the context of sharing political views [9], and location data or temporal information [12]. These base scenarios are built using a combination of arguments from position to know and cause to effect [17].

Each scenario is referred to as Xi with $1 \leq i \leq 7$ and designed as a set of propositions, where each proposition $r_j \in \mathcal{L}$. In order to collect a relatively large amount of data with less cognitive effort for the user, two propositions per base scenario are chosen and tested by a single subject within our experiments. We combine propositions and base scenario using the same notation, writing Xi_j, where $j = \{0, 1\}$ refers to the proposition being tested. For convenience, we call Xi_j a *scenario*. Given 7 base scenarios and 2 propositions, we obtain a total of 14 scenarios.

5.3   Interpretations

For each scenario, two interpretations can be made:

At: An interpretation building on the number of extensions in which the conclusion is acceptable (via an argument theory AT, with rules and contraries between propositions), in which a justification ratio $\mu(r_i)$ is associated with each proposition r_i.

Pt: A possible worlds based probabilistic interpretation, in which each proposition r_i is associated with a probability $p(r_i)$ of its information being verified.

We associate the justification ratio of a conclusion r_i as the outcome of the probability semantics, with the likelihood that that piece of information is verified (e.g., is shared). Given that both interpretations are based on the same set of propositions, the key design link is such that the justification ratio of r_i within At is the same as the probability of r_i in Pt. Given this equivalence, we refer to this ratio as $\tau = \mu(r_i) = p(r_i)$. With two interpretations per scenario, we obtain 28 experimental conditions α. The next factors are further characteristics of conditions α.

5.4   Fractions, Percentages, and Ratios

In Domain 1 (see Sect. 5.1), the ratio τ of a proposition is an irreducible fraction varied between $1/6$ and $2/3$. That is, we ensured that the conclusion occurred in τ of the extensions. Besides the main objectives of the experiments, we want to show two further properties: that the scenario has limited influence on the results, and that the ratio—rather than the number of extensions—is the key factor that influences user believability ratings. For demonstrating the latter, we introduce redundant equivalent fractions γ (e.g., $1/2, 2/4, 3/6$) corresponding to the ratios τ using experimental conditions with 2, 3, 4, or 6 extensions. Each scenario Xi_j is associated with a fraction γ.

In Domain 2, we maintain the same fractions γ but also introduce another value, ω, representing the likelihood of the event described within the content of a proposition. For example, a proposition r_a = "Joe is a Republican" becomes r_b = "There is 70% chance that Joe is a Republican". We vary ω between 20% and 80% and the overall ratio is given by the product $\tau = \gamma * \omega$. Fractions γ and percentages ω in a scenario are associated using different combinations of both low or high ω and γ, or high ω and low γ and vice-versa.

User experiment task

In this domain, we intend to identify whether Joe is a Republican. Please read the following information and inference rules representing the domain.

- Initial Information:
  - Joe was not at the Labor Union (LU) office last week
  - Joe was offered a job at the LU
  - Joe is a member of a political party

- Inference rules:
  - If Joe is a member of a political party, then Joe could be a Democrat
  - If Joe is a member of a political party, then Joe could be a Republican
  - If Joe is a Republican, then Joe would not believe in LUs
  - If Joe does not believe in LUs, then Joe could not have taken the job
  - If Joe was offered a job at the LU, then Joe could have taken the job
  - If Joe has taken the job, then Joe could have a job at the LU
  - If Joe was not at the LU office last week, then Joe could not have a job at the LU

From this representation we also obtain the following 3 possible worlds.

- Possible Worlds:
  1: (Joe is a Democrat, Joe has taken the job, Joe has got a job at the LU)
  2: (Joe is a Democrat, Joe does not have a job at the LU, Joe has taken the job)
  3: (Joe is a Republican++, Joe does not have a job at the LU, Joe does not believe in LUs)

Corresponding argument framework

Mapping:
- r_{office} = "Joe was at the LU office last week"
- r_{offer} = "Joe was offered a job at the LU"
- r_{party} = "Joe is a member of a political party"
- r_{dem} = "Joe is a Democrat"
- r_{rep} = "Joe is a Republican"
- r_{taken} = "Joe has taken the job"
- r_{job} = "Joe has got a job at the LU"
- r_{bel} = "Joe believes in Labor Unions"

Knowledge base:
- $K = \{\neg r_{office}; r_{offer}; r_{party}\}$
- $\mathcal{R} = \{r_{party} \Rightarrow r_{dem}; r_{party} \Rightarrow r_{rep};$
  $r_{rep} \Rightarrow \neg r_{bel}; r_{offer} \Rightarrow r_{taken};$
  $r_{taken} \Rightarrow r_{job}; \neg r_{office} \Rightarrow \neg r_{job}\}$
- $\bar{\ } = \{(\neg r_{bel}, r_{taken}); (r_{dem}, r_{rep}); (r_{rep}, r_{dem})\}$

Arguments:
- $A_1 : \neg r_{office}$
- $A_2 : r_{offer}$
- $A_3 : r_{party}$
- $A_4 : A_3 \Rightarrow r_{dem}$
- $A_5 : A_3 \Rightarrow r_{rep}$
- $A_6 : A_5 \Rightarrow \neg r_{bel}$
- $A_7 : A_2 \Rightarrow r_{taken}$
- $A_8 : A_7 \Rightarrow r_{job}$
- $A_9 : A_1 \Rightarrow \neg r_{job}$

Extensions:
- $\xi_1 = \{A_1, A_2, A_3, A_4, A_7, A_8\}$
- $\xi_2 = \{A_1, A_2, A_3, A_5, A_6, A_9\}$
- $\xi_3 = \{A_1, A_2, A_3, A_4, A_7, A_9\}$

Abstract framework:

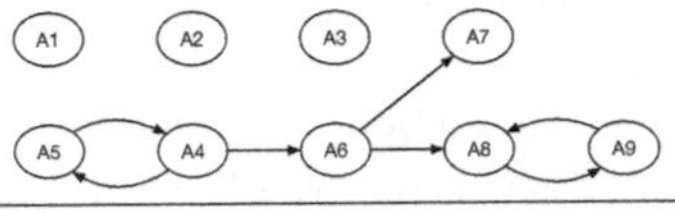

Fig. 2. User experiment argument interpretation & framework – Domain 1

Example 2. To obtain an argument theory based interpretation, one of our scenarios presented the user with a set of premises, and grounded defeasible rules from which arguments can be formed. We then only presented the conclusions of arguments from the preferred extensions which result from our framework. For example, Fig. 2 presents an example of the argument theory interpretation that is shown to the user during the experiment, and below its correspondent argument framework. In this scenario, 3 preferred extensions existed referred to as possible worlds, and the conclusion r_{rep} = "Joe is a Republican" is valid in one of these extensions. The experimental condition corresponding to this example is $\alpha_1 = \langle$Domain:1, Scenario:$X1\_1$, Proposition r_{rep}:"Joe is a Republican", Interpretation:At, Percentage ω:1, Fraction γ:1/3, Ratio τ:1/3$\rangle$. We then asked the user:

> *Given the 3 stated possible worlds, how likely is that you would believe that "Joe is a Republican"?*

The user's response to the question is recorded as $u_\mu(r_{rep}) = \{1, \ldots, 5\}$ on a 5-points Likert scale. Assuming that the extensions are equiprobable, we obtain: $P(\xi_1) = P(\xi_2) = P(\xi_3) = 1/3$ as shown in Fig. 2. The justification ratio for the tested proposition r_{rep} obtained is $\mu(r_{rep}) = 1/3$.

User experiment task

In this domain, we intend to identify whether Joe is a Republican. Assume that we have a stream of information composed by one or many copies of the following messages.

- Joe was not at the Labor Union office last week
- Joe was offered a job at the Labor Union
- Joe is a member of a political party
- Joe is a Democrat
- Joe is a Republican ++
- Joe has taken the job
- Joe has got a job at the Labor Union
- Joe does not have a job at the Labor Union
- Joe does not believe in Labor Unions

Fig. 3. User experiment probabilistic interpretation – Domain 1

To obtain a correspondent probabilistic interpretation, we presented the set of propositions to the user as a list of hypothetical messages, which included both premises and conclusions of the above argumentation framework in no particular order. In Fig. 3 we present the corresponding experimental scenario. The user was informed that a stream of information would release a number of messages from the list, and asked to comment on the likelihood that a message would state the tested proposition. In this scenario, we also informed the user that 1 out of 3 messages reported that "Joe is a Republican". The experimental condition corresponding to this example is $\alpha_2 = \langle$Domain:1, Scenario:$X1\_1$, Proposition r_{rep}:"Joe is a Republican", Interpretation:Pt, Percentage ω:1, Fraction γ:1/3, Ratio τ:1/3$\rangle$, where the only difference with α_1 is the interpretation. To determine $u_p(r_{rep})$, the user was asked the question:

If 3 messages are released, how likely is that a message would state that "Joe is a Republican"?.

For these scenarios, $\tau = 1/3$, and $\gamma = 1/3$. In the experiments we also tested for situations in which $\tau = 1/3$, and $\gamma = 2/6$ for example constructing a similar domain with 6 extensions, where r_{rep} was valid in only 2 of those. The justification ratio of a proposition in At corresponds to the probability in Pt in Domain 1 such that $p(r_{rep}) = \mu(r_{rep}) = \gamma = \tau$. In Domain 2, the proposition r_{prep} = "There is 90% chance that Joe is a Republican" is used instead, with $\omega = 0.9$ in both interpretations and $\omega * p(r_a) = \omega * \mu(r_a) = \omega * \gamma = \tau$. Figure 4 shows an example of the experiment scenario including r_{prep} for the argument interpretation At. The corresponding probabilistic interpretation Pt can be derived by extracting all the propositions from this scenario.

User experiment task

In this domain, we intend to identify whether Joe is a Republican. Please read the following information and inference rules representing the domain.

- Initial Information:
  - Joe was not at the Labor Union (LU) office last week
  - Joe was offered a job at the LU
  - Joe is a member of a political party

- Inference rules:
  - If Joe is a member of a political party, then there is 10% chance that Joe could be a Democrat
  - If Joe is a member of a political party, then there is 90% chance that Joe could be a Republican
  - If Joe is a Republican, then Joe would not believe in LUs
  - If Joe does not believe in LUs, then Joe could not have taken the job
  - If Joe was offered a job at the LU, then Joe could have taken the job
  - If Joe has taken the job, then Joe could have a job at the LU
  - If Joe was not at the LU office last week, then Joe could not have a job at the LU

From this representation we also obtain the following 3 possible worlds.

- Possible Worlds:
  1: (There is 10% chance that Joe is a Democrat
     Joe has taken the job,
     Joe has got a job at the LU)
  2: (There is 10% chance that Joe is a Democrat
     Joe does not have a job at the LU,
     Joe has taken the job)
  3: (There is 90% chance that Joe is a Republican++,
     Joe does not have a job at the LU,
     Joe does not believe in LUs)

Fig. 4. User experiment argument interpretation – Domain 2

6   Methodology and Results

We ran our experiments using Amazon Mechanical Turk[2], a web service that recruits participants to complete tasks. We recruited 420 participants for the experiment from the USA[3]. Data collection was performed with a questionnaire

---

[2] Amazon Mechanical Turk: https://www.mturk.com/.

[3] Ethical approval for these experiments was granted by the College Ethics Review Board of the University of Aberdeen on 10/08/2016.

including four experimental conditions, such that a participant would see two different scenarios, and respond to questions of both problems and interpretations. Initially participants were shown a training example for the argumentation theory to provide them with a basic understanding of argumentation. Each participant was then asked to respond to four combinations of different experimental conditions (α as described in Sect. 5).

- Domain 1: two questions within a scenario Xi, related to conditions Xi_0 and Xi_1 and an interpretation At (or Pt).
- Domain 2: two questions within a scenario Xj, where $i \neq j$, related to conditions Xj_0 and Xj_1, and an interpretation Pt (or At respectively).

Hence, no user would respond to an interpretation At and its corresponding interpretation Pt, and each user would see two different domains. We obtained 30 responses per condition α. In the remainder of the section, we detail they hypotheses associated with each type of problem, and describe our results.

6.1   Domain 1: Hypotheses

The aim of the first set of experiments is to understand whether people agree with the outcome of the probabilistic semantics when the propositions are purely qualitative. We study the believability rating of a proposition r_i in interpretation At as the outcome of the probabilistic semantics $u_\mu(r_i)$, and in the corresponding probabilistic interpretation Pt, $u_p(r_i)$. Our hypotheses are as follows.

H1.1: There is a correlation between the believability rating of At, $u_\mu(r_i)$, and the justification ratio of the conclusions, $\mu(r_i)$, obtained via the outcome of the probabilistic semantics.
H1.2: There is a correlation between the believability rating of Pt, $u_p(r_i)$, and the probability of the information being verified $p(r_i)$.
H1.3: The two correlations in At and Pt are similar.

We also test the following secondary hypotheses:

H1.4: The scenario does not influence the results: for any two scenarios with the same fraction γ there is no difference in the believability rating.
H1.5: The number of extensions does not influence the results: for any two scenarios with same τ but different γ there is no difference in the believability rating.

6.2   Domain 1: Results

Figure 5 presents the believability ratings $u_\mu(r_i)$ and $u_p(r_i)$ recorded for Domain 1. The horizontal axis is ordered according to the fraction γ associated with the experimental conditions. We also report the ratio τ corresponding to the fraction. For each scenario, $u_\mu(r_i)$ of At is shown besides $u_p(r_i)$ of Pt. The graph uses a divergent colour palette; the neutral rating is associated with the brightest colour, ratings below correspond to participants who consider the proposition unlikely, ratings above correspond to those who consider the conclusion likely. Moving from lower to higher γ (left to right), we observe that the darker area above the neutral bars increases for both At and Pt interpretations. Within each scenario, the neutral bar is approximately within the same range, with some exceptions. This provides some initial evidence that there is a correlation between the believability ratings and fractions γ.

A Spearman's rank-order correlation was run for each scenario Xi_j to determine the relationship between the believability ratings $u_\mu(r_i)$ in At and the justification ratios $\mu(r_i) = \gamma$, the outcome of the probabilistic semantics. This non-parametric test is used since the results are not normally distributed. The test showed a positive correlation value, rs, which was statistically significant $(rs(418) = .288, p \ll 0.001)$. This provides evidence for hypothesis H1.1—that there is a correlation between the probabilistic semantics and the user believability rating of a conclusion. A similar test determined that there is a statistically significant positive correlation between the believability ratings $u_p(r_i)$ in Pt and the probabilities $p(r_i) = \gamma$ $(rs(418) = .280, p \ll 0.001)$. This validates hypothesis H1.2; i.e., there is a correlation between the believability rating of a piece of information and its probability. A comparison between the two correlations was examined using a Fisher's r-to-z transformation. The overall z-score value (based on the difference between the correlations and their variance) was observed to be $z = 0.13$ with $p = 0.448$. Here, we accept the null hypotheses that the two

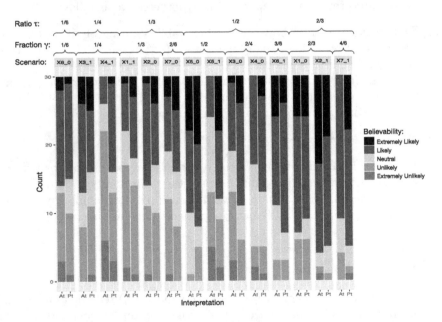

Fig. 5. Believability ratings $u_\mu(r_i)$ and $u_p(r_i)$ - Domain 1

Table 1. Mann-Whitney U tests on $u_\mu(r_i)$ vs. $u_p(r_i)$ within scenarios

Scenario	$X6\_0$	$X3\_1$	$X4\_1$	$X1\_1$	$X2\_0$	$X7\_0$	$X5\_0$
p-value	0.824	0.516	0.010*	0.265	0.888	0.247	0.744
Scenario	$X5\_1$	$X3\_0$	$X4\_0$	$X6\_1$	$X1\_0$	$X2\_1$	$X7\_1$
p-value	0.005*	0.015*	0.254	0.710	0.771	0.357	0.014*

correlations are not significantly different. This confirms hypothesis H1.3, and characterises how people interpret the outcome of the probabilistic semantics.

There are, however, some outliers that can be noticed in Fig. 5. This was investigated with a post-hoc analysis using a series of Mann-Whitney U tests for each scenario Xi_j comparing $u_\mu(r_i)$ and $u_p(r_i)$. Table 1 reports only the p-values, where we consider significance at $p < 0.001$. None of the comparisons shows a significant difference, however, for the three scenarios marked with a star (*), the p-value tends to be low indicating the outliers.

Similar tests are used for the two secondary hypotheses. H1.4 seeks to prove that given the same fraction γ (e.g. 1/3), there is no difference between the believability rate of different scenarios associated to that fraction (e.g. $X1\_1$ vs. $X2\_0$). In Table 2 we report the p-values of comparisons between different conditions, where significant values are highlighted in bold. Hypothesis H1.4 is only partially supported: the scenario tends not to influence the results in Pt, however, in At, the hypothesis is only supported in 3 out of 5 conditions.

Hypothesis H1.5 focussed on understanding the believability ratings in experimental conditions associated with different fractions γ but same ratio τ (e.g. 1/2 for $X5\_0$ vs. 3/6 for $X6\_1$). In Table 3 we report the p-values for comparisons between these conditions. H1.5 is mainly supported, with the exception of three cases in At. This provides partial evidence that it is the ratio rather than the fraction that influences the believability ratings among different conditions.

Table 2. Mann-Whitney U tests on At and Pt between scenarios with similar γ

Fraction γ	X_a	X_b	$u_\mu(r_i)$ vs. $\mu(r_i)$	$u_p(r_i)$ vs. $p(r_i)$
1/4	$X3\_1$	$X4\_1$	0.403	**0.000**
1/3	$X1\_1$	$X2\_0$	0.407	0.660
1/2	$X5\_0$	$X5\_1$	0.259	**0.000**
2/4	$X3\_0$	$X4\_0$	0.669	0.208
2/3	$X1\_0$	$X2\_1$	0.147	0.056

Table 3. Mann-Whitney U tests on At and Pt between scenarios with similar τ

Ratio τ	X_a	X_b	$u_\mu(r_i)$ vs. $\mu(r_i)$	$u_p(r_i)$ vs. $p(r_i)$
1/3	$X1\_1$	$X7\_0$	0.201	0.187
1/3	$X2\_0$	$X7\_0$	0.629	0.579
1/2	$X5\_0$	$X3\_0$	0.147	**0.001**
1/2	$X5\_0$	$X4\_0$	0.068	0.006
1/2	$X5\_0$	$X6\_1$	0.417	0.526
1/2	$X5\_1$	$X3\_0$	0.932	0.370
1/2	$X5\_1$	$X4\_0$	0.677	0.016
1/2	$X5\_1$	$X6\_1$	0.574	**0.000**
1/2	$X3\_0$	$X6\_1$	0.353	0.003
1/2	$X4\_0$	$X6\_1$	0.147	0.035
2/3	$X1\_0$	$X7\_1$	0.244	0.169
2/3	$X2\_1$	$X7\_1$	0.799	**0.001**

6.3   Domain 2: Hypotheses

The second problem focusses on understanding whether the outcome of the probabilistic semantics is a factor in assessing the believability of conclusions that are about event likelihood. We hypothesised that the product between the justification ratio of a conclusion and its likelihood influences people's believability ratings in the At interpretation and is comparable with the multiplication law in the probability interpretation Pt. We consider similar hypotheses as in Domain 1, with the difference that the believability rating is now tested for correlation with the product of the fraction γ and the likelihood w expressed within the content of a proposition ($\tau = \gamma * w$). Hypothesis H2.1 tests for correlation in the interpretation At where $\mu(r_j) = \tau$. Hypothesis H2.2 tests for correlation in Pt where $p(r_j) = \tau$ and H2.3 tests for similarity between the two correlations.

6.4   Domain 2: Results

Our initial tests study the correlation between the believability ratings and the fractions γ or the likelihood w alone. Statistical tests were performed using the Spearman's rank-order correlation, and similarity is tested using the Fisher's r-to-z transformation, with significance at $p < 0.001$. We observed no correlation for fraction γ in both interpretations At ($rs(418) = .59, p = 0.228$) and Pt ($rs(418) = .26, p = 0.596$). There is, instead, a low correlation with w in both At ($rs(418) = .193, p \ll 0.001$) and Pt ($rs(418) = .184, p \ll 0.001$) with high similarity ($z = 0.13, p = 0.448$). More interestingly, we found a correlation between the product of γ and w reflecting the multiplication law of probability in both At ($rs(418) = .293, p \ll 0.001$) and Pt ($rs(418) = .250, p \ll 0.001$) with similar behaviour ($z = 0.67, p = 0.251$). We now focus on this last result.

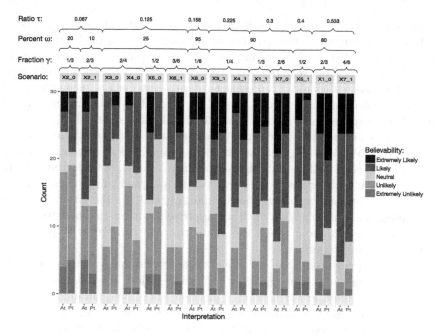

Fig. 6. Believability rating $u_\mu(r_i)$ and $u_p(r_i)$ - Domain 2

In Fig. 6, we present the believability rating $u_\mu(r_i)$ and $u_p(r_i)$ recorded for Domain 2. The horizontal axis is ordered according to $\tau = \gamma * \omega$. The results support hypothesis H2.1 for At: there is a positive correlation between the believability rating and the product of the likelihood expressed within a conclusion and the justification ratio due to the probabilistic semantics. The outcome of the probabilistic semantics is a factor required to interpret the believability ratings: the correlation with the likelihood expressed within a conclusion alone is low ($rs = .193$) and moderately improves when the product is used ($rs = .293$). Similar behaviour is observed in Pt supporting H2.2: there is a correlation between the believability rating and the product of the likelihood expressed within the proposition and its probability of occurring. This is stronger than the correlation with the former only ($rs = .250$ vs. $rs = .184$). Finally, H2.3 is supported as no significant difference between the two correlations values is observed.

7   Discussion

We have demonstrated that the outcome of Thimm's probabilistic semantics is an important factor in understanding the believability ratings of the conclusions, even in the case in which a proposition is about the likelihood of an event. The results indicate that people tend to agree with the outcome of the probabilistic semantics. Furthermore, our results confirm that the outcome of the probability semantics may be understood by people in a way similar to the understanding of probability. In the second problem, we showed that this similarity is due to a heuristic associating the product of probabilities to the believability of conclusions. Note that as discussed in Sect. 4, the multiplication law assumes that there is independence between the event reported by the proposition and it being inferred. We also tested for τ representing dependent events, using the law of conditional probability. The results showed no correlation with the believability ratings. Due to space constraints, we have omitted these results.

The results of our study are built on a standard (structured) approach to argumentation. While other techniques, such as weighted argumentation could have been used (and will be investigated as future work), we selected the approach used in this paper due to (1) the widely accepted and well understood nature of the standard argumentation semantics; and (2) the ease with which multi-extension semantics from such an approach can be mapped to a many worlds interpretation, from which the comparison to a frequentist probability interpretation can be performed.

The results presented here are—in a sense—preliminary. There are many aspects of this research that need further investigation. To name some, both correlation coefficients are significantly positive but show a moderate correlation between the degree of believability and the justification ratio or associated probability. This suggests that other factors need to be investigated further in the future. One of these aspects is the role of the domains used within the scenarios as we have shown that in the argumentation interpretation this has a more significant role than in the probabilistic view. From an argumentation perspective, further studies should focus on considering other semantics, such as the ranking-based semantics [3]. Further studies should also focus on understanding how people combine probabilities and on analysing human factors, for example, by considering the background of participants involved. We also wish to investigate how cycles and self attacks in the argument graphs, as well as the introduction of preferences may affect our results.

8   Conclusions

We investigated whether qualitative argumentation captures some notion of uncertainty by associating a degree of believability of conclusions with the number of preferred extensions. To do so, we examined whether people agree with the outcome of the probabilistic semantics. More broadly, our work can be seen to follow a strand of research similar to that of Cerutti et al. [5], aiming to study the alignment between argumentation semantics and human intuition. The novelty of our work is in that we focus on the particular role that multiple extensions play in evaluating the believability of a conclusion.

In this paper, we designed our experiments with a two-fold objective: to determine whether our claim was valid; and to investigate whether there is a similarity between probabilistic and argumentation-based reasoning. Our results show that people tend to agree with the outcome of the probabilistic semantics and that people employ a similar heuristic in understanding both preferred extensions and probabilities. Through our experiments, we obtained some initial promising insights into the use of probability within argumentation frameworks that may guide researchers in better supporting human reasoning in their work.

Acknowledgements. This work was partially funded by a grant to the University of Aberdeen made by the UK Economic and Social Research Council; Grant reference ES/MOO1628/1.

References

1. Bailin, S., Battersby, M.: Conductive argumentation, degrees of confidence, and the communication of uncertainty. In: van Eemeren, F.H., Garssen, B. (eds.) Reflections on Theoretical Issues in Argumentation Theory. AL, vol. 28, pp. 71–82. Springer, Cham (2015). https://doi.org/10.1007/978-3-319-21103-9_5
2. Bench-Capon, T.J.M.: Persuasion in practical argument using value-based argumentation frameworks. J. Logic Comput. **13**(3), 429–448 (2003)
3. Bonzon, E., Delobelle, J., Konieczny, S., Maudet, N.: A comparative study of ranking-based semantics for abstract argumentation. In: Proceedings of the 30th AAAI Conference on Artificial Intelligence, pp. 914–920 (2016)
4. Cayrol, C., Lagasquie-Schiex, M.C.: Graduality in argumentation. J. Artif. Intell. Res. **23**(1), 245–297 (2005)
5. Cerutti, F., Tintarev, N., Oren, N.: Formal arguments, preferences, and natural language interfaces to humans: an empirical evaluation. In: Proceedings of the 21st European Conference on Artificial Intelligence, pp. 207–212 (2014)
6. Croitoru, M., Vesic, S.: What can argumentation do for inconsistent ontology query answering? In: Liu, W., Subrahmanian, V.S., Wijsen, J. (eds.) SUM 2013. LNCS (LNAI), vol. 8078, pp. 15–29. Springer, Heidelberg (2013). https://doi.org/10.1007/978-3-642-40381-1_2
7. Dunne, P.E., Hunter, A., McBurney, P., Parsons, S., Wooldridge, M.: Weighted argument systems: basic definitions, algorithms, and complexity results. Artif. Intell. **175**(2), 457–486 (2011)
8. Haenni, R.: Probabilistic argumentation. J. Appl. Logic **7**(2), 155–176 (2009)
9. Heatherly, R., Kantarcioglu, M., Thuraisingham, B.: Preventing private information inference attacks on social networks. IEEE Trans. Knowl. Data Eng. **25**(8), 1849–1862 (2013)
10. Hunter, A., Thimm, M.: Probabilistic argument graphs for argumentation lotteries. In: Computational Models of Argument, Frontiers in Artificial Intelligence and Applications, vol. 266, pp. 313–324. IOS Press (2014)
11. Li, H., Oren, N., Norman, T.J.: Probabilistic argumentation frameworks. In: Modgil, S., Oren, N., Toni, F. (eds.) TAFA 2011. LNCS (LNAI), vol. 7132, pp. 1–16. Springer, Heidelberg (2012). https://doi.org/10.1007/978-3-642-29184-5_1
12. Mayer, J.M., Schuler, R.P., Jones, Q.: Towards an understanding of social inference opportunities in social computing. In: Proceedings of the 17th ACM International Conference on Supporting Group Work, pp. 239–248 (2012)
13. Modgil, S., Prakken, H.: The ASPIC+ framework for structured argumentation: a tutorial. Argum. Comput. **5**(1), 31–62 (2014)
14. Prakken, H.: An abstract framework for argumentation with structured arguments. Argum. Comput. **1**(2), 93–124 (2010)

15. Tang, Y., Cai, K., McBurney, P., Sklar, E., Parsons, S.: Using argumentation to reason about trust and belief. J. Logic Comput. **22**(5), 979 (2012)
16. Thimm, M.: A probabilistic semantics for abstract argumentation. In: Proceedings of the Twentieth European Conference on Artificial Intelligence, pp. 750–755 (2012)
17. Walton, D., Reed, C., Macagno, F.: Argumentation Schemes. Cambridge University Press, Cambridge (2008)
18. Zenker, F.: Bayesian Argumentation: The Practical Side of Probability. Synthese Library, vol. 362. Springer, Heidelberg (2013). https://doi.org/10.1007/978-94-007-5357-0. pp. 1–11

A Dynamic Model of Trust in Dialogues

Gideon Ogunniye[1]([✉]), Alice Toniolo[2], and Nir Oren[1]

[1] Department of Computing Science,
University of Aberdeen, Aberdeen, Scotland, UK
g.ogunniye@abdn.ac.uk
[2] School of Computer Science, University of St Andrews,
St. Andrews, Scotland, UK

Abstract. In human interactions, trust is regularly updated during a discussion. For example, if someone is caught lying, any further utterances they make will be discounted, until trust is regained. This paper seeks to model such behaviour by introducing a dialogue game which operates over several iterations, with trust updates occurring at the end of each iteration. In turn, trust changes are computed based on intuitive properties, captured through three rules. By representing agent knowledge within a preference-based argumentation framework, we demonstrate how trust can change over the course of a dialogue.

1 Introduction

Within a dialogue, participants exchange arguments, aiming to achieve some overarching goals. Typically, these participants have partial information and individual preferences and goals, and the parties aim to achieve an outcome based on these individual contexts. Importantly, some dialogue participants may be malicious or incompetent, and—to achieve desirable dialogical outcomes—the inputs from these parties should be discounted. In human dialogues, such participants are characterised by the lack of trust ascribed to them, and in this work we consider how such trust should be computed.

While previous work [12] has considered how the trust of participants should be updated *following* a dialogue, we observe that in long-lasting human discussions, trust can change during the dialogue itself. For example, within a courtroom, a witness who repeatedly appears to lie will not be believed even if they later act honestly. Trust can be viewed as making the arguments of more trusted agents be preferred—in the eyes of those observing the dialogue—to the arguments of less trusted agents. Importantly, there appears to be a feedback cycle at play within dialogue: low trust in a dialogue participant can lead to further reductions of trust as they are unable to provide sufficient evidence to be believed. To accurately model dialogue and reason about the trust ascribed to its participants, it is critical to take this feedback cycle between utterances and trust into account. This paper considers such a feedback cycle.

© Springer International Publishing AG, part of Springer Nature 2018
E. Black et al. (Eds.): TAFA 2017, LNAI 10757, pp. 211–226, 2018.
https://doi.org/10.1007/978-3-319-75553-3_15

The research questions we address in this work are as follows. (1) How should trust change during the course of a dialogue based on the utterances made by dialogue participants? (2) How should trust affect the justified conclusions obtained from a dialogue?

To answer these questions, we describe a dialogue model in which participants interact by exchanging arguments. Within this model, we define a trust relation for each participant with respect to other participants (encoded as a preference ordering over the participants), and describe how each participant updates its trust relation. In particular, each participant observes the behaviours of others and uses these observations as an input to update its trust relation (for the other participants) through a trust update function.

To compute the justified conclusions of a dialogue, we instantiate a preference-based argumentation framework (PAF) [1]. As a result, each participant can identify its own set of preferred conclusions, and a set of justified conclusions can be identified from these sets.

The proposed framework permits us to better represent the feedback relationship between trust and dialogue. The remainder of the paper is organised as follows: Sect. 2 recalls preference-based argumentation frameworks [1] and provides a brief overview of our notion of trust in dialogues. Section 3 describes our proposed dialogue model. Section 4 describes the trust update rules and the process we considered for dynamically updating trust within our dialogue model. Section 5 describes how the preference-based argumentation framework is instantiated in our model. Section 6 illustrates how trust update rules are applied through an example. Section 7 compares our approach with some existing works. Section 8 presents our conclusions and some directions for future work.

2   Background

Preference-based argumentation frameworks extend abstract argumentation frameworks [7], and we therefore begin by describing the former.

Definition 1. *An Argumentation Framework $\mathcal{F}$ is defined as a pair $\langle \mathcal{A}, \mathcal{R} \rangle$ where $\mathcal{A}$ is a set of arguments and $\mathcal{R}$ is a binary attack relation on $\mathcal{A}$.*

Extensions are sets of arguments that are, in some sense, justified. These extensions are computed using one of several *argumentation semantics*.

Preference-based argumentation frameworks [1] seek to capture the relative strengths of arguments and can be instantiated in different ways. In this paper, we will use preference-based argumentation frameworks to encode trust in other dialogue participants, allowing us to compute which arguments should, or should not be considered justified.

Within a preference-based argumentation framework, preferences are encoded through a reflexive and transitive binary relation $\geq$ over the arguments of $\mathcal{A}$. Given two arguments $\phi_1, \phi_2 \in \mathcal{A}$, $\phi_1 \geq \phi_2$ means that ϕ_1 is at least as preferred as ϕ_2. The relation $>$ is the strict version of $\geq$ i.e., $\phi_1 > \phi_2$ iff $\phi_1 \geq \phi_2$ but $\phi_2 \not\geq \phi_1$. As usual, $\phi_1 = \phi_2$ iff $\phi_1 \geq \phi_2$ and $\phi_2 \geq \phi_1$.

Given this, a preference-based argumentation framework is defined as follows.

Definition 2. *A Preference-based argumentation framework (PAF for short) [1] is a tuple $\mathcal{T} = \langle \mathcal{A}, \mathcal{R}, \geq \rangle$ where $\mathcal{A}$ is a set of arguments, $\mathcal{R} \subseteq \mathcal{A} \times \mathcal{A}$ is an attack relation and $\geq\; \subseteq \mathcal{A} \times \mathcal{A}$ is a (partial or total) preorder on $\mathcal{A}$. The extensions of $\mathcal{T}$ under a given semantics are the extensions of the argumentation framework $(\mathcal{A}, \mathcal{R}_r)$, called the* repaired *framework, under the same semantics with $\mathcal{R}_r = \{(\phi_1, \phi_2)|(\phi_1, \phi_2) \in \mathcal{R} \text{ and } (\phi_2 \not> \phi_1)\} \bigcup \{(\phi_2, \phi_1)|(\phi_1, \phi_2) \in \mathcal{R} \text{ and } \phi_2 > \phi_1\}$.*

Given a PAF, one can identify different sets of justified conclusions by considering different extensions. PAFs extend standard Dung argumentation frameworks with the addition of preferences between arguments to repair *critical attacks* and refine the extension of the repaired PAF. Therefore, we also define the semantics of standard argumentation frameworks, the notion of critical attacks and extension refinement. In this paper we will focus on the preferred semantics.

Definition 3. *Given $\mathcal{F} = \langle \mathcal{A}, \mathcal{R} \rangle$, a set of arguments $\mathcal{E} \subseteq \mathcal{A}$ is said to be* conflict-free *iff $\forall \phi_1, \phi_2 \in \mathcal{E}$, there is no $(\phi_1, \phi_2) \in \mathcal{R}$. Given an argument $\phi_1 \in \mathcal{E}$, $\mathcal{E}$ is said to defend ϕ_1 iff for all $\phi_2 \in \mathcal{A}$, if $(\phi_2, \phi_1) \in \mathcal{R}$ then there is a $\phi_3 \in \mathcal{E}$ such that $(\phi_3, \phi_2) \in \mathcal{R}$. $\mathcal{E}$ is* admissible *iff it is conflict-free and defends all its elements. $\mathcal{E}$ is a* complete extension *iff there are no other arguments which it defends. $\mathcal{E}$ is a* preferred extension *iff it is a maximal (with respect to set inclusion) complete extension.*

Preferred semantics admit multiple extensions; here, such an extension represents a potentially justified view (which conflicts with other views). If an argument is present in all extensions, then it is *sceptically* justified; while if it is present in at least one extension, it is *credulously* justified.

Definition 4. *(Critical attack) [1]. Let $\mathcal{F}$ be an argumentation framework and $\geq\, \subseteq \mathcal{A} \times \mathcal{A}$. An attack $(\phi_2, \phi_1) \in \mathcal{R}$ is* critical *iff $\phi_1 > \phi_2$.*

PAFs repair critical attacks on the graph of attacks by *inverting* the arrow of the attack relation (i.e., $(\phi_2, \phi_1) \in \mathcal{R}$ with $\phi_1 > \phi_2$ becomes $(\phi_1, \phi_2) \in \mathcal{R}$). This repair property ensures that arguments that are more preferred in an argumentation framework *defeat* arguments that are less preferred. An argument ϕ_1 defeats ϕ_2 iff $((\phi_1, \phi_2) \text{ or } (\phi_2, \phi_1)) \in \mathcal{R}$ and $\phi_1 > \phi_2$. For a symmetric attack relation, removing critical attacks gives the same results as inverting attacks. Extensions are then constructed from the corresponding repaired PAF using the semantics of $\mathcal{F}$. In addition, in PAFs, a refinement relation is used to refine the results of a framework by comparing its extensions.

Definition 5. *(Refinement relation) [1]. Let $(\mathcal{A}, \geq)$ be such that $\mathcal{A}$ is a set of arguments and $\geq\, \subseteq \mathcal{A} \times \mathcal{A}$ is a (partial or total) preorder. A refinement relation denoted by $\geq_r$, is a binary relation on $\mathcal{P}(\mathcal{A})^2$ such that $\geq_r$ is reflexive, transitive and for all $\mathcal{E} \subseteq \mathcal{A}$, for all $\phi_1, \phi_2 \in \mathcal{A}\backslash\mathcal{E}$, if $\phi_1 > \phi_2$ then $\mathcal{E} \bigcup \{\phi_1\} >_r \mathcal{E} \bigcup \{\phi_2\}$.*

Let Ags be a set of participants within a dialogue. We consider that each dialogue participant $Ag_{i \in Ags}$, for $i = 1, \ldots, n$, has an associated trust relation over other participants, encoded through a preference ordering $\succeq_{Ag_i}$.

Definition 6. *Let Ags be a set of dialogue participants. The trust relation of a given participant Ag_i over Ags is a preference ordering $\succeq_{Ag_i} \subseteq Ags \times Ags$. $Ag_j \succeq_{Ag_i} Ag_k$ denotes that Ag_i prefers (trusts) Ag_j to Ag_k.*

We consider the following properties for the trust relation:

- *Non-Symmetric:* if a participant Ag_i trusts another participant Ag_j, this does not imply that Ag_j trusts Ag_i.
- *Transitive:* Unlike some other works on trust [9,17], we assume that transitivity of trust (also known as *derived trust*) is not required in our model. As a result, we assume that a given participant has the ability to decide whether or not to trust another participant at any stage of the dialogue.

The trust relation represents the viewpoint of a given participant independently of the trust relations of other participants. Therefore, unlike the systems described in, for example, [9,17], there is no need to represent a 'global map' of trust relations—a trust network—in our model.

3   A Formal Dialogue Model

We consider a dialogue system where each participant Ag_i has two main components: a *knowledge base* (containing its trust relation over other participants, a set of arguments, and a set of attacks between arguments) and a *commitment store*. We follow Hamblin (as cited in [19]) in defining a commitment store as a "store of statements" that represents the arguments a participant is publicly committed to.

Definition 7. *The knowledge base of a participant $Ag_i \in Ags$ is a tuple $\mathcal{KB}_{Ag_i} = \langle A_{Ag_i}, R_{Ag_i}, \succeq_{Ag_i} \rangle$, where A_{Ag_i} is the set of arguments known by Ag_i (representing their own knowledge); $R_{Ag_i} \subseteq A_{Ag_i} \times A_{Ag_i}$ is a set of attacks where $(\phi_1, \phi_2) \in R_{Ag_i}$ iff $\phi_1 \in A_{Ag_i}$ and ϕ_2 is an argument provided by any participant Ag_j; and $\succeq_{Ag_i}$ is the trust relation (c.f., Definition 6) of Ag_i with regards to other participants.*

Each participant updates its knowledge base at the end of each iteration of a dialogue. Intuitively, an iteration represents a subdialogue, including an exchange of arguments arising from a participant's (potentially) controversial assertion. Unlike the knowledge base, the commitment store is updated after every dialogue move made by the participant.

Definition 8. *The commitment store of a participant $Ag_i \in Ags$ at iteration $t \in \{1 \ldots n\}$ is a set $CS_{Ag_i}^t = \{\phi_1, \ldots, \phi_n\}$ which contains arguments introduced into the dialogue by Ag_i at iteration t such that $CS_{Ag_i}^0 = \emptyset$.*

The union of the commitment stores of all participants is called the *universal commitment store* $\mathcal{UCS}^t = \bigcup_{Ag_i} CS^t_{Ag_i}$. An argument put forward by a participant may be attacked by an argument from another participant. Therefore, in our dialogue system, an argumentation framework $\langle \mathcal{UCS}^t, \mathcal{R} \rangle$ is induced by the set of arguments exchanged during dialogue in the universal commitment store and their respective attacking relationships as in [7]. Hence, $(\phi_1, \phi_2) \in \mathcal{R}$ if $(\phi_1, \phi_2) \in R_{Ag_i}$, $\phi_1 \in CS_{Ag_i}$ and $\phi_2 \in \mathcal{UCS}^t$. The universal commitment store can be viewed as the global state of the dialogue at a given iteration.

We now turn our attention to the dialogue game itself. A dialogue game like the one described in [13] specifies the major elements of a dialogue, such as its commencement, combination, and termination rules among others. Likewise, the system described in [11] specifies how the topic of discussion in a dialogue can be represented in some logical language. We are interested in how a participant updates its commitment store and its trust relation in a dialogue when it, or other participants, introduce arguments. We assume that at iteration t, a participant is allowed to add arguments to its commitment store if it is not already present within the store (and was not previously present), and retract arguments from its commitment store only if the argument was already present in the store.

3.1 Protocol Rules and Speech Acts

Protocol rules regulate the set of legal moves that are permitted at each iteration of a dialogue. In our framework, a dialogue consists of multiple discrete iterations t within which the moves are made. A dialogue move is referred to as M^t_x where $x, t \in \mathbb{N}$, denoting that a move with identifier x is made at iteration t. At its most general, a protocol identifies a legal move based on all previous dialogue moves.

Definition 9. *A dialogue D consists of a sequence of iterations such that $D = [[M^1_1, \ldots, M^1_x], \ldots, [M^t_1, \ldots, M^t_x]]$. The dialogue involves n participants $Ag_1, \ldots, Ag_n$ where $(n \geq 2)$. Within a dialogue D, iteration j consists of a sequence of moves $[M^j_1, \ldots M^j_x]$.*

A dialogue participant evaluates the set of arguments exchanged within an iteration to update its trust relation over other participants. Within each iteration, there is a claim to be discussed and arguments that attack or defend the claim. Note that a claim is abstractly represented as an argument. An iteration therefore represents a sub-discussion focused around a single topic of the overarching dialogue, which can be treated in an atomic manner with regards to trust.

The dialogue protocol is as described in Fig. 1. Each node—except the 'update' node (described in detail later)—represents a speech act, and the outgoing arcs from a node indicate possible responding speech acts. We consider four types of speech acts, denoted $assert(Ag_i, \phi, t)$, $contradict(Ag_i, \phi_1, \phi_2, t)$, $retract(Ag_i, \phi, t)$, and $exit$ respectively. A participant Ag_i uses $assert(Ag_i, \phi, t)$ to put forward a claim $\phi \in A_{Ag_i}$ at iteration t. A $contradict(Ag_i, \phi_1, \phi_2, t)$ move attacks a previous argument $\phi_1 \in A_{Ag_j}$ from another participant Ag_j by argument $\phi_2 \in A_{Ag_i}$ from participant Ag_i. A participant Ag_i uses $retract(Ag_i, \phi, t)$ to retract its previous argument. A participant uses $exit$ to exit an iteration. This move is made when a participant has no more arguments to advance within the iteration. When an iteration concludes (shown by the terminal *update* node in the figure), trust is updated. The dialogue then proceeds to the next iteration, or may terminate. A dialogue therefore consists of at least one, but potentially many more, iterations.

In addition to the constraints on the type of speech act that can be made in a dialogue, we also consider the *relevance* of a move. A move M_{x+i}^t, for $x, i \geq 1$ is *relevant* to iteration t if the argument of the move will affect the justification of the argument of the move M_x^t. Specifically, an argument ϕ_2 in move M_{x+i}^t affects the justification of an argument ϕ_1 in M_x^t if it attacks ϕ_1 (c.f., [14]). Relevance is defined from the second move of an iteration (i.e., when $x \geq 1$) because the first move is taken to introduce the claim to be discussed in the iteration. The protocol rules enforce that ϕ_2 is relevant to an iteration t if it affects the justification of ϕ_1 that has been previously moved in the iteration. However, if ϕ_1 is retracted in the iteration, ϕ_2 is no longer relevant and must be retracted except if it affects the justification of another argument ϕ_3. Furthermore, as the outgoing arcs in Fig. 1 depict, a move to exit an iteration is also considered relevant from the second move but a move to retract an argument is only considered relevant from the third move (i.e., when $x \geq 2$). These constraints help to prevent participants from making moves that are not relevant to the current iteration.

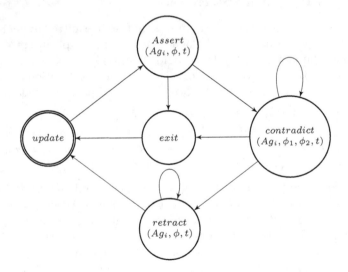

Fig. 1. Protocol rules

3.2 Commitment Rules

A participant's commitment store is revised throughout the dialogue as it advances arguments. Therefore, it is important to define how each of the proposed speech acts updates a participant's commitment store.

Definition 10. *The commitment store of a participant $Ag_i \in Ags$ is updated as follows:*

$$
CS_{Ag_i}^t = \begin{cases}
\emptyset & \text{iff } t = 0, \\
CS_{Ag_i}^{t-1} \bigcup \{\phi\} & \text{iff } m_x^t = assert(Ag_i, \phi, t), \\
CS_{Ag_i}^{t-1} \bigcup \{\phi_2\} & \text{iff } m_x^t = contradict(Ag_i, \phi_1, \phi_2, t) \\
CS_{Ag_i}^{t-1} \setminus \{\phi\} & \text{iff } m_x^t = retract(Ag_i, \phi, t) \\
CS_{Ag_i}^{t-1} & \text{iff } m_x^t = exit
\end{cases}
$$

4   Updating Trust

We now turn our attention to how trust should be updated as a dialogue progresses. We limit our focus to how the trust relation component of a participant's knowledge base ($\succeq_{Ag_i}$) is updated. A trust update function is used to perform this update when an iteration concludes, as represented by the 'update' node in Fig. 1.

As input, the trust update function takes a participant's *trust update rules* and its *preference on the trust update rules*. In the remainder of this section, we formalise both of these concepts.

Trust update rules describe the situations in which trust in a dialogue participant should change. In this paper, we consider the following trust update rules.

– A dialogue participant whose arguments are self-contradicting should be less trusted than a consistent participant.
– A dialogue participant who is unable to justify its arguments should be less trusted than one who can.
– A dialogue participant who regularly retracts arguments should be less trusted than one who does not.

These rules are similar to some of the properties that have been considered in the literature of ranking-based semantics for abstract argumentation (for a review on ranking-based semantics for abstract argumentation, see [3]). These rules are also supported by extension-based semantics (i.e., Dung's semantics [7]). For instance, the second rule could be represented as a participant having an argument ϕ in its commitment store, but not within an extension: $\phi \notin \mathcal{E}(\langle \mathcal{UCS}, \mathcal{R} \rangle)$[1]. We do not claim that the three trust update rules considered in this paper are exhaustive, and intend to investigate additional rules, taken from sources such as [3], in the future. We formalise the three trust update rules as follows.

Definition 11. *Self Contradicting Arguments (SC): A participant Ag_i is self contradicting if CS_{Ag_i} is not conflict free.*

---

[1] Here, $\mathcal{E}$ represents the extension(s) obtained on the argumentation framework $\langle \mathcal{UCS}, \mathcal{R} \rangle$.

Definition 12. *Lack of Justification (LJ): A participant Ag_i lacks justification for an argument ϕ_1 iff $\phi_1 \in CS_{Ag_i}$ and there is a $\phi_2 \in UCS \setminus CS_{Ag_i}$ such that ϕ_2 defeats ϕ_1.*

Defeats consider preferences among attacks and are defined in Sect. 2.

Definition 13. *Argument Retraction (AR): A participant Ag_i is inconsistent iff $\phi_1 \in CS_{Ag_i}$ and there is a $\phi_2 \in UCS \setminus CS_{Ag_i}$ such that ϕ_2 attacks ϕ_1 and Ag_i retracts ϕ_1 from CS_{Ag_i}.*

This rule also requires that if ϕ_2 attacks ϕ_1 and ϕ_1 is retracted by Ag_i, Ag_j is expected to retract ϕ_2 as enforced by the dialogue protocol without any loss of trust for Ag_j except if ϕ_2 attacks another argument ϕ_3 that is not retracted.

Given the three trust update rules considered, there are four possible combinations of these rules in an iteration. These possible combinations are given below.

- *(SC, LJ, AR):* This combination means all the three trust updates rules occur within a particular iteration under consideration.
- *(SC, LJ):* This combination means *self contradiction* and *lack of justification* occur within a particular iteration under consideration.
- *(SC, AR):* This combination means *self contradiction* and *argument retraction* occur within a particular iteration under consideration.
- *(LJ, AR):* This combination means *lack of justification* and *argument retraction* occur within a particular iteration under consideration.

Note that within an iteration, the arrangement of trust update rules in a combination is not important. For instance, *(SC, AR)* and *(AR, SC)* is considered to be the same combination.

Agents have preferences over trust update rules. For example, one may trust somebody who contradicts themselves much less than they trust someone who regularly retracts arguments. Such preferences on trust update rules are a partial order over trust update rules. This partial order specifies the order of importance a given participant attaches to the trust update rules.

Definition 14. *Let $TR_{Ags}^t = \{SC, LJ, AR\}$ be a set of trust update rules for the set of participants Ags at iteration t. A given participant's preference on TR_{Ags}^t is a partial ordering $\succeq_{Ag_{i(TR)}}^t$ such that for rules $X, Y \in TR_{Ags}^t$, $X \succeq_{Ag_{i(TR)}}^t Y$ denotes rule X has preference over rule Y in $\succeq_{Ag_{i(TR)}}^t$.*

Since we are concerned with the viewpoint of a given participant, dialogue participants may have varying preferences on trust update rules. Furthermore, such preferences may change from one iteration to another. For instance, in a particular iteration, a given participant may consider argument retraction as the least inconsistent behaviour if a target participant retracts an argument from its commitment store as a result of learning from the arguments of other participants that the retracted argument is inaccurate. This may not be the case if the target participant is forced to retract an argument from its commitment store as a result of its inability to advance other arguments to defend it.

If the preference on the trust update rules of a given participant Ag_i is $\succeq_{Ag_{i(TR)}} = (SC \succ_{Ag_{i(TR)}} LJ \succ_{Ag_{i(TR)}} AR)$, then, *self contradiction* is most important when updating the participant's trust relation, followed by *lack of justification* and *argument retraction* respectively.

Consider a dialogue participant Ag_i, with a trust update function denoted by $\mathcal{UF}$ at *iteration* t of a dialogue. The participant exchanges arguments with other participants in the dialogue through defined *speech acts* and *protocol rules*. It updates its *commitment store* $CS^t_{Ag_i}$ after each of its *moves* m^t_x in the dialogue. It observes some *trust updates rules* based on the observed behaviours of other participants in a particular iteration of the dialogue. As earlier stated, the commitment store of all dialogue participants is publicly observable. Ag_i updates its *trust relation* $\succeq_{Ag_i}$ over other participants based on its *trust update rules* and *preference on the rules* $\succeq^t_{Ag_{i(TR)}}$, repeating the process in the next iteration.

We formalise the trust update function as follows.

Definition 15. *Let $TR^t_{Ag_i}$ be the trust update rules of a given participant Ag_i; $\succeq^t_{Ag_{i(TR)}}$ be the participant's preference on the trust update rules; and $\succeq^t_{Ag_i}$ its trust relation over other participants at iteration $t \in \{1 \ldots n\}$. The trust update function $\mathcal{UF}$ is a function of the form $\mathcal{UF} \colon (TR^t_{Ag_i} \times \succeq^t_{Ag_{i(TR)}}) \to \succeq^t_{Ag_i}$ which takes in Ag_i's trust update rules and current trust preferences, and returns an updated set of trust preferences.*

A given participant's trust relation over other participants is updated via the trust update function. Such a relation provides the basis for computing what the participant deems justified in an iteration.

In the next section, we analyse how each participant computes extensions in their personalised preference-based argumentation frameworks.

5   Dialogue Outcome

Given an argumentation framework induced by the set of arguments exchanged during dialogue in the universal commitment store and their respective attacking relationships. Also, given a preference ordering over dialogue participants, we instantiate a PAF by providing a rational basis for the preferences between arguments. We prefer arguments $\phi_1 \geq \phi_2$ (or strictly prefer arguments $\phi_1 > \phi_2$) iff there are some dialogue participants Ag_i and Ag_j such that $\phi_1 \in CS_{Ag_i}, \phi_2 \in CS_{Ag_j}$ and $Ag_i \succeq Ag_j$ (respectively $Ag_i \succ Ag_j$). If there are critical attacks in $\langle \mathcal{UCS}, \mathcal{R} \rangle$, the attacks are repaired (c.f., Sect. 2). Moreover, the extensions generated from the $\langle \mathcal{UCS}, \mathcal{R} \rangle$ are refined as shown in Sect. 2.

Since the preference orderings over dialogue participants represent the viewpoint of a given participant in our model, it is possible to have as many preference orderings over participants as the number of participants in a dialogue. By implication, the notions of preferences between arguments; critical attacks; and argument defeat are relative to each participant. In what follows, we introduce the notion of a *participant* for a *PAF* similar to the notion of an *audience* in [2]. Participants are individuated by their preferences over other dialogue participants leading to their preferences between arguments. The arguments in the $\mathcal{UCS}$ will then be evaluated by each participant in accordance with its preferences between arguments. This leads to the following argument framework.

Definition 16. *Let Ags be a set of participants $\{Ag_1, \ldots, Ag_n\}$ then for $i = 1, \ldots, n$, the preference-base argumentation framework of participant Ag_i is a tuple $\mathcal{T}_{Ag_i} = \langle \mathcal{A}, \mathcal{R}, \succeq^{\mathcal{A}}_{Ag_i} \rangle$ where $\mathcal{A} \subseteq \mathcal{UCS}$ is a set of arguments, $\mathcal{R} \subseteq \mathcal{A} \times \mathcal{A}$ is an attack relation and $\succeq^{\mathcal{A}}_{Ag_i} \subseteq \mathcal{A} \times \mathcal{A}$ is a (partial or total) preorder on $\mathcal{A}$ according to Ag_i.*

An attack succeeds in the preference-based argumentation framework of a participant if it is not a critical attack or if the participant has no preference between the arguments. Thus, the set of *defeat relations* (attacks that succeed) in one participant's context may be different from the one in another participant's context. An argument $\phi_1 \in \mathcal{A}$ *defeats* another argument $\phi_2 \in \mathcal{A}$ iff $(\phi_1, \phi_2) \in \mathcal{R}$ and $\phi_2 \not\succeq^{\mathcal{A}}_{Ag_i} \phi_1$. Further, note that the *preferred semantics* of $\mathcal{T}_{Ag_i}$ may return a different refined preferred extension $\mathcal{E}_{Ag_i}$ to the *preferred semantics* of $\mathcal{T}_{Ag_j}$.

Definition 17. *A set of arguments $\mathcal{E}_{Ag_i}$ in a preference-based argumentation framework $\mathcal{T}_{Ag_i}$ is a preferred extension for a participant Ag_i if it is maximal (with respect to set inclusion) complete extension obtained from $\mathcal{T}_{Ag_i}$.*

To define the set of justified conclusions in our model, we borrow the notions of *objectively acceptable* and *subjectively acceptable* arguments from [2].

Definition 18. *Given a preference-based argumentation framework $\mathcal{T}_{Ags} = \langle \mathcal{A}, \mathcal{R}, \succeq^{\mathcal{A}}_{Ags} \rangle$ for some participants Ags, an argument ϕ is objectively acceptable iff for all $Ag_i \in Ags$, ϕ is in every $\mathcal{E}_{Ag_i}$. On the other hand, ϕ is subjectively acceptable iff for some $Ag_i \in Ags$, ϕ is in some $\mathcal{E}_{Ag_i}$.*

In the discussion thus far, we have shown that each dialogue participant computes its preferred extensions in a dialogue based on preference ordering (i.e., trust) over the other dialogue participants—leading to preference ordering over arguments. It then follows that out of the set of preferred extensions a given participant may have, the refined preferred extension is the extension whose arguments are more trusted than the other extensions in the set. Consequently, the set of objectively acceptable arguments is the set that the participants simultaneously considered as the most trusted set of arguments in the dialogue. We consider this set as the most justified conclusion of a dialogue similar to how the set of sceptically justified arguments is considered as the set of most justified arguments in standard argumentation frameworks and PAF. With this property, we show how trust can have an effect on the justified conclusions of a dialogue.

Next, we consider the notion of a cycle within the preference ordering.

Definition 19. *A preference-based argumentation framework $\mathcal{T}_{Ags} = \langle \mathcal{A}, \mathcal{R}, \succeq^{\mathcal{A}}_{Ags} \rangle$ for participants Ags has a cycle iff there are two arguments $\phi_1, \phi_2 \in \mathcal{A}$ such that $\phi_1 \succ^{\mathcal{A}}_{Ags} \phi_2$ and $\phi_2 \succ^{\mathcal{A}}_{Ags} \phi_1$.*

Proposition 1. *Assume preferred semantics, for any T_{Ag_i}, if $(\phi_1, \phi_2) \in \mathcal{R}$ and $\phi_2 \succ^A_{Ag_i} \phi_1$, then ϕ_1 is not accepted—$\phi_1 \notin \mathcal{E}_{Ag_i}$.*

Proof. For any T_{Ag_i} that is cycle free, there is a unique corresponding $\mathcal{F}$, $\mathcal{F}_{Ag_i} = \langle \mathcal{A}, \mathcal{R} \rangle$, such that an element of attack relation $(\phi_1, \phi_2) \in \mathcal{R}$ in $\mathcal{F}_{Ag_i}$ is an element of defeat relation $(\phi_1, \phi_2) \in \mathcal{R}$ in T_{Ag_i}. Therefore, the preferred extension of $\mathcal{F}_{Ag_i}$ will contain the same arguments as the preferred extension of T_{Ag_i}. If T_{Ag_i} is cycle free, it means there is a preference ordering $\succeq^A_{Ags}$ over $\mathcal{A}$. For $\phi_1, \phi_2 \in \mathcal{A}$, $(\phi_1, \phi_2) \in \mathcal{R}$ and $\phi_2 \succ^A_{Ag_i} \phi_1$. The attack from ϕ_1 to ϕ_2 will be inverted. Therefore, this attack will not appear in $\mathcal{F}_{Ag_i}$. Instead, an attack from ϕ_2 to ϕ_1 will appear and since attack from ϕ_1 to ϕ_2 is not in $\mathcal{F}_{Ag_i}$, ϕ_2 is accepted in a preferred extension of $\mathcal{F}_{Ag_i}$ and ϕ_1 rejected. This applies to T_{Ag_i} since T_{Ag_i} corresponds to $\mathcal{F}_{Ag_i}$.

Proposition 2. *Suppose T_{Ag_i} has a cycle between all arguments (i.e., $(\forall \phi_1, \phi_2 \in \mathcal{A})$ s.t.$(\phi_1, \phi_2) \in \mathcal{R}$, $\phi_1 =^A_{Ag_i} \phi_2$), then any extension of T_{Ag_i} is also an extension of Dung's framework $\mathcal{F} = (\mathcal{A}, \mathcal{R})$ and vice versa under the same semantics.*

Proof. This follows from Definition 2 and Proposition 1.

This property ensures that when Ag_i has equal or no preferences for some arguments in T_{Ag_i}, then there can be no critical attacks between these arguments and preferences play no role in the evaluation of this set of arguments.

Proposition 3. *If a set of arguments $S \in \mathcal{A}$ is objectively acceptable in all preferred extensions $\mathcal{E}_{Ags}$ of T_{Ags} for all the participants Ags in a dialogue, then the set S is the set of most trusted arguments in the dialogue.*

Proof. Since every $\mathcal{E}_{Ag_i}$ is conflict free as the preferred extensions of PAF and corresponding F are conflict free, it follows that in T_{Ag_i}, every $\phi_1 \in \mathcal{E}_{Ag_i}$ is either unattacked or attacked by some argument $\phi_2 \in \mathcal{A} \backslash \mathcal{E}_{Ag_i}$ such that $\phi_1 \succ^A_{Ag_i} \phi_2$. For the latter, we know that such attack is *critical* and is *repaired* such that $(\phi_2, \phi_1) \in \mathcal{R}$ becomes $(\phi_1, \phi_2) \in \mathcal{R}$. If ϕ_1 is objectively acceptable in all preferred extensions $\mathcal{E}_{Ags}$ of T_{Ags}, it follows that in all $T_{Ag_i} \subseteq T_{Ags}$, ϕ_1 is either unattacked or is attacked by some less preferred argument ϕ_2. Since, $\phi_1 \succ^A_{Ag_i} \phi_2$ denotes that ϕ_1 is more trusted (more preferred) than ϕ_2, it follows that the set of arguments $S \subseteq \mathcal{E}_{Ags} = \{\phi_1 | \nexists \phi_2 \in \mathcal{A} \backslash \mathcal{E}_{Ags}$ such that $(\phi_2, \phi_1) \in \mathcal{R}$ and $\phi_1 \succ^A_{Ag_i} \phi_2\}$ is the set of most trusted arguments.

6   Example

To illustrate how a participant updates its trust relation with regards to other participants, we provide an extended example, adapted from [16]. We connect the arguments in the dialogue to the participants that advance them as shown in the *Speech Acts* column of Table 1. The *Moves* column of the table shows that the dialogue has two iterations with five moves in the first iteration and four moves in the second iteration. Figures 2 and 3 show the argumentation frameworks derived from the dialogue by one of the participants Ag_k. $\mathcal{T}^t_{Ag_k}$ represents argumentation framework of Ag_k at iteration t where nodes are arguments and edges are attack relation. Let us consider that participant Ag_k evaluates $\mathcal{T}^1_{Ag_k}$ and $\mathcal{T}^2_{Ag_k}$.

1st *Iteration*: **Trust Update Rules** $TR^1_{Ag_k}$—In this iteration, Ag_k observes two trust update rules SC w.r.t Ag_i and LJ w.r.t Ag_j. Ag_k observes contradiction in the commitment store of Ag_i (i.e., ϕ_4 attacks ϕ_1 by defending ϕ_2 that attacks ϕ_1). Furthermore, Ag_k observes that Ag_j lacks justification for ϕ_2 as ϕ_5 defeats ϕ_2 (ϕ_2 is defeated by an undefeated argument ϕ_5). Note that the symmetric attack between ϕ_3 and ϕ_4 is obtained by the attack from ϕ_4 to ϕ_3 exchanged via the contradict move, while the ϕ_3 to ϕ_4 attack is known by Ag_k from its knowledge base $\mathcal{KB}_{Ag_k}$.

Preference on Trust Update Rules $\succeq^1_{Ag_{i(TR)}}$ —Let Ag_k's preference on the trust update rules be $LJ \succ^1_{Ag_{k(TR)}} SC \succ^1_{Ag_{k(TR)}} AR$.

Trust Update $\succeq^1_{Ag_k}$ —Given the trust update rules and Ag_k's preference on the rules, from Definition 15, we can infer that Ag_k prefers (i.e., trusts) Ag_i to Ag_j. Likewise, Ag_k prefers itself to Ag_i (i.e., $\succeq^1_{Ag_k} = Ag_k \succ^1_{Ag_k} Ag_i \succ^1_{Ag_k} Ag_j$).

Ag_k's Conclusion $\mathcal{E}_{Ag_k}$—In $\mathcal{T}^1_{Ag_k}$, Ag_k considers that ϕ_1 and ϕ_5 defeat ϕ_2, ϕ_3 defeats ϕ_4, and $\mathcal{E}^1_{Ag_k}$ is $\{\phi_1, \phi_3, \phi_5\}$.

Table 1. Example: Dialogue

Moves	Speech acts	Arguments
m^1_1	$assert(Ag_i, \phi_1, 1)$	ϕ_1: Death penalty is a legitimate form of punishment
m^1_2	$contradict(Ag_j, \phi_1, \phi_2, 1)$	ϕ_2: God does not want us to kill
m^1_3	$contradict(Ag_k, \phi_2, \phi_3, 1)$	ϕ_3: God does not exist
m^1_4	$contradict(Ag_i, \phi_3, \phi_4, 1)$	ϕ_4: Some people believe in God
m^1_5	$contradict(Ag_k, \phi_2, \phi_5, 1)$	ϕ_5: The legal status of the death penalty should not depend on some random people's belief
m^2_1	$assert(Ag_j, \phi_6, 2)$	ϕ_6: The state has no right to put its subjects to death
m^2_2	$contradict(Ag_i, \phi_6, \phi_7, 2)$	ϕ_7: If child rapists and murderers are put to death it will reduce the number of suicides by the survivors
m^2_3	$contradict(Ag_k, \phi_6, \phi_8, 2)$	ϕ_8: Majority opinion in some democratic countries favour death penalty
m^2_4	$contradict(Ag_j, \phi_7, \phi_9, 2)$	ϕ_9: There is no strong evidence that the death penalty makes victims of child abuse feel good

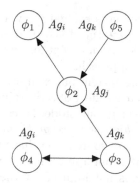

Fig. 2. $\mathcal{T}^1_{Ag_k}$ for 1st iteration

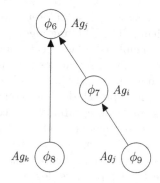

Fig. 3. $\mathcal{T}^2_{Ag_k}$ for 2nd iteration

2nd *Iteration*: **Trust Update Rules** $TR^2_{Ag_k}$—Ag_k observes that Ag_j lacks justification for ϕ_6 and Ag_i lacks justification for ϕ_7. Therefore, Ag_k observes one trust update rule LJ w.r.t to both Ag_i and Ag_j.

Preference on Trust Update Rules $\succeq^2_{Ag_{k(TR)}}$—Ag_k observes just one trust update rule. Therefore, preference over the trust update rules is not applicable in this iteration.

Trust Update $\succeq^2_{Ag_k}$—Note that, Ag_j has an undefeated argument ϕ_9 in this iteration while Ag_i has none. Therefore, Ag_k prefers Ag_j to Ag_i and itself to Ag_j (i.e., $\succeq^2_{Ag_k} = Ag_k \succ^2_{Ag_k} Ag_i \succ^2_{Ag_k} Ag_j$).

Ag_k's Conclusion $\mathcal{E}_{Ag_k}$—In $\mathcal{T}^2_{Ag_k}$, Ag_k considers that ϕ_8 defeats ϕ_6, ϕ_9 defeats ϕ_7, and $\mathcal{E}^2_{Ag_k}$ is $\{\phi_8, \phi_9\}$.

This example demonstrates how trust evolves in a dialogue and how such trust is used as a basis for expressing preferences between the arguments exchanged in the dialogue. In addition, the example illustrates how trust affects the justified conclusions obtained from a dialogue.

7   Related Work

Recent works on the integration of trust and argumentation has provided paradigms for handling inherent uncertainties in the interactions among agents in multi-agent systems. The importance of relating trust and argumentation was highlighted in [6]. In [10], arguments are considered as a separate source of information for trust computation.

There are four works in the literature which are closely related to the research described in this paper. The first is [12], where the authors propose a model of argumentation where arguments are related to their sources and a degree of acceptability is computed on the basis of the trustworthiness degree of the sources. The model also provides a feedback such that the final quality of the arguments influences the source evaluation as well. In this approach, different dimensions of trust are represented as graded beliefs ranging between 0 and 1 which change across different domains and arguments evaluated by a labelling algorithm. The labelling algorithm computes a fuzzy set of accepted arguments whose membership assigns to each argument a degree of acceptability unlike the extension-based semantics that we apply in our approach.

While related, the work of [12] differs from the current paper in several ways. First, the approach does not consider the cumulative effect of converging sources on argument acceptability. We consider this effect in our model by categorising accepted arguments into two categories namely *objectively acceptable* and *subjectively acceptable* extensions, based on the number of sources that have the arguments acceptable in their extensions. Second, unlike our approach, the evaluation of the trustworthiness degree of a target agent is not induced by the trusting agent's argumentation framework, but determined by the internal mechanism of the trusting agent. Third, [12] considers that in a dialogue, the final acceptability value of the arguments provides a feedback on the trustworthiness degree in the information source. In our approach, we observe that trust can change during the dialogue itself and as such the trust rating of a target participant should be updated at every stage (iteration) of a dialogue.

The works in [15,17] are closely related to ours. The authors present a framework which considers the source of arguments, and expresses a degree of trust in them. They define trust-extended argumentation graphs in which each premise, inference rule and conclusion of an argument is associated with the trustworthiness degree of the source proposing it. In this approach, the trust rating associated with the arguments and their sources does not change. In our approach, trust ratings associated with arguments and sources change between iterations. This notion of dynamic trust rating is captured by socio-cognitive models of trust [4] and other computational trust approaches [5,8].

Lastly, [18] models the connection between arguments about the trustworthiness of information sources and the arguments from the sources—as well as the attacks between the arguments. An information source is introduced into an argumentation framework as a meta-argument and an attack on the trustworthiness of the source is modelled as an attack on the meta-argument. A source is considered trustworthy if its meta-argument is accepted. Like us, [18] model the feedback from sources to arguments and vice-versa. However, like [12], they do not consider how trust evolves in the course of a dialogue.

8   Conclusions

This paper describes how trust changes during argumentation-based dialogues and how such change affects the justified conclusion of the dialogue. In particular, as arguments are exchanged in a dialogue, we formalise a number of trust update rules that a given participant can take into consideration for updating its trust relation over other target participants. The first contribution of our approach is that it captures how trust is dynamically updated in dialectical argumentation and how trust can affect the set of justified conclusions.

It is worth mentioning that the semantics of abstract argumentation frameworks have only focused on identifying which points of view are defensible and preference-based argumentation frameworks have extended these semantics to deal with preferences between arguments. However, they do not describe *why* one argument should be preferred over another. In our approach, the trust rating of the sources of arguments provides such a basis.

As future work, we intend to find out how change in trust in dialectical argumentation can affect the goals and argumentative strategies of participants. In addition, change in trust during a dialogue may require less trusted participants to present more evidence for their arguments to be believed, while the burden of proof reduces on more trusted participants. This is also an issue for future work. Finally, we are investigating an orthogonal approach to modelling changes in trust within an ongoing dialogue through the use of meta-argumentation. Doing so will eliminate the need for discrete iterations as used in the current work, and an empirical evaluation of the two approaches with regard to human intuitions will allow us to determine which approach is more realistic and useful.

References

1. Amgoud, L., Vesic, S.: Rich preference-based argumentation frameworks. Int. J. Approx. Reason. **55**(2), 585–606 (2014)
2. Bench-Capon, T.J.: Persuasion in practical argument using value-based argumentation frameworks. J. Log. Comput. **13**(3), 429–448 (2003)
3. Bonzon, E., Delobelle, J., Konieczny, S., Maudet, N.: A comparative study of ranking-based semantics for abstract argumentation. In: Proceedings of the 30th AAAI Conference on Artificial Intelligence, pp. 914–920 (2016)
4. Castelfranchi, C., Falcone, R.: Trust Theory: A Socio-Cognitive and Computational Model, vol. 18. Wiley, Hoboken (2010)
5. da Costa Pereira, C., Tettamanzi, A.G., Villata, S.: Changing ones mind: erase or rewind? Possibilistic belief revision with fuzzy argumentation based on trust. In: Proceedings of the 22nd International Joint Conference on Artificial Intelligence (2011)
6. Dix, J., Parsons, S., Prakken, H., Simari, G.: Research challenges for argumentation. Comput. Sci.-Res. Dev. **23**(1), 27–34 (2009)
7. Dung, P.M.: On the acceptability of arguments and its fundamental role in non-monotonic reasoning, logic programming and n-person games. Artif. Intell. **77**(2), 321–357 (1995)
8. Fullam, K.K., Barber, K.S.: Dynamically learning sources of trust information: experience vs. reputation. In: Proceedings of the 6th International Conference on Autonomous Agents and Multiagent Systems, pp. 164:1–164:8 (2007)
9. Jøsang, A., Keser, C., Dimitrakos, T.: Can we manage trust? In: Herrmann, P., Issarny, V., Shiu, S. (eds.) iTrust 2005. LNCS, vol. 3477, pp. 93–107. Springer, Heidelberg (2005). https://doi.org/10.1007/11429760_7
10. Matt, P.A., Morge, M., Toni, F.: Combining statistics and arguments to compute trust. In: Proceedings of the 9th International Conference on Autonomous Agents and Multiagent Systems, pp. 209–216 (2010)
11. McBurney, P., Parsons, S.: Games that agents play: a formal framework for dialogues between autonomous agents. J. Log. Lang. Inf. **11**(3), 315–334 (2002)
12. Paglieri, F., Castelfranchi, C., Pereira, C.D.C., Falcone, R., Tettamanzi, A., Villata, S.: Trusting the messenger because of the message: feedback dynamics from

information quality to source evaluation. Comput. Math. Organ. Theory **20**(2), 176 (2014)

13. Panisson, A.R., Meneguzzi, F., Vieira, R., Bordini, R.H.: Towards practical argumentation-based dialogues in multi-agent systems. In: Proceedings of the IEEE/WIC/ACM International Conference on Web Intelligence and Intelligent Agent Technology, vol. 2, pp. 151–158. IEEE (2015)

14. Parsons, S., McBurney, P., Sklar, E., Wooldridge, M.: On the relevance of utterances in formal inter-agent dialogues. In: Proceedings of the 6th International Conference on Autonomous Agents and Multiagent Systems, pp. 240:1–240:8 (2007)

15. Parsons, S., Tang, Y., Sklar, E., McBurney, P., Cai, K.: Argumentation-based reasoning in agents with varying degrees of trust. In: Proceedings of the 10th International Conference on Autonomous Agents and Multiagent Systems, pp. 879–886 (2011)

16. Spanring, C.: Conflicts in abstract argumentation. Cardiff Argumentation Forum (2016)

17. Tang, Y., Cai, K., Sklar, E., McBurney, P., Parsons, S.: A system of argumentation for reasoning about trust. In: Proceedings of the 8th European Workshop on Multi-Agent Systems, Paris, France (2010)

18. Villata, S., Boella, G., Gabbay, D.M., Van Der Torre, L.: A socio-cognitive model of trust using argumentation theory. Int. J. Approx. Reason. **54**(4), 541–559 (2013)

19. Walton, D., Krabbe, E.C.: Commitment in Dialogue: Basic Concepts of Interpersonal Reasoning. SUNY Press, Albany (1995)

Author Index